MODERN ELECTRONIC STRUCTURE THEORY AND APPLICATIONS IN ORGANIC CHEMISTRY

T0338556

MODERN ELECTRONIC STRUCTURE THEORY AND APPLICATIONS IN ORGANIC CHEMISTRY

Editor

Ernest R Davidson

Indiana University

World Scientific

Singapore • New Jersey • London • Hong Kong

Published by

World Scientific Publishing Co. Pte. Ltd.

P O Box 128, Farrer Road, Singapore 912805

USA office: Suite 1B, 1060 Main Street, River Edge, NJ 07661

UK office: 57 Shelton Street, Covent Garden, London WC2H 9HE

Library of Congress Cataloging-in-Publication Data
Modern electronic structure theory and applications in organic
  chemistry / [edited by] Ernest R. Davidson.
    p.   cm.
  Includes index.
  ISBN 9810231687
  1. Chemistry, Physical organic.   I. Davidson, Ernest R.
QD476.M62   1997
547'.128--dc21   97-27046
                                       CIP

British Library Cataloguing-in-Publication Data
A catalogue record for this book is available from the British Library.

This book is printed on acid-free paper.

Printed in Singapore by Uto-Print

PREFACE

This volume on *Modern Electronic Structure Theory and Application in Organic Chemistry* focusses on the use of quantum theory to understand and explain experiments in organic chemistry. High level *ab initio* calculations, when properly performed, are useful in making quantitative distinctions between various possible interpretations of structures, reactions and spectra. Chemical reasoning based on simpler quantum models is, however, essential to enumerating the likely possibilities. The simpler models also often suggest the type of wave function likely to be involved in ground and excited states at various points along reaction paths. This preliminary understanding is needed in order to select the appropriate higher level approach since most higher level models are designed to describe improvements to some reasonable zeroth order wave function.

In the first chapter in this volume, Zimmerman discusses a wide variety of thermal and photochemical reactions of organic molecules. Quantum theory is used mostly as a qualitative tool to explain facts taken from experiments. The next two chapters focus in greater depth on two particular classes of reactions. Gronert discusses the use of *ab initio* calculations and experimental facts in deciphering the mechanism of β-elimination reactions in the gas phase. He gives a detailed review of the experimental facts, the important questions they raise, and the answers obtained from high level calculations. Bettinger *et al.* focus on carbene structures and reactions with comparison of the triplet and singlet states. This chapter also contrasts the results found by many of the standard *ab initio* methods with the large body of experiments on carbenes. Next, Hrovat and Borden discuss more general molecules with competitive triplet and singlet contenders for the ground state structure. Again, the focus is on experimental facts, qualitative reasoning and quantitative calculations with the most appropriate *ab initio* methods.

None of the preceding chapters describe the theoretical methods in detail. All of the methods used have already been treated in a more elaborate fashion in the book, *Modern Electronic Structure Theory*, edited by David R. Yarkony. These chapters illustrate the use of these standard *ab initio* methods, coupled with reasoning based on simple quantum models, to treat problems in organic chemistry.

The final three chapters are somewhat different in nature. Treatment of the large variety of possible wave functions for excited states is not a routine problem. Reliable "black-box" methods do not yet exist. Cave explains the difficulties and considerations involved with many of the methods and illustrates the difficulties by comparing with the UV spectra of short polyenes.

Jordan *et al.* discuss long-range electron transfer using model compounds and model Hamiltonians. For compounds of this size, *ab initio* calculations are still not feasible for studying the mechanism of electron transfer. Hence, the approach follows the long tradition of semi-empirical theory applied to organic molecules.

Finally, Hiberty discusses the breathing orbital valence bond model as a different approach to introducing the crucial $\sigma\pi$ correlation that is known to be important in organic reactions (see, e.g. W.T. Borden and E.R. Davidson, *Acc. Chem. Res.* **29**, 67 (1996)). This concept is illustrated by looking at the bond energies of very small inorganic and organic molecules.

Ernest R. Davidson

CONTENTS

SOME THEORETICAL APPLICATIONS TO ORGANIC CHEMISTRY

HOWARD E. ZIMMERMAN

Chemistry Department, University of Wisconsin

Madison, WI 53706

The material in this chapter is a survey of some theoretical concepts, ideas and methods introduced in our research. Most of the theory has been in conjunction with experimental results. Thus, theory has proven especially useful in understanding ground state chemistry and much more essential in interpreting excited state behavior.

1 Ground State Examples

1.1 The Birch Reduction

The basic mechanism of the Birch Reduction,[1] as postulated by Birch, involved a one-electron reduction to give the radical-anion, followed by subsequent protonation of this species by an alcohol, introduction of a second electron, and finally a second protonation step. The kinetics reported by Krapcho and Bothner-by[2] established that the rate-limiting step is the protonation of the radical-anion **2**. However, what was controversial was the regioselectivity of that initial protonation step. The original Birch

mechanism suggested, on a qualitative basis, that the site of greatest negative charge and thus the initial protonation was at the meta position in alkoxylated and alkylated benzenes (e.g. anisole, compound 8 in Scheme I). However, in our very early studies,[3] simple Hückel calculations were the best available and these were used to probe the electron densities for a variety of alkoxylated aromatics. It was found that it was invariably the ortho position which was the site of highest density in the radical-anion. This became a controversial topic, and the views on the two mechanisms shown in Scheme I oscillated over the years. Finally, in more recent research[4] we obtained experimental evidence indicating that ortho protonation was preferred seven to one. Since more modern computational methodology was available, Gaussian90[5a] and Gamess[5b] (ROHF/3-21G) were used with the result that the electron density at the ortho position was found to be 1.302 versus 1.259 at the meta one. Interestingly, the frontier MO coefficients (i.e. the singly occupied antibonding MO) did not correlate with experiment. Additionally, the energy of the ortho-protonated species proved to be lower than that of the meta-protonated counterpart. Similar results were obtained with alkyl substitution.

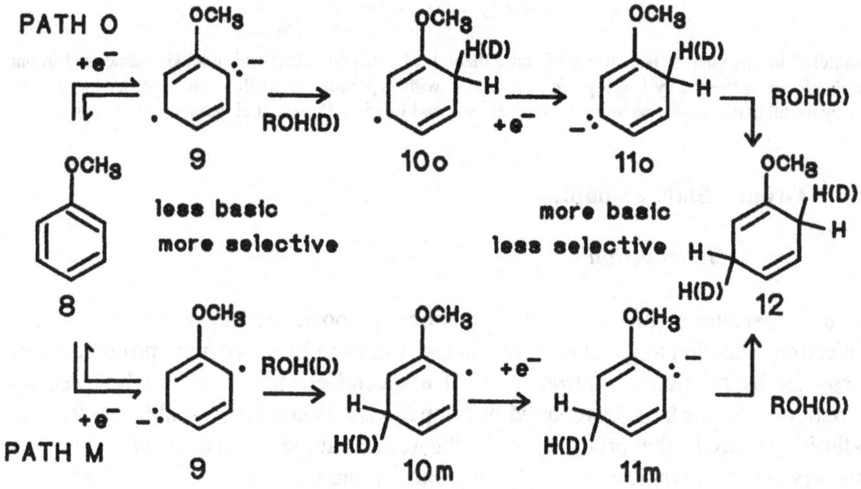

PATH O

9 → 10o → 11o → 12

less basic
more selective

more basic
less selective

PATH M

8 → 9 → 10 m → 11m

Scheme I: Possible mechanisms for the Birch reduction of anisole.

     With regard to the second protonation step in the Birch reduction, wherein the central site of a cyclohexadienyl carbanion is protonated at the central carbon giving the unconjugated Birch product, we note that this is a general phenomenon. The highest electron density is obtained at the central carbon. This has precedent in the case of the dienolates obtained from enones[6] where kinetic protonation of the dienolate affords the thermodynamically less stable β,γ-enone.

     Several points should be noted. First, use of frontier MO reasoning runs a risk of giving incorrect predictions. Secondly, Hückel computations are generally of real value. Hückel theory is synonymous with graphic theory and reflects topology. While more sophisticated computations are needed, in many cases Hückel results give insight to the electronics of molecular systems.

1.2   1,2-Carbanion Rearrangements

While 1,2 carbon to carbon shifts of alkyl and aryl groups are common for carbocations, the corresponding odd-electron and carbanions case are more rare. The electronic rationale for this behavior was noted by the present author in 1961.[7] Interestingly, this paper presented the first correlation diagram for an organic reaction; note Scheme II. A group was considered to migrate from carbon-1 to carbon-2, and the computations used a truncated basis set. This set was comprised of two p-orbitals and an sp^2 hybrid for the alkyl migration and a total of nine localized basis orbitals for the phenyl migration where there were six p-orbitals in the aromatic ring. One in-plane sp^2-hybrid at the ring carbon originally bonded to carbon-1 and one localized orbital was taken at each of C-1 and C-2 as shown in Figure 1. The correlation diagram considered the starting MO's and those of the half-migrated species, either with an alkyl group (e.g. methyl) half-migrated from C-1 to C-2 or with the phenonium intermediate being formed. The solid dots indicate electron occupation in the MO's and the asterisk represents zero electrons for the carbocation case, one electron for the free radical situation, or two electrons for a carbanion rearrangement.

It is seen that for a carbanion, the two non-bonding electrons taken as at an energy defined as zero, become badly anti-bonding where an alkyl group as methyl is migrating. However, for the phenyl migration, the rise in energy is much smaller. This was shown to be a general situation with further variation in substitution [also see Ref 8 for more details].

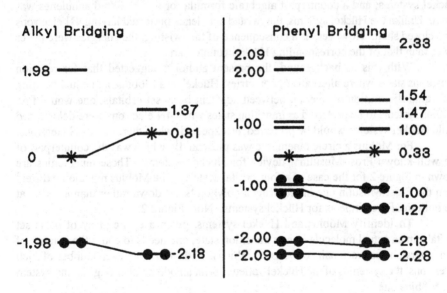

Scheme II: Correlation diagrams for alkyl and phenyl migration; reactant and half-migrated species given.

4

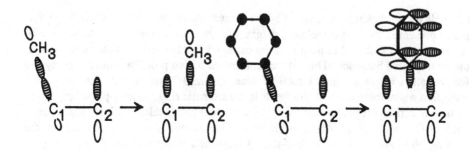

Figure 1: Basis Set Orbitals for Methyl and Phenyl 1,2-Migrations.

1.3 The Möbius-Hückel Concept

The idea of allowedness vs. forbiddenness, or aromaticity vs anti-aromaticity, of pericyclic reactions was advanced by Woodward and Hoffmann[9] and Longuet-Higgins and Abrahamson.[10] Edgar Heilbronner[11] had published a paper noting that a twisted, Möbius cyclic annulene would have a set of MO's which differed from that cyclic Hückel systems; and a counterpart algebraic formula for these twisted annulenes was given. Unlike the Hückel systems, the twisted annulenes preferred having 4N electrons for a closed shell. However, as a consequence of the twisting, the energies were never lower than that of the corresponding Hückel counterpart.

With this as background, the present author[12] suggested that for reaction transitions states where alternative geometries, Hückel and Möbius, had equal twisting and equal diminution of overlap between adjacent basis set orbitals, one would find Möbius systems to be preferred as transition states when 4N electrons were delocalized while Hückel systems would be preferred, as expected, when there were 6N electrons.

For Möbius a circle mnemonic was presented[12] which was the counterpart of the well-known Frost-Musulin[13] device for Hückel systems. These mnemonics are shown in Figure 2 for the case of a four orbital system. The Möbius mnemonic differs from the Frost-Musulin one in requiring a polygon 'side' down rather than a vertex at the bottom of the circle as for Hückel systems. Note Figure 2.

To identify Möbius and Hückel systems, given a cyclic array of basis set orbitals, in an actual molecule or in a transition state, one needs to count the number of sign inversions between pairs of basis set orbitals. With zero or an even number of such inversions, the system is of the Hückel variety. With an odd number (e.g. 1), the system is a Möbius one.

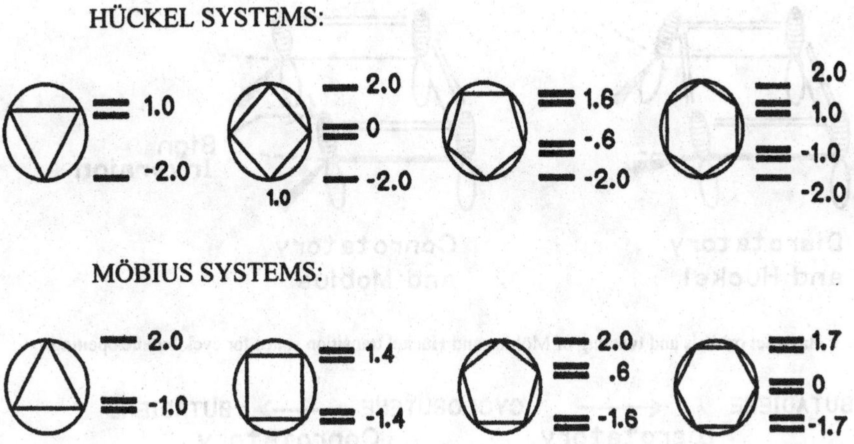

HÜCKEL SYSTEMS:

MÖBIUS SYSTEMS:

Figure 2: The Möbius - Hückel circle mnemonics.

For consideration of allowedness versus forbiddenness, one can write the cyclic array for a given pair of pericyclic reaction transition states, populate the MO's, and then compare energies of the Hückel and the Möbius counterparts. The lower energy one is the allowed reaction pathway. More interestingly, one can develop the correlation diagram for the reactions without the use of symmetry. This is done by recognizing that for each degenerate pair in the transition state, there is a crossing of a set of MO's as one proceeds from reactant to product.

Figures 3 and 4 give one example. This shows that the opening of cyclobutene to give butadiene in the ground state will prefer a conrotatory pathway as opposed to a disrotatory one (Note Figure 4), in which a Möbius transition state is employed. The correlation diagram for both conrotatory and disrotatory reactions is readily drawn by obtaining the transition state degeneracies from the circle mnemonics and recognition that each degeneracy gives rise to crossing of two MO's. The conrotatory mechanism leaves all of the bonding MO's as bonding while the disrotatory mechanism converts a bonding MO to an antibonding one. Thus if one were to picture the first-order situation of an adiabatic reaction (i.e. before any configuration interaction), a ground state reactant proceeding in a disrotatory fashion would afford a very high energy doubly excited state of product.

The Möbius-Hückel concept applies to more than the transition states discussed above. There are many Möbius molecules already known. One is barrelene. This molecule has three ethylene bridges. One may start with the bonding MO's of ethylene

6

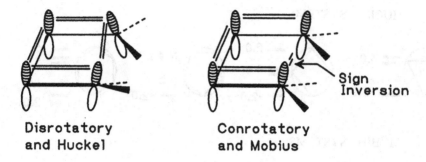

Disrotatory
and Huckel

Conrotatory
and Mobius

Sign
Inversion

Figure 3: Basis set orbitals and twisting of Möbius and Hückel transition states for cyclobutene opening.

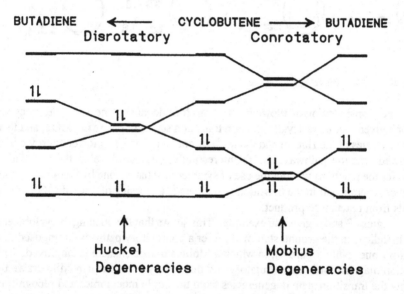

BUTADIENE ← CYCLOBUTENE → BUTADIENE
Disrotatory   Conrotatory

Huckel
Degeneracies

Mobius
Degeneracies

Figure 4: Correlation diagram interconverting cyclobutene and butadiene by conrotatory and disrotatory twisting.

on each bridge and note that these interact in a Möbius fashion. At a simple Hückel level, the bonding MO of ethylene is at -1, the energy scale using units of the absolute value of β. The transannular overlap is less by a factor (ϵ) than that for two adjacent p-orbitals. The Möbius splitting of the three bonding MO's thus results in a degenerate pair of MO's at -1 -ϵ and a single MO at -1 +2ϵ. This is illustrated in Figure 5. Similar consideration of the antibonding ethylenic orbitals as a basis set, leads to a similar splitting about the antibonding ethylenic MO at +1. Note Figure 5.

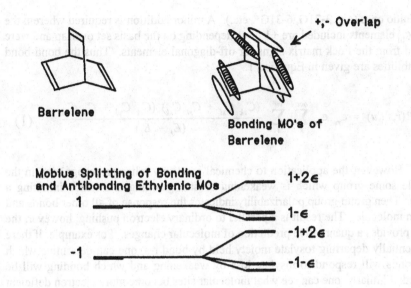

Figure 5: Admixture of three ethylene bridges of barrelene.

1.4   Prediction of Organic Reactions Using Group-Group Polarizability

In Hückel theory bond-bond, atom-bond, bond-atom, and atom-atom polarizabilities give the change in one part of the molecule as another is modified. For example, in bond-bond polarizability, one obtains the change in a bond order (e.g. r-s) as the resonance integral β is changed for another pair of atoms (e.g. t-u). In the case of atom-atom polarizability, one obtains the change in charge at one center as the Coulomb integral at another atom is modified. The atom-bond and bond-atom parameters are defined similarly.

While these polarizabilities[14] are of intellectual interest, they have had severe limitations. (a) They have been limited to use with π-systems (e.g. aromatics), (b) the neglect of overlap approximation has assumed different p-orbitals do not overlap, (c) and the approximate nature of Hückel wavefunctions is generally a problem.

It was recognized by the author[15] that the mutual-polarizability concept could be applied to organic molecules in general, including those having both sigma and pi bonds. Further, the most modern self-consistent field methods could be employed, not only the more approximate semi-empirical wavefunctions, but also ab initio ones. This was made possible by the realization that Weinhold "natural hybrid orbitals" (i.e. NHO's)[16] comprise an orthogonal basis unlike a Hückel set. Additionally, the NHO basis orbitals are readily available from both semi-empirical computations (e.g. AM1)

8

and ab initio ones (e.g. 3-21G, 6-31G*, etc.). A minor addition is required wherein the ϵ_{rs} and ϵ_{tu} elements included are +1 or -1 depending on the basis set overlap and were obtained from the Fock matrix signs for off-diagonal elements. Thus the bond-bond polarizabilities are given in Equation 1.

$$P(r,s,t,u) = \epsilon_{rs}\, \epsilon_{tu} \sum_{k}^{occ} \sum_{l}^{vir} \frac{(C_{rk}\, C_{sl} + C_{rl}\, C_{sk})\, (C_{tk}\, C_{ul} + C_{tl}\, C_{uk})}{(E_l - E_k)} \qquad (1)$$

However, the application to chemical reactivity relies on recognizing in the molecule some group which is weakening or strengthening and hence initiating a reaction. Then group-group polarizability indicates the response of all other bonds and atoms in molecule. The result is a parallel to ordinary electron pushing; however, the method provides a quantitative prediction of molecular changes. For example, if there is a potentially departing tosylate moiety held by bond r-s, one can determine which other bonds will respond to tosylate loss by weakening and which bonding will be enhanced. Similarly, one can see what molecular sites become more electron deficient in such a case.

While a number of examples were given, one representative example is the solvolysis of exo- and endo-norbornyl chloride.[17] It is well known that the exo stereoisomer solvolyses in polar media much more rapidly than the endo isomer. Also, there is evidence for participation of the π-bond.

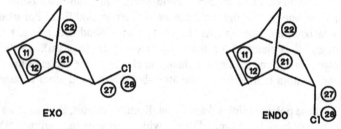

EXO   ENDO

Scheme III: The exo and endo norbornenyl systems with labeled orbitals.

We see interaction between the π-bond 11-12 and the C-Cl sigma bond with a positive polarizability (0.00575) in the case of the exo stereoisomer compared with essentially zero for the endo isomer. The positive value for the exo isomer indicates withdrawal of electron density and weakening of C=C π-bond 11-12 as the C-Cl sigma bond (27-28) is broken. The difference between the exo and endo isomers, again, correlates with the experimental behavior of the two compounds on solvolysis. One

Table 1: Exo and endo dehydronorbornyl bond-bond polarizabilities.

Exo Norbornenyl Polarizabilities

r	s	t	u	Orb r	Orb s	Orb t	Orb u	Polariz-ability
27	28	11	12	C 5(Cl 8)	Cl 8(C 5)	C 2(C 3)	C 3(C 2)	0.005753
27	28	21	22	C 5(Cl 8)	Cl 8(C 5)	C 4(C 7)	C 7(C 4)	0.000723
28	28	11	12	Cl 8(C 5)	Cl 8(C 5)	C 2(C 3)	C 3(C 2)	-0.000429
28	28	22	22	Cl 8(C 5)	Cl 8(C 5)	C 4(C 7)	C 7(C 4)	-0.000547

Endo Norbornenyl Polarizabilities

r	s	t	u	Orb r	Orb s	Orb t	Orb u	Polariz-ability
27	28	11	12	C 5(Cl 8)	Cl 8(C 5)	C 2(C 3)	C 3(C 2)	-0.000092
27	28	21	22	C 5(Cl 8)	Cl 8(C 5)	C 4(C 7)	C 7(C 4)	0.011229
28	28	11	12	Cl 8(C 5)	Cl 8(C 5)	C 2(C 3)	C 3(C 2)	0.000158
28	28	22	22	Cl 8(C 5)	Cl 8(C 5)	C 4(C 7)	C 7(C 4)	-0.008572

interesting facet is the heavy electron withdrawal from the sigma bond 21-22 in the endo stereoisomer (i.e. the chlorine is endo, note 27-28:21-22). Additionally, the sigma bond anti-coplanar with the C-Cl bond of the exo-stereoisomer exhibits an appreciably large interaction with that bond.

While it is true that the polarizabilities computed are related to an "overlap path" of conjugating basis orbitals between interacting groups, the role of electron affinity and availability at the various centers plays a large role as well. But then these are the factors which control reactivity.

Further, polarizabilities represent a precise counterpart to electron pushing and something which has not been considered in predicting reactivity of sigma systems. Beyond this, polarizabilities represent just one case where one can extend the Hückel treatment to sigma systems as a consequence of the orthogonality of the basis orbitals (i.e. NHO's) used.

1.5 Electron Density and Central Protonation of Mesomeric Anions

In the Birch Reduction discussed above, the final step was protonation of a mesomeric carbanion. Since we know that conjugated dienes are lower in energy than the unconjugated ones, formation of the unconjugated reaction product cannot be controlled by relative thermodynamic stability of the product. While in some reactions, the final product energy is seen sufficiently in the transition state, that the more stable product is the one formed. Not so here. It is observed as a general phenomenon that kinetic protonation of anions of the type C=C-C=C-$\underline{C}$ ↔ C=C-$\underline{C}$-C=C ↔ $\underline{C}$-C=C-C=C occurs at the central carbon. The same is true of dienolates such as C=C-C=C-O⁻.

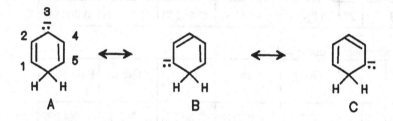

Figure 6: Three resonance structures for the cyclohexadienyl carbanion.

Simple Hückel calculations on the pentadienyl type anions predict an equal distribution of negative charge at atoms 1,3 and 5 (i.e. 0.333 e at each of the three centers) in agreement with the appearance of three resonance structures shown in Figure 6 for the example of the cyclohexadienyl carbanion. But in looking at the three resonance structures we note that two of the three structures have double bonds at C1-C2 and C4-C5 but single bonds at C2-C3 and C3-C4. Thus this qualitative resonance reasoning suggests that bond 1-2 and 4-5 should be shorter. In fact, if one adjusts the resonance integrals to reflect this, with the absolute value of the C1-C2 and C4-C5 values being larger, the electron density shifts towards C3. This is precisely what was done using the Mulliken-Wheland-Mann modification[18] of Hückel theory in which the resonance integral β for bond r-s is given by Equation 2.

$$\beta_{rs} = \beta_0[.08(P_{rs}(pi)+P_{rs}(sigma)) + 0.115]/0.276 \qquad (2)$$

$P_{rs}(sigma)$ was taken as unity while $P_{rs}(pi)$ is obtained from one iteration and then used in Equation 2 for the next iteration. The constants were selected so that with a π-bond order of unity, β_{rs} becomes equal to the standard β_0. Although not here, in other applications, the Coulomb integral was also varied. Interestingly, Equation 2 is philosophically related to the equations of SCF theory which is also iterative and where each successive iteration relies on the LCAO MO coefficients of the previous iteration.

More modern inspection of the mesomeric anions using the semiempirical AM1 and the ab initio Gaussian92 (and Gaussian94), is thus of real interest. Two alternative approaches have been considered.[19] One is computation of the isolated mesomeric carbanions and the other is completion with the sodium cation included to give an ion-pair. Interestingly, both AM1[20] and Gaussian[5] predict the highest electron density at the central atom for the ion pair in accord with experiment as well as for the simple carbanion. For the ion pair, one starts with the sodium cation positioned above either the central or the terminal carbon of the pentadienyl moiety. Geometry optimization moves the sodium cation until it is directly over the central carbon. These results are given in the first four columns of Table II.

Table II: Charge distribution in cyclohexadienyl anions (with and without the sodium cation present) and the cyclohexadienolate anion.

	Cyclohexadienyl		Na_cyclohexadienyl		Dienolate
	AM1	3-21G	AM1	3-21G	3-21G
1	-0.3669	-0.42178	-0.1723	-0.35756	-0.768332 (O)
2	-0.0487	-0.16345	-0.1192	-0.21866	0.520120
3	-0.3891	-0.57926	-0.0978	-0.70493	-0.460252
4	-0.0487	-0.16346	-0.1550	-0.21934	-0.169823
5	-0.3671	-0.42185	-0.1188	-0.35600	-0.330626

Also included are computations for the dienolate system. Experimentally, it is observed as a general phenomenon[6] that protonation and alkylation bond to the carbon α to the carbonyl group ultimately engendered. While in this case the computations reveal the dienolate oxygen to be most electron rich, protonation at oxygen to afford the dienol is reversible and we need only consider the competition between (e.g.) protonation α to the incipient carbonyl group (i.e. numbered as C-3 below) versus γ to the carbonyl (i.e. at C-5); but C-3 is most electron rich, thus accounting for preferential protonation and alkylation at this site.

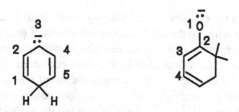

Figure 7: Cyclohexadienyl and dienolate anions.

2 Excited State and Open Shell Examples

2.1 n-π* Reactivity

It was in 1961 that the present author suggested[21] that n-π* singlets and triplets exhibit two types of photochemical reactivity and presented a general theory of excited state reactivity of organic species. This provided a basis for relating the three-dimensional electronic structure of a given excited state to possible photochemical transformations.

Of the two types of excited state reactions of n-π* states, one arises from the singly occupied p_y (or "n") orbital and the other from the π-system which has one extra, antibonding electron. The excitation process is shown in Scheme IV for a simple, unconjugated ketone. Since the p_y orbital is only singly occupied, it is highly electrophilic and quite willing to attack π-bonds and to abstract hydrogen atoms and

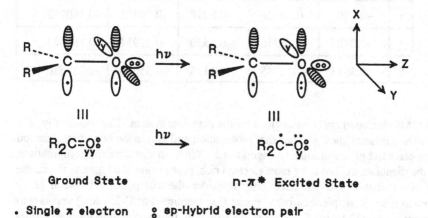

Scheme IV: n-π* excitation. Three-dimensional version above and Lewis structure equivalent below.

attack π-bonds. The hydrogen abstraction capability was independently reported by Kasha.[22]

The p_y orbital of the n-π* excited state also has an "electron hole" which can be distributed in any coplanar σ-system. Thus, in the n-π* excited state this p-orbital is close to coplanar with the σ-bonds bonding alkyl groups to the carbonyl carbon. It was noted in our early work[21] that delocalization of the "electron hole" leads to weakening of the sigma bond and thus accounts for the Norrish Type I cleavage which leads to an acyl and alkyl radical pair.

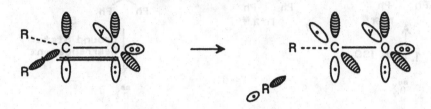

Scheme V: A basis orbital picture of the Norrish Type I cleavage reaction.

The second type of reactivity exhibited by n-π* excited states arises from the electron rich π-system which is isoconjugate with a radical anion. One example of a reaction of this class which was presented in that early study is the "Type A" Cyclohexadienone Rearrangement.[21] The mechanism is outlined in Scheme VI. In this case, and in many others, there are four basic steps involved in the photochemistry. The first step is excitation and intersystem crossing (in this triplet example); intersystem crossing of ketones is a very rapid process. Step 2 is some molecular bond alteration; in the case of the Type A Cyclohexadienone Rearrangement, the molecular change involves π-π bonding between the carbons β to the carbonyl group. Step 3 is radiationless decay, here by intersystem crossing to ground state; this step leads to a zwitterion. The last step 4 is a ground state process characteristic of the S_o species engendered by Step 3. Here it is a rearrangement known[23] to occur from such zwitterion species.

While this is just one case of π-system reactivity of an n-π* excited state, we might stop and inspect some of the facets encountered. The first question is why the triplet excited state undergoes β-β bonding. Scheme VII gives one view.[24] We see that in promotion of one electron from the p_y (or "n") orbital to MO 4, one has an excitation from an MO with a zero β-β bond order to MO 4 which has an appreciable β-β bond order. Thus n-π* excitation enhances the β-β bond order. This is not true of promotion from MO 3 to MO 4 (i.e. a π-π* excitation) where both MOs are weighted positively at the β carbons and thus have positive bond orders. Our early computations used an

SCF-CI approach with a limited basis set comprised of the p_y and the π-system orbitals. These revealed that, of the triplet configurations, only the n-π^* one is β-β bonding. Interestingly S_0 is also bonding.

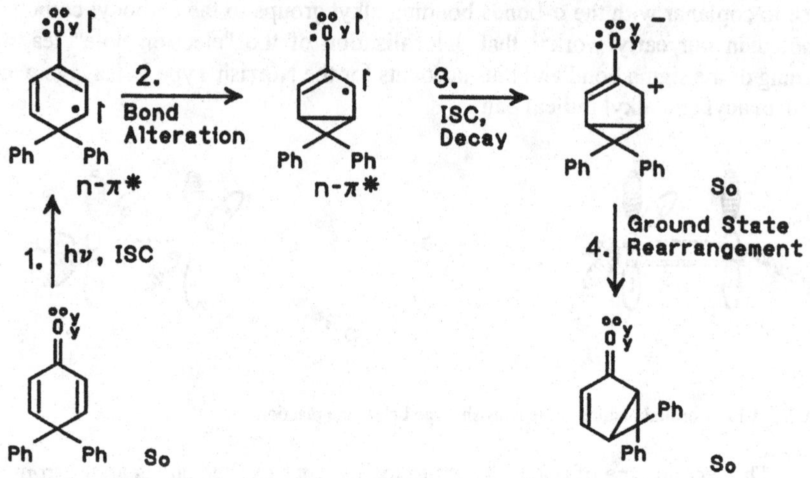

Scheme VI: Type A rearrangement mechanism.

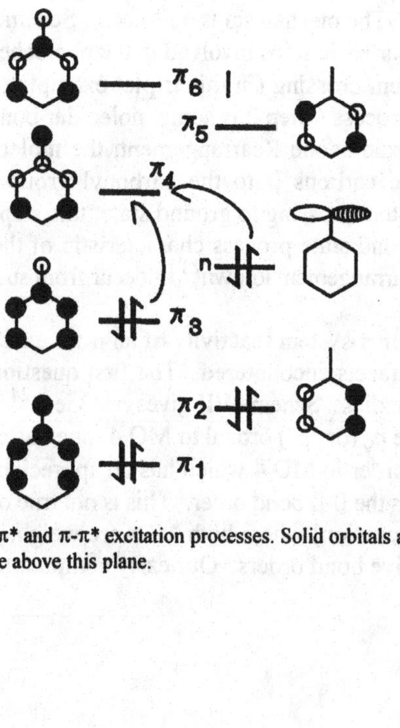

Scheme VII: n-π^* and π-π^* excitation processes. Solid orbitals are positive above the paper plane and white ones are negative above this plane.

2.2 π-π* Reactivity

Of π-π* excited state reactivity, the Di-π-Methane Rearrangement[25] provides one of the most common and most illustrative examples. The rearrangement was encountered in the rearrangement of barrelene to semibullvalene in 1965[25] but the correct mechanism and the generality of the reaction was realized only the next year.[26]

The basic mechanism is depicted in Equations 3 and 4. The only requirement for the reaction is having two π groups bonded to a central sp³ hybridized carbon. In Equation 3, both groups are taken as alkenyl and in Equation 4 one is pictured as an aryl group. Some examples of the Di-π-Methane Rearrangement proceed via a singlet excited state and some make use of the triplet.

An interesting case is the photochemical conversion of barrelene to semibullvalene as pictured mechanistically in Scheme VIII. It is seen that there are two bridgehead sp³ hybridized carbons, each of which satisfies the requirement for two

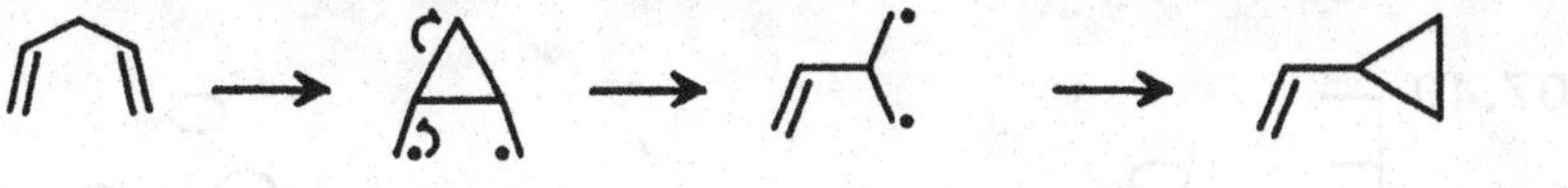

(3)

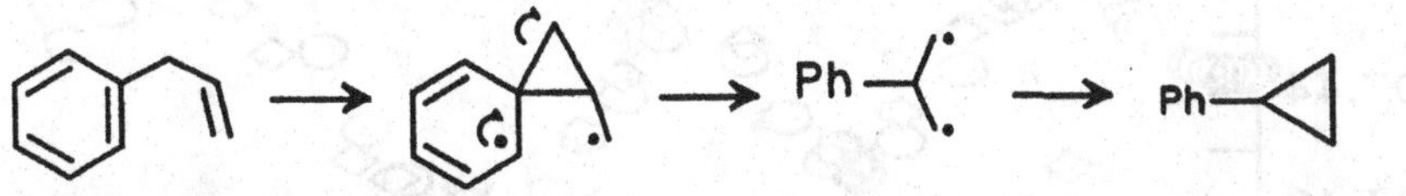

Di-π-methane reactant   "Diradical 1"   "Diradical 2"   Photoproduct

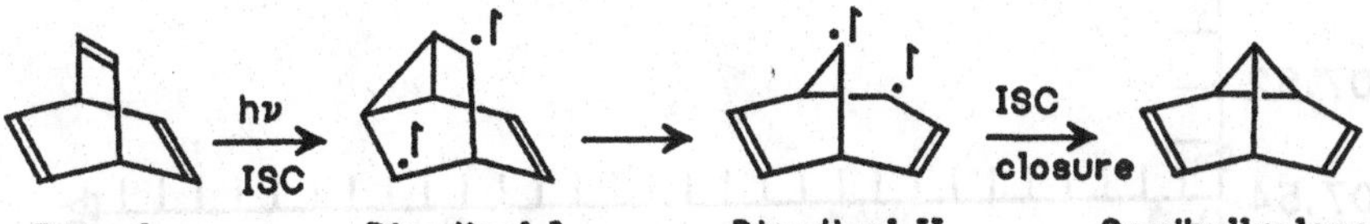

(4)

Barrelene   Diradical I   Diradical II   Semibullvalene

Scheme VIII: The mechanism of the barrelene to semibullvalene conversion.

attached π-groups, here vinyls. In this case the experimental evidence is that the reaction proceeds via a triplet excited state. Also, at the outset there was evidence that "Diradical II" is an intermediate with a lifetime sufficient for conformational equilibration. Further, in this early work,[26a,b] a reaction profile was obtained quantum mechanically by use of Extended Hückel computations.[27] Cartesian coordinates were

16

obtained for the reactant, product and diradical species from Dreiding molecular models taped to a piece of graph paper. The computations revealed both Diradical I and Diradical II to be energy minima and provided a limited hypersurface. A comparable computation for the originally assumed mechanism led to a vertical excited state in a deep minimum and faced with a large energy barrier.

However, much more recently with the availability of Gaussian90, G92 and G94, a more reliable computation was possible. This was done at the ROHF/6-31G* level for the triplet, RHF/6-31G* and then at the CASSCF(6,6)/6-31G* level in order to inspect more closely the effect of configuration interaction; this was done for both singlets and triplets.[26c] Note Figures 8 and 9.

Several points are noteworthy. One is that despite the qualitative nature of the very early results, the vastly improved computations still confirmed the presence of two triplet energy minima with the structures of Diradicals I and II. The second point of interest is the near degeneracy of the triplet states (i.e. note the CASSCF results) of the

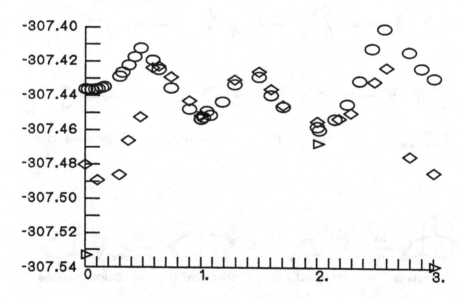

Figure 8: Triplet hypersurface for conversion of barrelene to semibullvalene including selected singlet points. The ordinate is in Hartrees. The abscissa is 0 for barrelene, 1 for diradical I, 2 for diradical II, and 3 for semibullvalene. The ellipsoids are for the triplet, the diamonds for the singlet with triplet geometry, the triangles for S0.

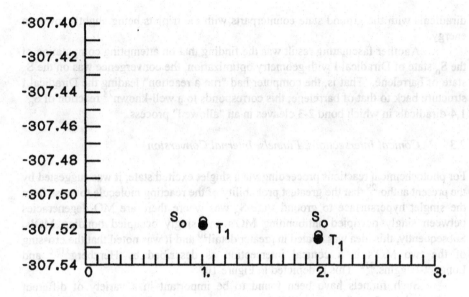

Figure 9: So-T1 near degeneracies for diradicals I and II.

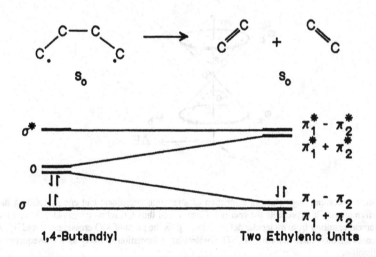

Scheme IX: Facile allowed fragmentation of S_0 1,4-Diyls. In this scheme, the MO counterpart correlation diagram is depicted with an S_0 configuration.

18

diradicals with the ground state counterparts with the triplets being slightly lower in energy.

Another fascinating result was the finding that on attempting computation of the S_0 state of Diradical I with geometry optimization, the convergence was on the S_0 state of barrelene. That is, the computer had "run a reaction" leading the Diradical I structure back to that of barrelene; this corresponds to a well-known[28] reaction of S_0 1,4-diradicals in which bond 2-3 cleaves in an "allowed" process.

2.3   Conical Intersections, Funnels, Internal Conversion

For photochemical reactions proceeding via a singlet excited state, it was suggested by the present author[29] that the greatest probability of the reacting molecule to cross from the singlet hypersurface to ground state S_0 was where there are MO degeneracies between singly occupied antibonding MO's and singly occupied bonding MO's. Subsequently, this idea was treated in greater detail[30] and it was noted that the crossing of the two MO's is a conical intersection as described by Herzberg[31a,c] and Longuet-Higgins.[31b] This is depicted in Figure 10.

Such funnels have been found to be important in a variety of different transformations. Such funnels providing routes to ground state have been noted by Michl as well.[32]

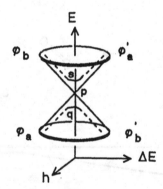

Figure 10. MO's becoming degenerate as a function of a reaction coordinate and energy splitting due to molecular deformation. `k` represents the energy difference and thus a reaction coordinate, `h` is a matrix interaction element resulting from molecular deformation. `p` is the point of MO crossing. `s` and `q` show the consequences of molecular deformation. The molecular deformation is likely as a consequence of a Jahn-Teller distortion.

In our own research, the role of these funnels in controlling the course of photochemical reactions has been noted in a variety of examples. One fascinating case is given in Scheme X.[33] This involves the photochemical singlet interconversion of three compounds. This provides an example of our Bicycle Rearrangement in which a carbon, part of a three-membered ring, moves stereospecifically along the π-surface of a molecule. The intriguing aspect of the case is presented in Scheme X. Here there are three compounds of interest, termed "Di-π", "Bicyclic" and "Spiro" in Scheme X. "Spiro" is photochemically unreactive. However, "Di-π" and "Bicyclic" react by processes whose gross mechanisms are quite clear from the structural standpoint. The reactions beginning with "Di-π" are depicted with bold arrows while those which begin with "Bicyclic" are shown with dashed arrows. The paradox arises when we see that both the "Di-Pi" and the "Bicyclic" reactions need to proceed via diradical "D"; yet the reaction product depends on which reactant is employed. How can one intermediate (i.e. diradical "D") give rise to different behavior depending on its source?

While there are chemical momentum and trajectory considerations which, a priori might be involved, in this case there is another factor. Here two different states of diradical "D" proved to be involved. Those reactions beginning with reactant "Di-π" are indicated with solid arrows while those pathways initiated from reactant "Bicyclic" are shown with a dashed arrow.

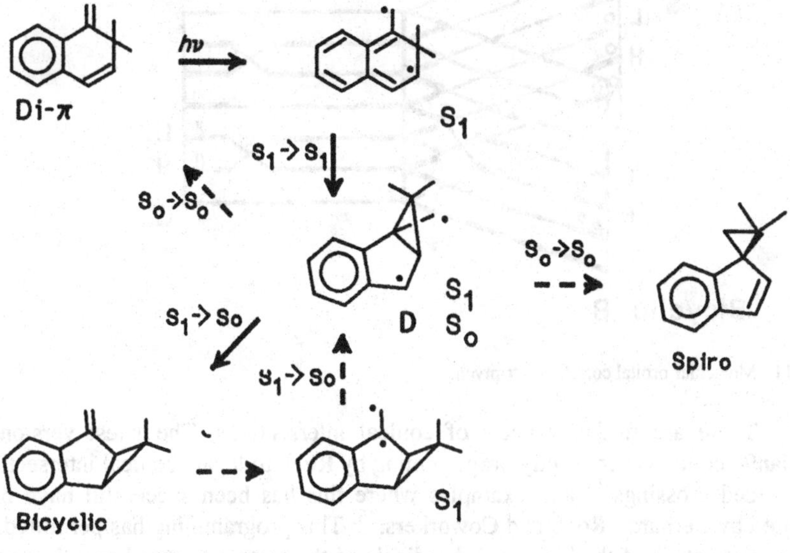

Scheme X: Interconversion of three compounds, "Di-π", "Bicyclic", and "Spiro".

20

The corresponding correlation diagram is given in Figure 11 which makes use of a triptych. Inspection of the correlation diagram reveals that the "Di-π" approach to Diradical "D" (i.e. using triptych branch A) involves no opportunity for loss of excitation, and this diradical is reached with an S_1 configuration. Conversely, the approach to Diradical "D" from "Bicyclic" reactant (i.e. using triptych branch B) affords a HOMO-LUMO crossing and thus an opportunity for radiationless decay to S_0. The S_0 diradical "D" has allowed pathways leading to photoproducts "Di-π" and "Spiro". Finally, the excited state of "Spiro" can be seen to lead (adiabatically) to a doubly excited "D". Configuration Interaction leads us to a description in terms of states rather than MO's and this, too, has been carried out.[33] Where there are MO crossings, avoided crossings were obtained. Each such funnel affords an opportunity for radiationless decay to ground state. Thus whether one considers the problem in terms of MO's or states, the conclusion is the same. This treatment of the role of radiationless decay has been applied in a number of examples.[34] Additionally, related is the suggestion that a funnel may be canted and thus lead to a preference in the direction of motion on the ground state surface.[33,34]

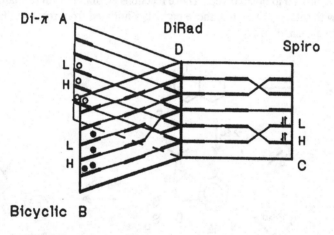

Figure 11: Molecular orbital correlation triptych.

There are further aspects of conical intersections. The latest version of Gaussian94 contains very pretty programming by Robb to locate conical intersections and avoided crossings;[35] and examples where this has been successful have been presented by Bernardi, Robb and Coworkers.[36] This programming has permitted the study of the nature of the cations and radicals of the meta and para benzylic system discussed below in context of the "meta" effect.

2.4   The Meta Effect

It was observed originally by Havinga[37] that in the dark, meta-nitrophenyl phosphate hydrolyzed in water exceptionally slowly while the para isomer was quite reactive in giving rise to p-nitrophenol and phosphate; this is totally expected behavior as the p-nitrophenolate anion is much more delocalized and of lower energy than the meta counterpart. However, in light, the meta isomer hydrolyzed while the para counterpart was no more reactive than in the dark. This was said to be inexplicable in terms of ordinary resonance reasoning. This prompted the present author to make use of the then available Hückel computations and to determine the generality of the phenomenon. First note Scheme XI which depicts the facile expulsion of the trityl cation in the photolysis of m-nitrophenyl trityl ether but the much lower efficiency when the nitro group is para. In this solvolysis reaction, the trityl cation picks up either water or methanol depending on the reaction solvent.[38] For the electron donor counterpart, an example studied both experimentally and theoretically[39] was the solvolysis of meta-methoxy-phenyl acetate, para-methoxyphenyl acetate, and 3,5-dimethoxyphenyl acetate; note Scheme XII. The observation was homolysis to give radical type products from the paraisomer, a mixture of ionic and radical products from the meta isomer, and only ionic solvolysis photoproducts from the 3,5-dimethoxybenzyl acetate. Since the early computations were of the Hückel variety, it seemed proper to determine how different more modern

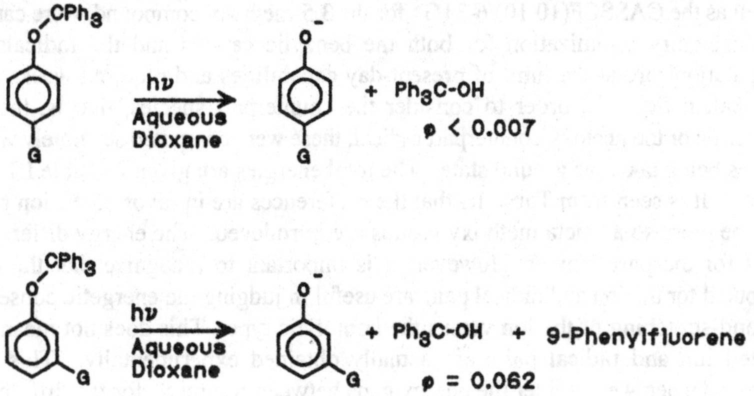

Scheme XI. The photochemical solvolytic behavior of the nitrophenyl trityl ethers. $G = NO_2$.

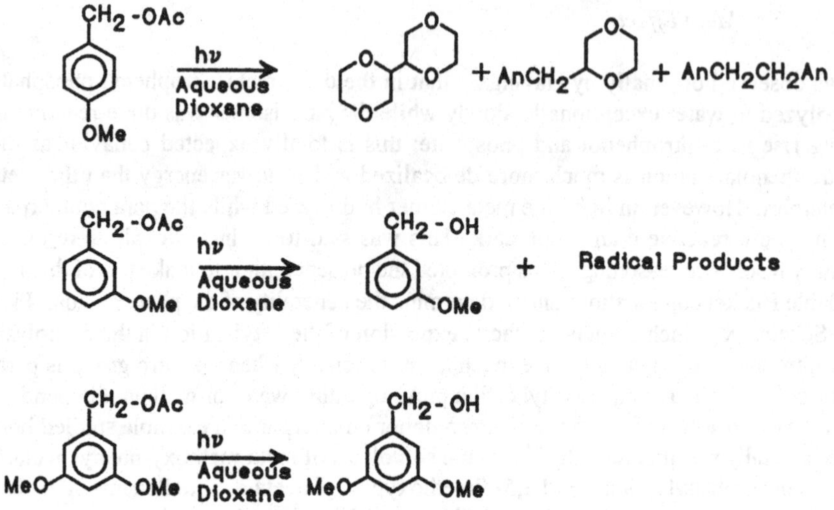

Scheme XII: The solvolytic behavior of the methoxybenzyl acetates.

computational technique would prove. Thus, Gaussian94 was employed to obtain the energies of the p-methoxylbenzyl, the m-methoxybenzyl and 3,5-dimethoxylbenzyl cations and radicals in their first excited states.[40] To this end, CASSCF(8,8)/6-31G*, as well as the CASSCF(10,10)/6-31G* for the 3,5-methoxy compound, were carried out with geometry optimization for both the benzylic cations and the radicals. Such computations are at the limit of present-day capabilities and required weeks for each such calculation. In order to consider the counterpart species, that is, the acetate counterion or the acetoxy counterpart radical, these were computed separately with these species being taken as ground state. The total energies are given in Table III.

It is seen from Table III that the preferences are in favor of the ion pairs and become more so as meta methoxy groups are introduced. The energy differences are small for the para isomer. However, it is important to recognize that the energies computed for the ion and radical pairs are useful in judging the energetic consequences of bond-stretching of the ion versus the homolytic type. This does not mean that the excited ion and radical pairs are actually obtained experimentally. This point is addressed when we consider the energy gaps between S_0 and S_1 for the different pairs. gaps are much larger, meaning that selectively for the meta ion pairs the excited surface has come closer to ground state.

Table III. CASSCF(8,8) energies of ion and radical pairs.

	Ion Pair	Radical Pair	Δ Rad vs Cation	Cation Δ	Radical Δ
Mono-meta	-610.14687	-610.11433	0.03254	0.01580	0.09161
Para	-610.12340	-610.11704	0.00635	0.05534	0.09438
Di-Methoxy	-724.05620	-724.00483	0.05137	0.00250	0.09395

Cation and radical pair energies are for S_1. The energy units are hartrees (627.5 kcal/mole per hartree). Column 3 gives the energy favoring ionic versus homolytic dissociation. Columns 4 and 5 give the S0-S1 energy gaps.

Further, this led to a rather exciting result. Gaussian94 contains methodology for computing conical intersections and avoided crossings. The case of the dimethoxybenzylic cation was studied using a CASSCF(8,8)/6-31G* conical computation which led to a geometry with a minimum energy separation of 0.15 kcal/mole. The separation for the mono methoxylated cation was slightly larger (circa 0.3 kcal/mole). These computations are of the "state averaging" type and lead to final energies for the S_0-S_1 combination which are about 1/100th of a Hartree higher in energy than the optimized S_1. This signifies that the geometry found for the conical intersection (or avoided crossing) is slightly different and displaced from the S_1 energy minimum.[41]

2.5   The Large K - Small K Concept Controlling Multiplicity Dependent Reactions

Often in organic photochemistry one encounters reactants which afford one product from the singlet excited state and a different one from the triplet. There is a generalization which we uncovered some years ago.[42] The basic idea is that the singlet-triplet splitting in a photochemical transition state may vary widely. For a single configuration involving singly occupied HOMO and LUMO MO's, the splitting is given by twice the exchange integral, G_{kllk} or K, between these two MO's [note Ref. 43, Chap. 5 for a discussion]. This splitting of 2K may be small or large. The two extremes are qualitatively depicted in Figure 12. Assuming that the center of gravity of the MO's does not differ appreciably, it can be seen that the T1 and S1 states will be positioned energetically similarly to that in Figure 12. It is also seen that if one can categorize two competing reactions as "small K" and "large K" reactions, then a triplet will find the lower energy hypersurface and path on the "large K" route. Conversely, a singlet will make use of the lower energy "small K" route. The problem is to categorize reactions in terms of the exchange integrals on the excited state hypersurfaces. For simplicity, if

24

one uses the zero differential overlap (i.e. ZDO) approximation, one can write the

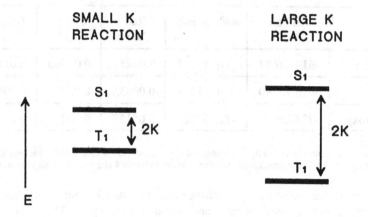

Figure 12: Large K - small K energetics.

exchange integral as the summation in Equation 5.

$$K = \sum_{r,s} [C_{rH} C_{rL} C_{sH} C_{sL} \gamma_{rs}] \tag{5}$$

or in matrix notation:

$$K = \omega \Gamma \omega$$

Here the ω vector and its transpose ω have columns and rows, respectively, consisting of the terms $C_{iL} C_{iH}$. L and H refer to HOMO and LUMO MO's and r and s refer to basis hybrid or atomic orbitals. The Γ matrix has the elements γ_{rs}, that is, the repulsion integrals between basis orbitals r and s. Thus it is the magnitude of such triple products occurring in the summation of Equation 5 and these, in turn, depend on the proximity of basis orbitals r and s as well as the LCAO MO coefficients at centers r and s. Where HOMO and LUMO are heavily weighted at the same sites, and thus HOMO and LUMO "match" the exchange integral tends to be large. For highly polarized species, HOMO and LUMO tend not to have heavy weightings at the same centers, and a small K results. Also, for 4N and photochemically allowed pericyclic reactions, HOMO and LUMO for symmetry reasons, there is poor matching and a small K results. Diradical processes of non-polar species usually have the relevant MO's similar distributed and large K values result.

A very simple example is that of butadienes which undergo an electro-cyclic ring closure to cyclobutenes from the singlet while the triplet excited state preferentially undergoes cis-trans isomerization about the π-bonds. A more complex case is the Di-π-Methane Rearrangement of the tetraphenyldiester shown in Scheme XIII. Here the usual cyclopropyldicarbinyl diradical is formed in the first step of the reaction. However, there are two alternative ways for this species to "unzip" and proceed onwards to photoproduct. The singlet, generated by "direct irradiation" (i.e. without addition of a sensitizer), leads to the product with carbomethoxy groups on the three-membered ring while the triplet, generated by use of a triplet sensitizer, affords the photoproduct with the ester groups on the double bond of product. The pathway leading onward to the singlet photoproduct retains the carbomethoxy groups at an odd-electron center as the three-ring opens. At this stage of the transformation, the diradicaloid species then is polarized and prefers a small K pathway from the singlet. However, in the case of the triplet, the alternative opening does not develop electron density on the electronegative ester group and prefers this large K pathway.

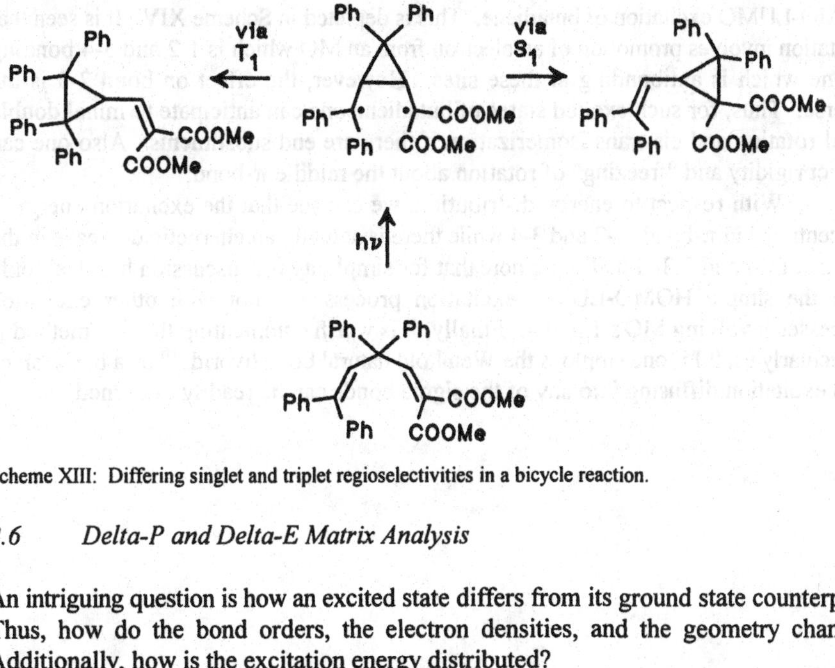

Scheme XIII: Differing singlet and triplet regioselectivities in a bicycle reaction.

2.6   Delta-P and Delta-E Matrix Analysis

An intriguing question is how an excited state differs from its ground state counterpart. Thus, how do the bond orders, the electron densities, and the geometry change? Additionally, how is the excitation energy distributed?

These questions have been addressed[44] with the definition of a ΔP matrix. This matrix is defined as the difference between an excited state bond order or density matrix and the corresponding ground state one. Note Equation 6.

$$\Delta P = P* - P_0 \qquad (6)$$

The basic idea is that on excitation most of the bond orders tend to decrease, although some do increase. These changes are seen in the off-diagonal elements of the ΔP matrix. The diagonal elements give the changes in electron density resulting from excitation.

Thus, in a molecule such as n-octylnaphthalene one expects excitation to be heavily localized in the naphthyl chromophore and only slightly delocalized into the octyl side-chain. Thus, the ΔP matrix has appreciably sized elements corresponding to the atoms and bonds of the naphthyl group and only slightly in the octyl side chain. The method is applicable to any computations which affords Mulliken population analysis matrices or a density matrices.

As noted, most off-diagonal elements of the ΔP matrix tend to be negative and these elements correspond to the distribution of excitation energy in making these bonds more antibonding. A simple case to consider in understanding the idea, is that HOMO-LUMO excitation of butadiene. This is depicted in Scheme XIV. It is seen that excitation involves promotion of an electron from an MO which is 1-2 and 3-4 bonding to one which is antibonding at these sites. However, the effect on bond 2-3 is the reverse. Thus, for such excited states of butadiene one can anticipate terminal double bond rotation and cis-trans isomerization if there are end substituents. Also one can expect rigidity and "freezing" of rotation about the middle π-bond.

With respect to energy distribution, we can see that the excitation energy is concentrated in π-bonds 1-2 and 3-4 while there is actually an energetic decrease in the energy of π-bond 2-3. Finally, we note that for simplicity the discussion has dealt only with the simple HOMO-LUMO excitation process and not with other excitation processes involving MO's 1 and 4. Finally, it is worth commenting that the method is particularly useful if one employs the Weinhold natural bond hybrids[16] as a basis, since then excitation diffusing into any of the sigma bonds can be readily discerned.

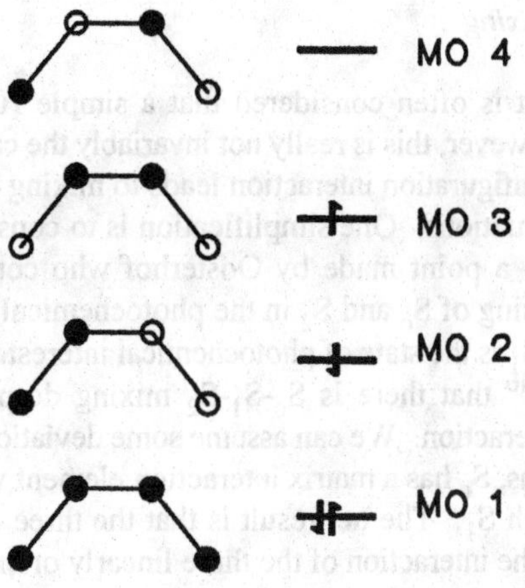

Scheme XIV: HOMO-LUMO butadiene excitation bond order effects.

Beyond this, if one is computing the hypersurface for a reaction, the ΔP treatment may be applied at the various points of interest on the hypersurface. For example, the Di-π-Methane Rearrangement of the diphenyldimethyl-1,4 pentadiene in Equation 7 provides one interesting case.[44c] As might be anticipated application of the ΔP matrix idea to the initial excited state, a singlet here, revealed that excitation is heavily localized in the diphenyl-vinyl moiety; this is the lowest energy chromophore in the molecule. However, as one moves along the reaction coordinate to the cyclopropyldicarbinyl species, the largest matrix elements correspond to bonds of the three-membered ring. In accord with this, inspection of the eigenvectors of the dicarbinyl diradical reveals typical benzhydryl radical type weightings, and it is the cyclopropyl dicarbinyl group which has been perturbed. Of course, it is unlikely that the molecule will proceed adiabatically to excited state of photoproduct, and along the hypersurface one expects decay to the S_0 surface. In any case, the migration of excitation during an excited state transformation is general.

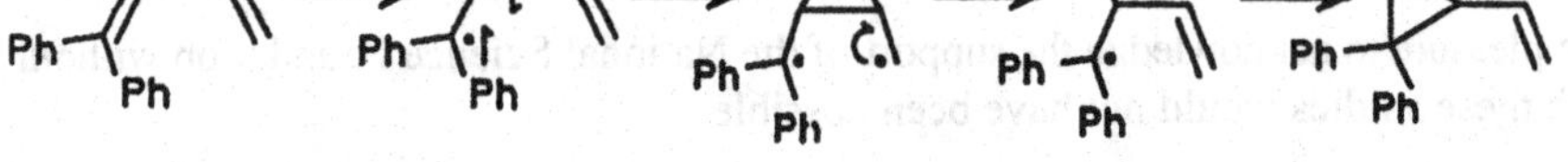

(7)

2.7   The Role of So-S1-S2 Mixing

In discussions of excited states it is often considered that a simple HOMO-LUMO excitation process is involved. However, this is really not invariably the case, and in any event is an oversimplification. Configuration interaction leads to mixing of a variety of excited configurations in most situations. One simplification is to consider just three configurations. This is related to a point made by Oosterhof who commented that, although there is an avoided crossing of S_0 and S_2 in the photochemical cyclization of butadiene to cyclobutene and yet S_1 is the state of photochemical interest.[45] It has been shown by the present author[44f,46] that there is S_0-S_1-S_2 mixing despite Brillouin's theorem which precludes S_0-S_1 interaction. We can assume some deviation from perfect symmetry (e.g. vibrationally). Thus, S_0 has a matrix interaction element with S_2, which is not precluded from mixing with S_1. The net result is that the three configurations interact in a fashion analogous to the interaction of the three linearly oriented p-orbitals of an allylic species (i.e. here S_0-S_2-S_1). However, in this case, unlike the interaction of the three atomic orbitals, the off-diagonal matrix elements are positive. Thus, the nodal character of the "linear" combination of the three configurations is inverted from that of the p-orbital case. The lowest energy resulting state has the weighting of S_0 - S_2 + S_1 and the highest energy state has all configurations weighted positively.

      In the same work it was found that the Möbius-Hückel approximation still operates correctly in the ZDO-SCF approximation, thus justifying our earlier efforts based on Hückel theory (i.e. topology).

Conclusion

This survey has selected a limited number of situations where a theoretical analysis has been of value in the author's ground-state and photochemical research. A general conclusion is that both experimental exploration of chemical phenomena and theoretical interpretation of the observations, together, provide a satisfying mode of doing mechanistic research.

Acknowledgment

It is a pleasure to acknowledge the support of the National Science Foundation without which these studies would not have been possible.

References

1.   A.J. Birch and D. Nasipuri, *Tetrahedron*, 148 (1959).
2.   (a)A.P. Krapcho and A.A. Bothner-By, *J. Am. Chem. Soc.* **81**, 3658 (1959).
3.   *Tetrahedron* **16**, 169 (1961).
4.   (a)H.E. Zimmerman and P.A. Wang, *J. Am. Chem. Soc.* **112**, 1280 (1990).
        (b)H.E. Zimmerman and P.A. Wang, *J. Am. Chem. Soc.* **115**, 2205 (1993).
5.   (a)Gaussian 92, Revision A. M.J. Frisch, G.W. Trucks, M. Head-Gordon, P.M.W. Gill, M.W. Wong, J.B. Foresman, B.G. Johnson, H.B. Schlegel, M.A. Robb, E.S. Replogle, R. Gomperts, J.L. Andres, K. Raghvachari, J.S. Binkley, C. Gonzalez, R.L. Martin, D.J. Fox, D.J. Defrees, J. Baker, J.J.P. Stewart, J.A. Pople, Gaussian, Inc., Pittsburgh PA, 1992.
        (b)Gaussian 90, Revision J, M.J. Frisch, M. Head-Gordon, G.W. Trucks, J.B. Foresman, H.B. Schlegel, K. Raghavachari, M. Robb, J.S. Binkley, C. Gonzalez, D.J. Defrees, D.J. Fox, R.A. Whiteside, R. Seeger, C.F. Melius, J. Baker, R.L. Martin, L.R. Kahn, J.J.P. Stewart, S. Topial, J.A. Pople, Gaussian Inc., Pittsburgh, PA, 1990.
        (c)QPCE Program No. QG01, Quantum Chemistry Program Exchange, Indiana University; M.W. Schmidt, K.K. Baldridge, J.A. Boatz, J.H. Jensen, S. Koseki, M.S. Gordon, K.A. Nguyen, T.L. Windus, S.T. Elbert, QPCE Bull., August 1990, p. 10.
        (d)Gaussian94, Revision. D.
6.   H.E. Zimmerman in *Molecular Rearrangements*, ed. P. DeMayo (Interscience, New York, 1963), pp. 347-349; ibid, idem, pp. 346-347.
7.   H.E. Zimmerman and A. Zweig, *J. Amer. Chem. Soc.* **83**, 1196 (1961).
8.   H.E. Zimmerman, *Quantum Mechanics for Organic Chemists* (Academic Press, New York, 1975).
9.   R.B. Woodward and R. Hoffmann, *J. Am. Chem. Soc.*, 395 (1965).
10.   H.C. Longuet-Higgins and E.W. Abrahamson, *J. Am. Chem. Soc.*, 2045 (1965).
11.   E. Heilbronner, *Tetrahedron Lett.*, 1923 (1964).
12.   H.E. Zimmerman, *J. Amer. Chem. Soc.* **88**, 1564 (1966).
13.   A.A. Frost and B. Musulin, *J. Chem. Phys.* **21**, 572 (1953).
14.   For a discussion and original references see, A. Streitwieser, *Molecular Orbital Theory For Organic Chemists* (Wiley, New York, 1961).
15.   H.E. Zimmerman and F. Weinhold, *J. Am. Chem. Soc.* **116**, 1579 (1994).
16.   (a)J.P. Foster and F. Weinhold, *J. Am. Chem. Soc.* **102**, 7211 (1980).
        (b)A. Reed, L.A. Curtiss and F Weinhold, *Chem. Rev.* **88**, 899 (1988).
17.   (a)S. Winstein, H.M. Walborsky and K. Schreiber, *J. Am. Chem. Soc.* **72**, 5795 (1950).

(b)S. Winstein and D.S. Trifan, *J. Am. Chem. Soc.* **74**, 1154 (1952).

18. (a)G.W. Wheland and D.W. Mann, *J. Chem. Phys.* **17**, 264 (1949).
     (b)G. Wheland, *J. Am. Chem. Soc.* **64**, 900 (1942).

19. H.E. Zimmerman, unpublished.

20. MOPAC Vers. 6.1, QCPE #455.

21. (a)H.E. Zimmerman, Seventeenth National Organic Symposium of the Amer. Chem. Soc., Bloomington, Indiana, (1961), pgs. 31-41.
     (b)H.E. Zimmerman and D.I. Schuster, *J. Amer. Chem. Soc.* **83**, 4486 (1961).
     (c)H.E. Zimmerman and D.I. Schuster, *J. Amer. Chem. Soc.* **84**, 4527 (1962)
     (d)H.E. Zimmerman, in *Advances in Photochemistry*, Eds. A. Noyes, Jr., G. S. Hammond and J. N. Pitts, Jr. (Interscience, Vol. 1, 1963), pp. 183-208.

22. M. Kasha, 1960 Radiation Res., Suppl. 2, 243; M. Kasha, in *Light and Life*, Eds. W.D. McElroy and B. Glass (Johns Hopkins Press, Baltimore, 1961).

23. (a)H.E. Zimmerman, D. Döpp and P.S. Huyffer, *J. Amer. Chem. Soc.* **88**, 5352 (1966).
     (b)H.E. Zimmerman, D.S. Crumrine, D. Döpp and P.S. Huyffer, *J. Amer. Chem. Soc.* **91**, 434 (1969).

24. H.E. Zimmerman, R.W. Binkley, J.J. McCullough and G.A. Zimmerman, *J. Amer. Chem. Soc.* **89**, 6589 (1967).

25. H.E. Zimmerman and G.L. Grunewald, *J. Amer. Chem. Soc.* **88**, 183 (1966).

26. (a)H.E. Zimmerman, R.W. Binkley, R.W. Givens and M.A. Sherwin, *J. Amer.Chem. Soc.* **89**, 3932 (1967).
     (b)H.E. Zimmerman, R.W. Binkley, R.S. Givens, M.A. Sherwin and G.L. Grunewald, *J. Amer. Chem. Soc.* **91**, 3316 (1969).
     (c)H.E. Zimmerman, H.M. Sulzbach and M.B. Tollefson, *J. Am. Chem. Soc.* **115**, 6548 (1993).

27. R. Hoffmann and W.N. Lipscomb, *J. Chem. Phys.* **36**, 3489 (1962).

28. H.E. Zimmerman, D. Armesto, M.G. Amezua, T.P. Gannett and R.P. Johnson, *J. Am. Chem. Soc.* **101**, 6367 (1979).

29. H.E. Zimmerman, *J. Amer. Chem. Soc.* **88**, 1566 (1966).

30. H.E. Zimmerman, *Accounts of Chem. Res.* **5**, 393 (1972).

31. (a)H.A. Jahn and E. Teller, *Proc. Roy. Soc. Ser. A* **161**, 220 (1937).
     (b)G. Herzberg and Longuet-Higgins, *Discuss. Faraday Soc.* **35**, 77 (1963).
     (c)E. Teller, *J. Phys. Chem.* **41**, 109 (1937).

32. J. Michl, *Mol. Photochem.* 1972, 243-255, 257-286, 287-314.

33. H.E. Zimmerman and R.E. Factor, *J. Amer. Chem. Soc.* **102**, 3538 (1980).

34. (a)H.E. Zimmerman, *Topics in Current Chemistry* **100**, 45 (1982) Springer-Verlag, Heidelberg-New York.
     (b)H.E. Zimmerman and R.E. Factor, *Tetrahedron* **37**, 125 (1981), Supplement 1.

(c)H.E. Zimmerman and T. P. Cutler, *J. Org. Chem.* **43**, 3283 (1978).

35.   See Reference 5. Revision D of Gaussian94 contains the conical code.

36.   (a)M.A. Robb, F. Bernardi and M. Olivucci, *Pure App. Chem.* **67**, 783 (1995).
(b)M. Olivucci, In. Ragazos, F. Bernardi and M.A. Robb, *J. Am. Chem. Soc.* **115**, 3710 (1993).

37.   E. Havinga, R.O. De Jong and W. Dorst, *Rec. Trav. Chim.* **75**, 378 (1956).

38.   H.E. Zimmerman and S. Somasekhara, *J. Amer. Chem. Soc.* **85**, 922 (1963).

39.   H.E. Zimmerman and V.R. Sandel, *J. Amer. Chem. Soc.* **85**, 915 (1963).

40.   H.E. Zimmerman, *J. Am. Chem. Soc.* **117**, 8988 (1995).

41.   H.E. Zimmerman, unpublished results.

42   (a)H.E. Zimmerman, J.H. Penn and M.R. Johnson, *Proc. Natl. Acad. Sci. USA*, **78**, 2021 (1981).
(b)H.E. Zimmerman, D. Armesto, M.G. Amezua,T.P. Gannett and R.P. Johnson, *J. Amer. Chem. Soc.* **101**, 6367 (1979).
(c)H.E. Zimmerman and R.E. Factor, Ref. 34b.
(d)H.E. Zimmerman, J.H. Penn and M.R. Johnson, *Proc. Natl. Acad. Sci. USA*, **78**, 2021 (1981).
(e)H.E. Zimmerman and G.-S. Wu, *Canadian J. Chem.* **61**, 866 (1983).

43.   H.E. Zimmerman, *Quantum Mechanics for Organic Chemists* (Academic Press, New York, 1975).

44.   (a)H.E. Zimmerman, W.T. Gruenbaum, R.T. Klun, M.G. Steinmetz and T.R. Welter, *J.C.S. Chemical Communications* (1978) 228-230.
(b)H.E. Zimmerman and T.R. Welter, *J. Amer. Chem. Soc.* **100**, 4131 (1978).
(c)H.E. Zimmerman and R.T. Klun, *Tetrahedron* **43**, 1775 (1978).
(d)H.E. Zimmerman and M.G. Steinmetz, *J.C.S. Chemical Communications*, (1978), 231-232.
(e)See References 33c and 42c.
(f)H.E. Zimmerman, *Accts. of Chem. Research*, **10**, 312 (1982).
(g)H.E. Zimmerman, J.H. Penn and C.W. Carpenter, *Proc. Natl. Acad. Sci. USA*, **79**, 2128 (1982).

45.   W.Th.A.M. van der Lugt and L Oosterhoff, *J. Am. Chem. Soc.* **91**, 6042 (1969).

46.   H.E. Zimmerman, *Tetrahedron* **38**, 753 (1982).

AB INITIO STUDIES OF ELIMINATION
REACTION MECHANISMS

SCOTT GRONERT

Department Of Chemistry and Biochemistry, San Francisco State University,
San Francisco, CA 94132, USA.

Abstract: In an effort to build a better understanding of the mechanisms of β-elimination reactions, *ab initio* calculations have been used to investigate several gas-phase systems. Many of the studies focus on the regio- and stereoselectivity of the process, but competition with substitution and 1,4 elimination also has been evaluated. The results are compared to relevant gas-phase experiments and provide a foundation for interpreting the diverse observations from condensed phase studies.

1. Introduction

Base-induced, β–eliminations (Eq. 1) have fascinated organic chemists for the last half century and have played an important role in the development of modern physical organic chemistry.[1-8] Although the overall elimination process is simply loss of HX, the mechanism is complex and involves four distinct bonding changes: (1) formation of a new Y-H bond, (2) cleavage of the C_β-H_β bond, (3) formation of a C-C π-bond, and (4) cleavage of the C_α-X bond. These bonding changes can be concerted (E2 mechanism) or step-wise (E1 mechanisms) and can occur synchronously or non-synchronously.

$$Y^- + \underset{\beta}{\overset{H}{\underset{|}{C}}}-\underset{\alpha}{\overset{|}{\underset{|}{C}}}-X \longrightarrow YH + \overset{\diagup}{\underset{\diagup}{C}}=\overset{\diagdown}{\underset{\diagdown}{C}} + X^- \qquad (1)$$

Most of the data on eliminations has been derived from experimental studies in solution. Pioneering work in many laboratories has led to a basic understanding of the mechanism as well as the development of concepts such as concertedness and synchronicity. In addition, the regio- and stereoselectivity of the process has been explored with a variety of substrate, base, and solvent combinations. To some extent, the interpretation of experimental data has been limited by the confounding effects of solvation and ion pairing. It is often difficult to ascertain the specific role that the solvent or counterion plays in determining the selectivity of an elimination reaction so broad generalizations have been made in many cases. As a result, solvation and ion-pairing add two, poorly-defined variables to an already complex reaction system.

An obvious way to avoid the complications of solvation and ion-pairing is to study elimination reactions in the gas phase. Beyond simplifying the problem, there is a fundamental logic to this approach. When organic chemists write a mechanism, they generally do not include the solvent because there simply is not enough information to do so in a meaningful way (*i.e.*, the orientation of a complete solvent shell). In some cases, a solvent molecule is included as a proton donor or acceptor, but like Eq. 1, most mechanisms are written for the gas phase. Solvation and ion-pairing effects are viewed as a perturbation to this solvent-free mechanism. Therefore, organic chemists naturally think in terms of gas-phase mechanisms despite the fact that most experimental data comes from condensed-phase studies. As a result, there is intrinsic merit in studying reactions in the gas phase. Moreover, gas-phase studies are useful in the interpretation of data from solution because differences in reactivity are clear evidence of solvation and ion pairing effects. Therefore, it is the combination of condensed and gas-phase studies that leads to the most comprehensive understanding of a mechanism.

In the last twenty years, two important technological advances have allowed chemists to investigate the gas-phase mechanisms of base-induced elimination reactions. First, developments in mass spectrometry have opened the door to gas-phase studies of organic reactions. This work is leading to important new insights into elimination mechanisms, and already it is clear that effects that had been attributed to substrate reactivity may instead be artifacts of solvation phenomena. Although mass spectrometry is a powerful tool, it has several limitations including the inability to identify the neutral products of ionic reactions.

The second important development has been modern computational chemistry. With faster computers and more efficient algorithms, it is now possible to study realistic organic reactions with reasonable accuracy. Computational studies are particularly well suited for the study of eliminations because so many of the key questions focus on the structure of the transition state. Unlike experimental studies which are generally restricted to rate and product determinations, *ab initio* calculations can directly probe the transition state. As a result, geometric properties such as conformation, concertedness and synchronicity can be addressed in a straightforward manner. With this in mind, it is not surprising that computational chemistry has had an important impact on our understanding of β–eliminations.

This chapter focuses on recent computational studies of elimination reactions and is limited to gas-phase systems involving anionic bases. The following organization is used. First, important mechanistic aspects will be outlined. In this section, the results of classic experimental studies will be included to illustrate key points and provide a foundation for the computational studies. Next, important gas-phase experimental studies will be reviewed. Finally, the results of our computational studies will be presented and their importance in understanding elimination mechanisms will be discussed. Some of these studies have yet to be published.

2. Mechanistic Aspects

2.1 Concertedness and Synchronicity

Because base-induced β–eliminations involve four distinct bonding changes, a wide range of transition states is possible depending on the timing of events. For the sake of convenience, the four bonding changes are often combined together to give two dominate processes, proton transfer and leaving group expulsion. At the two ends of the transition state spectrum for β–eliminations are non-concerted mechanisms where the proton transfer and leaving group expulsion processes occur as separate steps and a stable intermediate is formed.   If leaving group expulsion occurs first (Eq. 2), a carbocation intermediate is formed and the mechanism is considered to be E1(elimination, unimolecular). This mechanism does not play a role in the gas phase eliminations of neutral substrates because the initial dissociation is very unfavorable in the absence of solvation or ion pairing (dissociation energies are generally > 300 kcal/mol).   If proton transfer occurs first (Eq. 3), a carbanion intermediate is formed and the mechanism is designated as E1cb (where cb indicates that the conjugate base of the substrate is an intermediate).

$$-\underset{|}{\overset{|}{C}}-\underset{|}{\overset{H}{C}}-X \longrightarrow X^- + -\underset{|}{\overset{H}{C}}-\overset{+}{C} \overset{Y^-}{\longrightarrow} \underset{/}{\overset{\backslash}{C}}=\underset{\backslash}{\overset{/}{C}} + YH \qquad (2)$$

$$-\underset{|}{\overset{H}{C}}-\underset{|}{\overset{|}{C}}-X \overset{Y^-}{\longrightarrow} YH + -\underset{|}{\overset{|}{C}}-\underset{|}{\overset{|}{C}}-X \longrightarrow \underset{/}{\overset{\backslash}{C}}=\underset{\backslash}{\overset{/}{C}} + X^- \qquad (3)$$

Between these extremes are the concerted, E2 (elimination, bimolecular) mechanisms.   Although E2 reactions are concerted by definition, they are not necessarily synchronous and therefore the transition states can vary in terms of the relative extent of proton transfer vs. leaving group expulsion.   Transition states where leaving group expulsion dominates are referred to as E1-like and those where

36

proton transfer dominates are referred to as E1cb–like.[9] At the center of the spectrum is the perfectly synchronous E2 transition state. In gas-phase studies, one does not expect E1-like transition states to play an important role because partial cleavage of the C-X bond increases the extent of charge separation in the system (the α-carbon has a partial positive charge and the X group has a partial negative charge). Of course, charge separation has a high energetic cost in a medium with a low dielectric (gas-phase). On the other hand, E1cb-like mechanisms are likely because proton transfer simply shifts the charge from the base to the β–carbon and the transition state does not suffer the same electrostatic consequences.

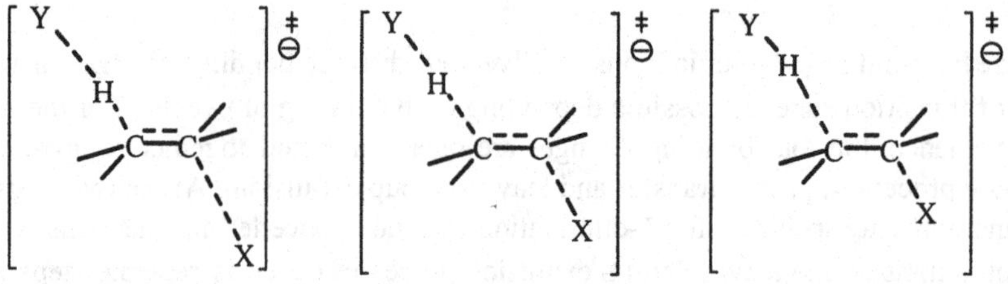

E1-like   synchronous E2   E1cb-like

2.2 Regioselectivity

In β–eliminations, there is often the opportunity to form more than one alkene product because the substrate has more than one type of β–carbon. For example, in a simple substrate such as a 2-halo-2-methylbutane, there is the possibility of two products:

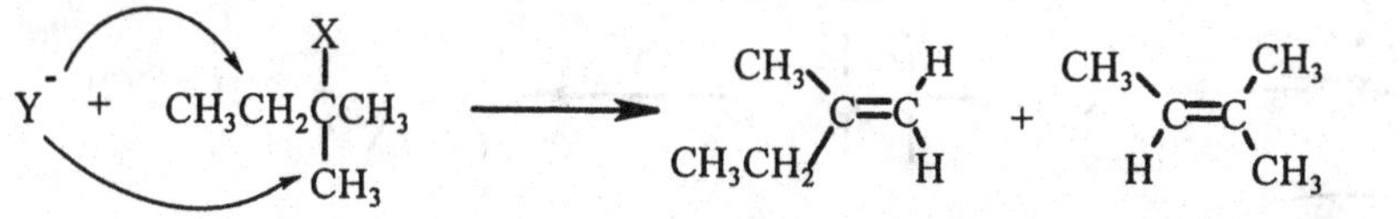

$$Y^- + CH_3CH_2\overset{X}{\underset{CH_3}{\overset{|}{C}}}CH_3 \longrightarrow$$

2-methyl-1-butene   2-methyl-2-butene

(4)

Attack at carbon 1 gives 2-methyl-1-butene and attack at carbon 3 gives 2-methyl-2-butene. The regioselectivity depends on a number of factors including the stability of the product alkene (Saytzev Rule)[10] as well as the acidity (accessibility) of the β–hydrogen (Hofmann Rule)[11]. In addition, the reaction conditions (solvent, base, etc.) have a significant effect. Even limiting the discussion to anionic leaving groups, the results from condensed phase work present a complex picture involving a subtle interaction of a number of factors. For example, the reaction of the 2-halo-2-methylbutanes with ethoxide in ethanol can give predominantly the 1-butene or the 2-butene product depending on the nature of the leaving group (Table 1).[12]

Table 1. Product Ratios from the Elimination Reactions of 2-Halo-2-Methylbutanes with Ethoxide in Ethanol.[12]

Halogen	Ratio of 2-Butene/1-Butene
Fluoride	0.4
Chloride	1.2
Bromide	1.6

There is a clear trend in these results with regards to leaving group ability. With a poor leaving group, one expects more E1cb-character and consequently proton transfer dominates the transition state. As a result, the reaction favors removal of the most acidic or accessible proton (carbon 1 => 1-butene). With better leaving groups, there is more π-bonding in the transition state and the more stable alkene (2-butene) is formed in the greatest yield. It should be noted that others have argued for an alternative explanation based on the size of the leaving group.[13]

The regioselectivity also depends on the nature of the base and solvent. For example, Brown and co-workers have investigated the eliminations of 2-bromo-2-methylbutane in a variety of alkoxide/alcohol solvent systems (Table 2).[14] As the size of the base increases, the amount of the 1-butene product also increases. This relationship points to the accessibility of the proton playing an important role in determining the regioselectivity because formation of the 2-butene product requires removal of the more hindered proton (2°) and is disfavored as the base becomes more sterically demanding. Across this series, base strength increases with more extensive alkyl substitution and this factor should also increase the yield of the 1-butene.

Table 2. Product Ratios from the Elimination Reactions of 2-Bromo-2-Methylbutane in RO⁻/ROH.[14]

RO⁻	Ratio of 2-Butene/1-Butene
$CH_3CH_2O^-$	2.3
$(CH_3)_3CO^-$	0.4
$(CH_3CH_2)_3CO^-$	0.1

For example, Froemsdorf and Robbins have completed studies in DMSO that show that base strength has a strong influence on regioselectivity (Table 3).[15] Using *sec*-butyl tosylate as a substrate, they found that the strongest base, *t*-BuO⁻, gave the greatest amount of 1-butene and that the weaker bases, phenoxides, favored the 2-butenes (E + Z). There may be some size dependence built into these results, but the decrease in 1-butene yield caused by p-nitro substitution on the phenoxide clearly points to a basicity effect. Of course as the strength of the base increases, the extent of proton transfer in the transition state also should increase and a more E1cb-like mechanism should result. Consequently, the less substituted product will dominate (see above).

Table 3. Product Ratios from the Elimination Reactions of *sec*-Butyl Tosylate in DMSO.[15]

Base	Ratio of 2-Butenes/1-Butene
$(CH_3)_3CO^-$	0.6
$CH_3CH_2O^-$	0.8
$C_6H_5O^-$	2.2
$p\text{-}NO_2\text{-}C_6H_5O^-$	5.2

2.3 Stereoselectivity

Stereochemistry plays a role in elimination reactions at two levels. The most obvious stereochemical aspect of eliminations is the formation of E and Z isomers. For example, the β–elimination of 2-halobutanes leads to a mixture of E-2-butene and Z-2-butene as well as 1-butene. With respect to the E and Z stereoisomers, generally the most stable alkene product dominates; however, there are many exceptions and it has been difficult to explain all the results. For example, using *t*-BuO⁻/*t*-BuOH as a reaction medium, *sec*-butyl tosylates give mainly Z-2-butene, but using *t*-BuO⁻/DMSO, E-2-butene is favored.[15] The addition of a crown ether to the *t*-BuO⁻/*t*-BuOH system leads to a reverse in the observed selectivity— the E-product is favored.[16] These results indicate that in the absence of strong ion pairing (DMSO or crown ether system), there is a preference for the more stable, trans product, but that solvation effects are significant and can actually reverse the selectivity in a less polar medium (*t*-BuOH).

A more subtle level of stereochemical control in base-induced eliminations centers on the relative orientation of the departing groups. It has been assumed that eliminations would adopt transitions states with periplanar conformations in order to maximize the π-overlap between the α and β carbons (see Section 2.7); however, periplanarity can be attained with either an anti or syn orientation of the departing groups.

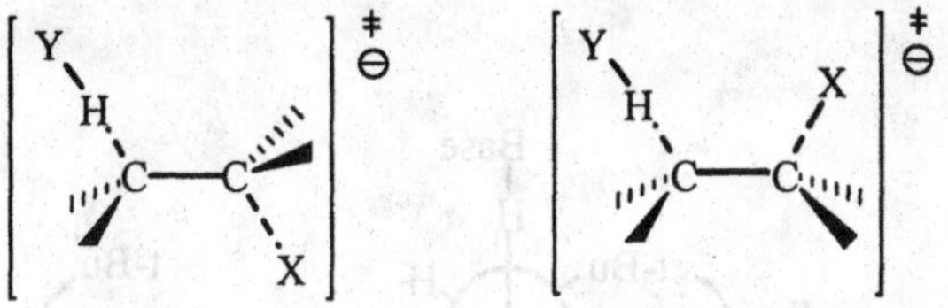

antiperiplanar   synperiplanar

Under most circumstances, it is the antiperiplanar arrangement which is preferred. For example, in the alkoxide-induced eliminations of diastereomeric 2-halo-1,2-diphenylpropanes, the products are consistent with a preference for anti elimination (Scheme 1).[17]

Scheme 1

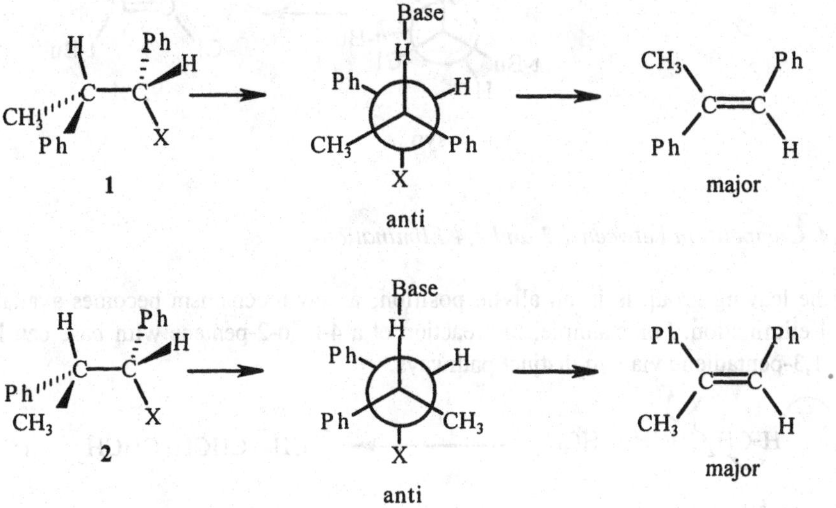

40

The preferential formation of the less stable, Z-1,2-diphenylpropene from substrate **2** suggests a strong preference for anti elimination. Syn elimination has been observed, but generally is restricted to cases where a an antiperiplanar transition state is precluded by steric strain or a rigid ring structure. Schlosser (Scheme 2) has shown that **3** gives predominately syn elimination presumably because of severe steric strain in the antiperiplanar transition state.[18]

Scheme 2

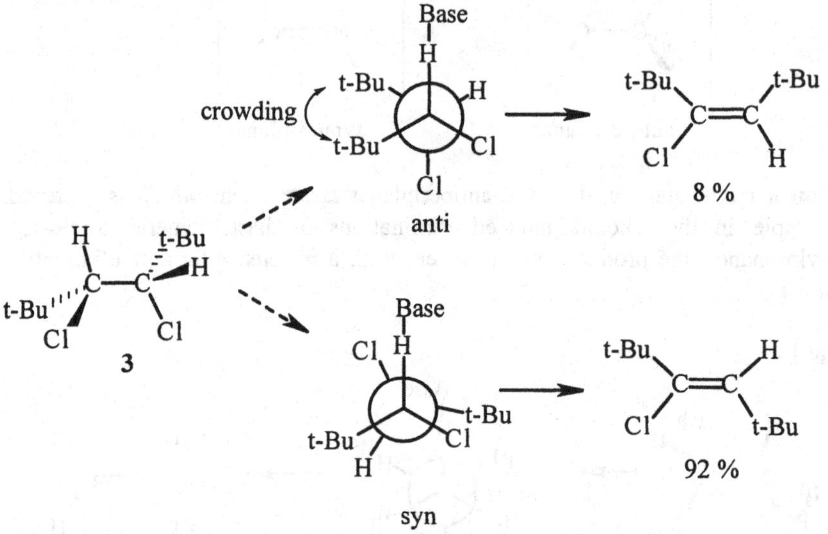

2.4 Competition between 1,2 and 1,4 Eliminations

If the leaving group is in an allylic position, a new mechanism becomes available, 1,4 elimination. For example, the reaction of a 4-halo-2-pentene with base can lead to 1,3-pentadiene via two distinct pathways.

$$Y^- \quad H\text{-}CH_2CHCH\text{=}CHCH_3 \quad \longrightarrow \quad CH_2\text{=}CHCH\text{=}CHCH_3 \quad (5)$$

$$CH_3CHCH\text{=}CHCH_2\text{-}H \quad Y^- \quad \longrightarrow \quad CH_3CH\text{=}CHCH\text{=}CH_2 \quad (6)$$

Attack at carbon 5 leads to a conventional β–elimination (Eq. 5) whereas attack at carbon 1 gives the 1,4 elimination (Eq. 6). In this case the products are indistinguishable, but with less symmetric substrates, the mechanisms lead to different products. Examples are available which demonstrate pure 1,4 or pure 1,2 elimination. Rickborn[19] has shown that the reaction of an allylic epoxide with $LiN(CH_2CH_3)_2$ proceeds exclusively *via* 1,4-elimination (Eq. 7). In this case, the 1,4-elimination has an advantage because the proton that is removed is especially acidic (allylic). Schlosser[20] gives an example where both the 1,2- and 1,4-elimination channels involve activated hydrogens (Eq. 8). Here, the reaction gives exclusively the 1,2-elimination product. These and other condensed phase results indicate that allylic activation is important and that 1,4 elimination will dominate unless the 1,2 pathway involves an activated proton. When both pathways involve activated protons, either may be preferred.

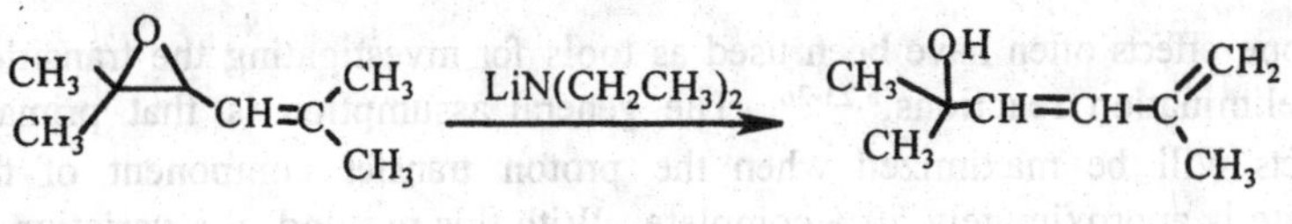

$$
\text{(7)}
$$

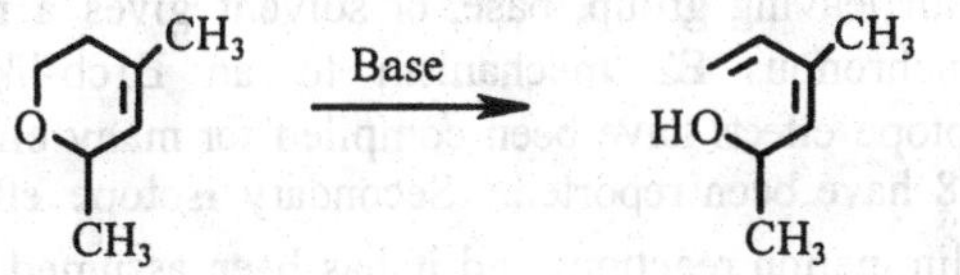

$$
\text{(8)}
$$

2.5 Competition Between Elimination and Substitution

Fundamentally, elimination and substitution reactions share the same set of reactants: a nucleophile (base) and a substrate with a good leaving group.

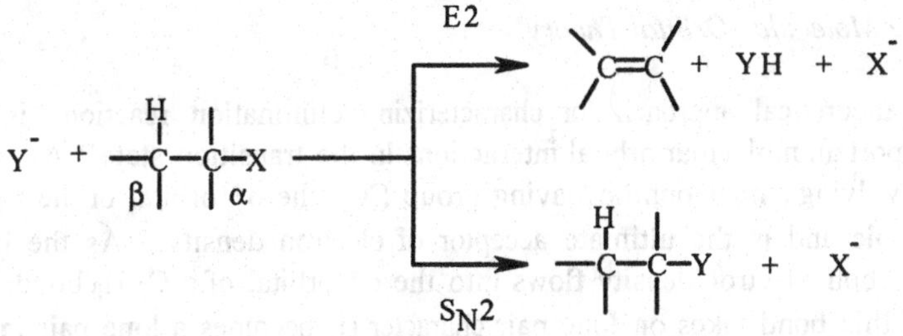

$$
\text{E2} \qquad \text{C=C} + YH + X^- \qquad \text{(9)}
$$

$$
Y^- + \underset{\beta}{\text{C}}\text{H}-\underset{\alpha}{\text{C}}\text{X}
$$

$$
S_N2 \qquad \text{C}-\text{C}-Y + X^- \qquad \text{(10)}
$$

As a result, elimination and substitution often compete, and depending on whether the base attacks the β or α carbon, alkenes (Eq. 9) or substituted alkanes (Eq. 10) can be formed. The competition between elimination and substitution is an important consideration in synthetic chemistry and has been the subject of numerous mechanistic studies. There is far too much data to adequately survey in this section and much of it goes beyond the scope of this chapter. Only a few generalizations will be stated. First, eliminations are favored by strong, localized bases such as alkoxides. Second, S_N2 substitution is retarded by steric hindrance at either the α or β carbon. Finally, bulkiness in the base (nucleophile) tends to favor elimination over substitution. More details on this subject can be found in authoritative sources.[21,22]

2.6 Isotope Effects in β–Eliminations

Kinetic isotope effects often have been used as tools for investigating the transition states of β–elimination reactions.[5,23-26] The general assumption is that primary isotope effects will be maximized when the proton transfer component of the transition state is approximately 50% complete. With this in mind, the variation in primary isotope effect with leaving group, base, or solvent gives a measure of the progression from a synchronous E2 mechanism to an E1cb-like or E1-like mechanism. Primary isotope effects have been compiled for many eliminations and k_H/k_D values as high as 8 have been reported. Secondary isotope effects have also been used in the study elimination reactions and it has been assumed that the larger the effect, the greater the extent of bond cleavage. Since deuterium can be substituted at either the α or β carbon, this technique can be used to assess the extent of proton transfer as well as the extent of leaving group cleavage. Finally, heavy atom isotope effects have also been investigated for β–eliminations. In gas-phase studies, some primary deuterium isotope effects have been reported (see Section 3.4).

2.7 Qualitative Molecular Orbital Theory

The simplest theoretical approach for characterizing elimination reactions is to identify the important molecular orbital interactions in the transition state. As with any reaction involving a nucleophilic leaving group (X^-), the σ* orbital of the C_α-X bond plays a role and is the ultimate acceptor of electron density. As the base attacks the β–carbon, electron density flows into the σ* orbital of a C_β-H_β bond and the σ orbital of this bond takes on lone pair character (it becomes a lone pair in an

E1cb mechanism). This electron flow also leads to two new bonding interactions: (1) the σ orbital of the C_β-H_β bond combines with the σ* orbital of the C_α-X bond to give a π-bond; (2) the interaction of the base (Y⁻) with the σ* orbital of a C_β-H_β bond leads to a new Y-H bond. An outline of important bonding interactions is given in Scheme 3. The Y-H σ and C_β-H_β σ* orbitals are omitted for clarity.

Scheme 3

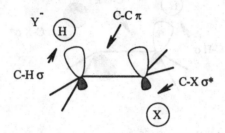

The σ* orbital of the C_α-X bond has its major lobe on the back-side of the C-X bond. X is more electronegative than carbon so the σ orbital of the C-X bond is polarized towards X and orbital orthogonality requirements force the opposite to be true for the σ* orbital. The major lobe of C_β-H_β σ orbital naturally is directed along the bond axis. As we will see later in this section, the polarization of these orbitals has an important effect on the stereochemistry of elimination reactions. It should be noted that a valence bond analysis of eliminations has been presented by Pross and Shaik.[27]

The orbital interactions in Scheme 3 are strongest when there is a periplanar arrangement for the H_β-C_β-C_α-X framework and π-overlap between the α and β carbons is maximized. Of course an orthogonal arrangement (H_β-C_β-C_α-X dihedral angle = 90°) would preclude π-bonding in the transition state and limit the electron flow between the α and β carbons. The stabilizing effect of partial π-bonding in the transition state explains the preference for periplanarity in β–eliminations.

This picture has been used by Bach[28] to rationalize the anti preference in E2 eliminations. To attain a syn transition state, one must rotate the system 180° around the C-C bond. In the resulting orbital picture (Scheme 4), the π-overlap between the α and β carbons is less satisfactory because there is no direct interaction between the major lobes of the carbon π-orbitals. As a result, π-bonding is less effective in the synperiplanar orientation and consequently, the transition state is less stable. The diminished π-overlap also makes syn eliminations more susceptible to asynchronous mechanisms because the electron flow between the α and β carbons is limited and therefore the coupling of the proton transfer and leaving group expulsion processes is reduced.

Scheme 4

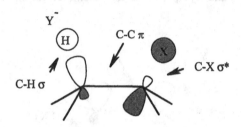

2.8 Gas-phase Potential Energy Surfaces

In gas-phase reactions involving a charged reactant, a long range, ion-dipole attraction plays an important role in determining the nature of the potential energy surface. As an anion approaches a neutral substrate, the energy of the system drops as a result of this ion-dipole attraction and a loosely bound complex is formed. Ion-dipole complexes are generally 5 to 20 kcal/mol more stable than the separated reactants,[29] but when strong hydrogen bonding is possible, complexation energies can exceed 40 kcal/mol. From the complex, the energy of the system rises to a transition state as the reaction occurs (*i.e.*, elimination) and then drops as a complex is formed between the products. The product complex is also bound by ion-dipole forces. Finally, the energy rises as this complex breaks down to give the observed products. The net result is a double-well potential (Figure 1). If the reaction barrier is smaller than the initial complexation energy, the transition state is more stable than the separated reactants and a negative activation barrier is observed (Figure 1a). This is common for ion-molecule reactions because complexation energies are quite large. In contrast, if the barrier is larger than the complexation energy, the transition state is less stable than the reactants and a positive activation energy results (Figure 1b). In reactions with more than one barrier, more complex surfaces are observed (*e.g.*, triple-wells). Typical experimental apparatus do not allow for long reaction times or high reagent concentrations; therefore, only fast gas-phase reactions may be studied. Generally speaking, mass spectrometric studies are limited to systems with activation energies less than 5 kcal/mol. In fact, most gas-phase studies have focused on systems with negative activation energies. It should be noted that despite the low activation energies, high selectivities are often seen in gas-phase reactions (see Section 3).

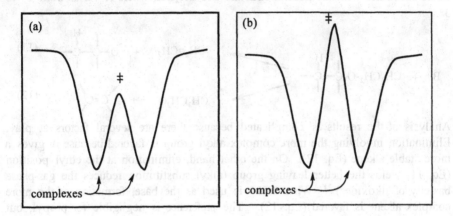

Figure 1. Typical potential energy surfaces for gas-phase ion-molecule reactions.

3. Gas-phase Experimental Studies

The development of the flowing afterglow (FA), ion cyclotron resonance spectrometers (ICR), and high-pressure mass spectrometers (HPMS) has allowed for detailed studies of organic reactions in the gas phase. In a few cases, these techniques have been applied to β–eliminations and the results provide important insights into the intrinsic mechanism (solvent-free). In the sections that follow, key results from these studies are outlined.

3.1 Regioselectivity

Although gas-phase experimental studies have made important contributions to our understanding of elimination reactions, they have had limited success in investigating the regiochemistry of the process. Mass spectrometry can only detect ionic products so the alkene cannot be identified directly. As a result, regiochemistry can only be determined in special cases where deuterium labeling or secondary reactions identify the alkene product. There are no reports for simple alkyl halides, but DePuy has presented some relevant data on the eliminations of unsymmetric dialkyl ethers.[30] Using strong bases, eliminations were observed for a series substrates with the general structure CH_3CH_2OR, where R varies from 1° to 3°.

$$B^- + CH_3CH_2\text{-}O\text{-}\overset{\overset{\displaystyle H}{|}}{\underset{|}{C}}\text{-}\overset{|}{\underset{|}{C}}\text{---}$$

$$CH_2=CH_2 \ + \ ^-O\text{-}\overset{\overset{\displaystyle H}{|}}{\underset{|}{C}}\text{-}\overset{|}{\underset{|}{C}}\text{---} \qquad (11)$$

$$CH_3CH_2O^- \ + \ \overset{/}{\underset{/}{C}}{=}\overset{\backslash}{\underset{\backslash}{C}} \qquad (12)$$

Analysis of the results is complicated because there are several factors at play. Elimination involving the more complex alkyl group is favored because it gives a more stable alkene (Eq. 12). On the other hand, elimination at the ethyl position (Eq. 11) yields the better leaving group (alkyl substitution reduces the gas-phase basicity of alkoxides).[31] When NH_2^- is used as the base, formation of the more complex alkene is favored (Eq. 12). The preference is negligible for propyl, but significant for t-butyl. Apparently leaving group ability has only a modest effect on the selectivity and an E1cb-like transition state is likely. With HO^- a similar trend is observed, but elimination at the ethyl group is more favorable. Overall, these results point to selectivity in gas-phase eliminations, but they do not give a clear picture of the transition state.

Table 4. Selectivity in the Gas-phase Eliminations of Unsymmetric Dialkyl Ethers.[30]

	% Ethene Formed (Eq. 11)	
Substrate	NH_2^-	HO^-
n-Pr-O-Et	45	80
i-Bu-O-Et	21	86
s-Bu-O-Et	17	73
t-Bu-O-Et	3	3

3.2 Stereoselectivity

For the reasons mentioned in the previous section, it is difficult to determine the stereochemistry of alkenes formed in gas-phase eliminations. However Kass and Rabasco have used careful deuterium labeling studies to distinguish between syn and anti pathways in the eliminations of substituted 3-methoxycyclohexenes.[32] For substrates such as **4**, reactions with base give mainly two ionic products that are the result of secondary reactions, the deprotonated diene (Eq. 13) and the leaving group complexed to the protonated base (Eq. 14). If carbon 4 is stereoselectively labeled with deuterium, both types of products identify the stereochemistry of the mechanism (anti or syn). After correcting for deuterium isotope effects, anti/syn ratios can be determined.

$$
\begin{array}{c} \text{t-Bu} \\ \text{D} \quad\quad \text{H(D)} \\ \text{[ring structure]} \quad + \text{CH}_3\text{OH} \quad (13) \\ \text{CH}_3 \quad \text{CH}_3 \end{array}
$$

$$
\begin{array}{c} \text{t-Bu} \quad \text{OCH}_3 \\ \text{D} \quad\quad \text{D} \\ \text{[ring structure]} \\ \text{CH}_3 \quad \text{CH}_3 \\ \textbf{4} \end{array}
\xrightarrow{\text{B}^-}
\begin{array}{c} \text{t-Bu} \\ \text{D} \quad \text{H(D)} \\ \text{[ring structure]} + \text{CH}_3\text{O}^- + \text{BH(D)} \\ \text{CH}_3 \quad \text{CH}_3 \end{array}
$$

$$\left[\text{CH}_3\text{O}^- \quad \text{H(D)B} \right] \quad (14)$$

The results in Table 5 indicate that gas-phase β–eliminations have a preference for anti elimination. The preference is surprisingly small and seemingly insensitive to base strength.

Table 5. Stereoselectivity Derived from the Gas-phase, Base-Induced Eliminations of **4** and Related Substrates.[32]

Base	P A	% Anti
(CH$_3$)$_2$N$^-$	396	57
HO$^-$	391	75
CH$_3$O$^-$	381	57
F$^-$	371	55

PA = gas-phase proton affinity of the base in kcal/mol.[31]

Rabasco[33] points out that in this sterically congested system, an antiperiplanar orientation may be difficult to achieve and therefore anti elimination is less favorable. Rabasco and Kass have also looked at systems where the β–carbon is activated (allylic).[32] In these systems, anti elimination is favored with the exception of reactions with very weak bases where formation of the complexed product, [CH$_3$O$^-$ HB], is the only exothermic channel. Under these conditions, an interaction between the leaving group and the protonated base maybe a requirement for a successful elimination and therefore syn stereochemistry is preferred. However, it should be noted that contrary to earlier assumptions,[34,35] complexed products, [CH$_3$O$^-$ HB], can be formed in either anti or syn eliminations.

3.3 1,2 vs. 1,4 Elimination

Rabasco and Kass have also used their 3-methoxycyclohexene system to investigate the competition between 1,2 and 1,4 elimination.[36,37] Using substrates such as **5**, they found that 1,4 elimination was heavily favored when strong bases such as HO$^-$ and NH$_2$$^-$ are used, but that 1,2 elimination is more competitive with weaker gas-phase bases such as fluoride. Nibbering and co-workers have studied the reactions of strong bases with 2-butenyl ethyl ethers and have observed almost exclusive 1,4 elimination.[38]

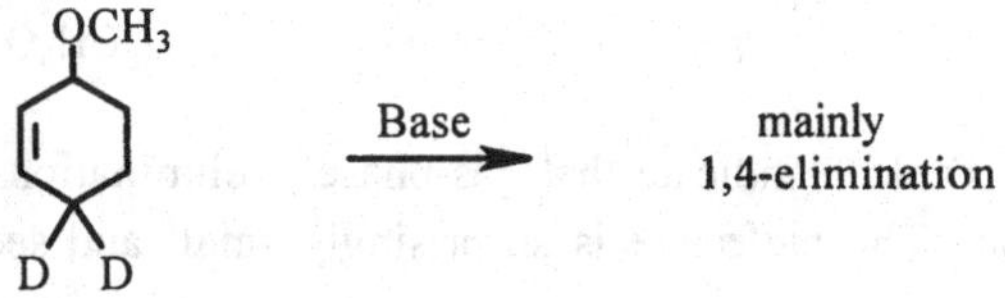

mainly
1,4-elimination

5

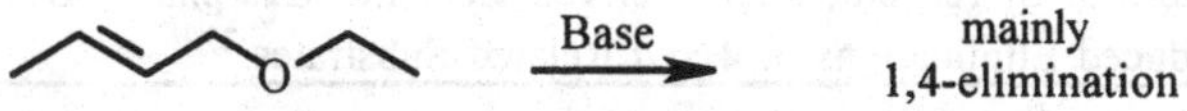

2-butenyl ether

mainly
1,4-elimination

Both of these examples are, to some extent, biased towards 1,4 elimination. In **5**, 1,4 elimination involves removal of an allylic proton which is much more acidic than the protons at carbon 4 (1,2 elimination). In the 2-butenyl ether system, 1,4 elimination gives a conjugated diene (butadiene) whereas 1,2-elimination gives ethene. Rabasco and Kass have designed a system which is more evenly balanced (each pathway involves the removal of an allylic proton).[36] When deuterium labeled derivatives of **6** are treated with strong bases, 1,2 elimination dominates, but with weak bases, 1,4 elimination is favored. It should be noted that this system slightly biased towards 1,2 elimination because this process leads to a conjugated triene.

$$6$$

Taken as a whole, the gas-phase results point to a very delicate balance between 1,2 and 1,4 elimination and suggest that subtle factors probably determine the preference.

3.4 Elimination vs. Substitution

As noted earlier, most gas-phase approaches to studying elimination reactions are unable to directly identify the neutral products. As a result, simply distinguishing between elimination and substitution is not a straightforward task because both reactions give the same ionic product. For example, the reaction of CH_3O^- with n-propyl bromide could lead to either propene (E2) or methoxy propane (S_N2), but each pathway produces bromide as the sole ionic product (Eq.'s 15,16). Therefore as the reaction progresses in a mass spectrometer, one simply sees the formation of Br^- and gains no information as to the mechanism.

$$CH_3CH_2CH_2Br + CH_3O^- \xrightarrow{\text{E2}} CH_3CH=CH_2 + CH_3OH + Br^- \quad (15)$$

$$\xrightarrow{S_N2} CH_3CH_2CH_2OCH_3 + Br^- \quad (16)$$

An obvious solution to this problem is to collect the neutral products and analyze them. Jones and Ellison[39] have undertaken this challenging task for the reaction of CH_3O^- with n-propyl bromide in a flowing afterglow. The only product they detected was propene suggesting that an E2 mechanism dominates. Although this technique gives direct results, it suffers from many technical difficulties and has not been widely used. For this reason, the gas-phase competition between elimination and substitution often has been evaluated on the basis of indirect methods such as reactivity patterns.[40] For example, DePuy and coworkers have observed that dimethyl ether does not react with strong bases in the gas phase, but that other ethers such as diethyl ether react readily to give alkoxide products.[30] Given that methyl should yield the most facile S_N2 reaction, DePuy reasoned that the other ethers must be undergoing an E2 rather than an S_N2 process. DePuy has also examined the gas-phase reactions of simple alkyl halides with nucleophiles.[41] The results in Table 6 point to a clear difference in reactivity for oxygen- and sulfur-centered nucleophiles. An interesting comparison can be made for $CF_3CH_2O^-$ and H_2NS^- because they have almost identical gas-phase proton affinities. With trifluoroethoxide, the rate of reaction increases with branching at the α-carbon, a result that is completely counter to what one would expect for an S_N2 reaction. In fact, the fastest rate is for t-butyl chloride, a substrate effectively incapable of an S_N2 reaction. The reaction with CH_3Cl indicates that S_N2 reactions are viable for trifluoroethoxide, but the relatively low rate constant suggests that elimination probably dominates in the systems with a β–hydrogen (i.e., the higher alkyl halides). In contrast, the reactivity pattern for H_2NS^- points to an S_N2 mechanism. A modest rate constant is observed for CH_3Cl and the 1° substrates, but no reactivity is observed for the 2° and 3° alkyl chlorides. This is the expected pattern for an S_N2 process and apparently the steric hindrance of a 2° system is sufficient to shutdown the substitution process. The absence of E2 reactivity suggests that sulfur is a poor base for eliminations. A similar pattern is seen for HS^-. It should be noted that contrary to conclusions based on condensed phase work, sulfur is not an exceptionally good nucleophile for gas-phase S_N2 reactions. In fact, it gives somewhat slower reactions than oxygen nucleophiles of similar basicity. The enhanced nucleophilicity of sulfur nucleophiles in solution therefore is probably the result of solvation phenomena.

Table 6. Rate Constants for the Gas-phase Reactions of Nucleophiles with Alkyl Chlorides.[41]

Alkyl Halide	$CF_3CH_2O^-$	H_2NS^-
CH_3-Cl	2.2	1.5
CH_3CH_2-Cl	2.5	0.38
$CH_3CH_2CH_2$-Cl	3.2	0.85
$(CH_3)_2CH$-Cl	4.3	<0.001
$(CH_3)_3C$-Cl	6.1	<0.001

Rate coefficients in 10^{-10} cm^3 molecule^{-1} s^{-1}

Further evidence for this reactivity pattern is seen in kinetic deuterium isotope effects. When substrates are labeled with deuterium at the β–carbon, significant kinetic isotope effects are observed for oxygen-centered nucleophiles.[42] The k_H/k_D values reach 4.7 and are consistent with a primary deuterium isotope effect in a mechanism involving proton transfer (E2). For the sulfur-centered nucleophiles, the k_H/k_D values are near unity suggesting a mechanism that does not involve β–proton transfer (S_N2).

4. Computational Studies

4.1 Questions to be Addressed by Computational Studies

From the previous sections, it is clear that there are several areas where computational chemistry could make an impact on the study of β–eliminations. First, there is still much to learned about the factors that affect the competition between anti and syn elimination. Work in solution has been clouded by solvation effects and gas-phase studies have been limited to a few systems. Second, the competition between Saytzev and Hofmann elimination has in no way been resolved by condensed phase studies where there seem to be too many experimental variables to clearly identify the controlling factors. Third, the competition between 1,2 and 1,4 elimination has been studied in solution and in the gas phase, but there are still many details to be explored particularly concerning the stereochemistry of these processes. Fourth, although isotope effects have been used in many mechanistic studies, the connection between transition state geometry and kinetic isotope effect has not been firmly established. Finally, although the competition between elimination and substitution has been well studied in solution, recent gas-phase results suggest that there are new insights to be gained with respect to the electronegativity of the nucleophile. In the following sections, recent computational efforts to address these questions will be presented.

4.2 Theoretical Approach

Of the β–eliminations that have been studied computationally, the reaction of fluoride with ethyl fluoride has received the greatest attention.[43-46] This is mainly because it is the smallest, realistic system and therefore would seemingly require the least computational resources. However, it has proven to be a difficult system because the high charge density of fluoride is not easily characterized. As a result, this system provides a demanding test of theoretical approaches for studying elimination reactions. The fluoride + ethyl fluoride system has been examined with a variety of methodologies including semi-empirical, density functional theory (DFT), and a broad spectrum of conventional *ab initio* approaches. To date, the most demanding calculations have been at the G2+ level and these can be used as a standard for comparison.[45] The G2+ level involves geometry optimizations at the MP2/6-31+G(d,p) level followed by a series of single point energy calculations on these geometries. An additivity scheme is used to approximate the QCISD(T)/6-311+(3df,2p) level of theory.[47,48]

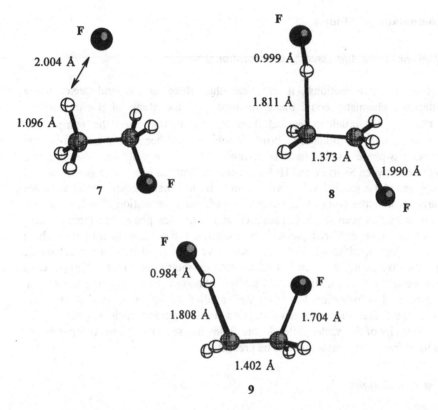

Figure 2. Structures in the reaction of fluoride with ethyl fluoride (MP2/6-31+G(d,p) level).

In Figure 2, an ion-dipole complex for the reaction of fluoride with ethyl fluoride is given. As in all the figures, heteroatoms are labeled, hydrogen is white, and carbon is gray. The ion-dipole complex (7) is characterized by a loose interaction between the fluoride and a β–hydrogen (F-H$_\beta$ = 2.004 Å) and is 16.5 kcal/mol more stable than the separated reactants (Table 7). The transition states for the E2(anti) and E2(syn) reactions are also shown in Figure 2. This β–elimination is endothermic in the gas phase and not surprisingly, the transition states occur late on the reaction coordinate. For the anti transition state (8), this leads to exceptionally long bond distances to the leaving group (C$_\alpha$-F = 1.990 Å) and the departing proton (C$_\beta$-H$_\beta$ =

1.811 Å). At the G2+ level of theory, this transition state is 3.1 kcal/mol more stable than the separated reactants. As a result, the barrier to the anti elimination is 13.4 kcal/mol with respect to the ion-dipole complex. The syn transition state (9) has more E1cb-character as indicated by a shorter C_α-F distance (1.704 Å), but a similar C_β-H_β distance (1.808 Å). The syn transition state is 4.2 kcal/mol less stable than the reactants or 7.3 kcal/mol less stable than the anti transition state. In the next few paragraphs, the G2+ results will be compared with those from other, less-demanding computational approaches.

Table 7. Results of Calculations on the Reaction of Fluoride with Ethyl Fluoride.[a]

	Complex		E2 (Anti)			E2 (Syn)		
Method	C_β-H_β	E	C_β-H_β	C_α-F_α	E	C_β-H_β	C_α-F_α	E
G2+	1.096	-16.5	1.811	1.990	-3.1	1.808	1.704	+4.2
AM1	1.75	-50.0	2.31	2.42	-1.6	3.60	2.05	-14.2
HF	1.084	-13.8	1.896	1.833	+9.6	2.050	1.653	+14.8
MP2	1.096	-15.0	1.811	1.990	-2.4	1.808	1.704	+5.3
B-VWN5	1.138	-13.3	1.721	1.968	-4.3	1.623	1.730	+3.9
B3-LYP	1.136	-15.6	NA	NA	NA	1.781	1.758	+4.3
X(α)	1.394	-10.6	2.334	2.055	-9.4	1.934	1.675	-0.5

[a]Energies in kcal/mol. G2+ uses MP2/6-31+G(d,p) level for optimizations and an additivity scheme that approximates QCISD(T)/6-311+G(3df,2p).[47] HF uses 6-31+G(d,p) basis set. MP2 uses 6-311+G(d,p) basis set. B-VWN5 and B3-LYP use aug-cc-pVDZ basis set. [45,46] X(α) uses DZP basis set for optimizations and LDA/NL/TZPP level for energies.[44] See references for details. NA indicates that transition state could not be located.

In modern computational chemistry, semi-empirical calculations (such as AM1[49]) provide the fastest and least demanding quantum calculations, but are inflexible and offer limited accuracy and reliability. For eliminations involving anionic bases, poor results are expected because the semi-empirical calculations are limited to a minimal basis set and do not contain the diffuse sp functions that are required to adequately characterize small anions. This should be particularly disastrous considering the high charge density of fluoride. In fact, AM1 overestimates the gas-phase basicity of fluoride by over 50 kcal/mol. The only hope for semi-empirical calculations in the β-elimination of fluoride + ethyl fluoride is that there could be a fortuitous cancellation of errors because the method equally overestimates the basicity of the nucleophile and leaving group (both are fluoride). At the AM1 level, the ion-dipole complex has a bizarre structure with a very long

C_β-H_β distance (1.75 Å) that suggests a completed proton transfer (Table 7). This is a direct result of AM1 overestimating the basicity of F^-. The ion-dipole complex also is predicted to be much too stable compared to the reactants (~ -50 kcal/mol). The anti transition state has bond lengths that are too long, but fortuitously, the transition state energy (-1.6 kcal/mol) is near that of the G2+ level. Finally, the syn transition state has a structure where the C_β-H_β bond is fully cleaved and the H-F fragment is interacting with the leaving group. At the AM1 level, the syn transition state appears to be more stable than the anti transition state by 12.6 kcal/mol. Obviously, these results indicate that AM1 calculations are unable to characterize the structures involved in this β–elimination. It should be noted that better results have been observed in a less demanding system (ClO^- + ethyl chloride),[50] but the errors are still significant (particularly the comparison to an S_N2 path), and therefore the use of semi-empirical calculations in the study of base-induced eliminations is probably risky.

Until recently, a standard approach for studying reaction mechanisms had been to complete optimizations at the Hartree-Fock level. This differs from the G2+ method where correlation corrections are included in the optimizations (MP2). The bond lengths in Table 7 indicate that the Hartree-Fock level is capable of producing reasonable geometries for the elimination. This is noteworthy given that this system is known for difficult optimizations on a flat potential energy surface.[45,46] Consequently, the common approach of employing Hartree-Fock geometries for higher-level calculations has merit in these systems; however, correlation has a profound effect on energies (see below) and this approach has given errant results in certain cases.[51] The values in Table 7 show that the elimination barriers are significantly overestimated at the Hartree-Fock level. This has been seen in a number of systems and therefore Hartree-Fock calculations are not appropriate for determining reaction energies.[51-56] It is not surprising that a system involving six electrons in a concerted reaction would require some level of electron correlation for reliable results.

An inexpensive approach for including correlation is to apply corrections with Moeller-Plesset perturbation theory. At its lowest level (MP2), this method is practical for moderate-size systems (up to 10 heavy atoms). The results in Table 7 indicate that the MP2 level can be useful for studying elimination reactions. The energies are very similar to those from the much more computationally demanding G2+ scheme and deviations of only about 1 kcal/mol are observed for the transition states. As a result, MP2 calculations offer an accurate and efficient approach that is tractable for larger systems.

Recently density functional theory (DFT) has become popular in the study of organic reaction mechanisms. Various levels have been applied to the reaction of F- with ethyl fluoride. Recent work using extended basis sets has shown promise in this elimination system.[45,46] For example, Kass has shown that the Becke88 exchange functional[57] with the Vosko, Wilk, and Nusair correlation functional[58] (B-VWN5) gives reasonable geometries for the anti and syn transition states as well as the ion-dipole complex. It is important to note that the anti transition state could not be located with common basis sets (*e.g.*, 6-31+G(d)) and therefore more complete basis sets were required for the study. The energies in Table 7 indicate that the B-VWN5 approach is capable of reasonable results. Although the complexation energy is underestimated, the transition state energies are within about 1 kcal/mol of the G2+ values. Results with other functional combinations indicate that good energies can be obtained with extended basis sets; however, the anti transition state cannot be located in many cases (the energy smoothly rises as the reaction progresses to products). For example, the popular combination of the Becke 3-parameter exchange functional[59] with the Lee, Yang, and Parr correlation functional[60] (B3-LYP) gives good energies and geometries for the syn transition state, but does not yield an anti transition state. Consequently, high-level DFT calculations are useful in this E2 system, but their generality is limited by problem optimizations. When DFT calculations have been applied to this system with less sophisticated functionals $(X(\alpha))$[61] and small basis sets (no diffuse functions),[44] the results have been less satisfactory. Specifically, the ion-dipole complex exhibits far too much extension in the C_β-H_β bond and the anti transition state appears to be too late on the reaction coordinate. In addition, higher-level calculations based on these geometries underestimate the transition state energies. These results reaffirm the need for extended basis sets in the study of elimination reactions.

In summary, some correlation is required to accurately characterize elimination reactions. It appears that the MP2 level is adequate and gives results that are near the G2+ values. DFT calculations are also useful, but they suffer from problem optimizations for the anti transition state in this system. As a result, MP2 calculations are probably the most versatile and reliable alternative to large scale calculations (G2+). Methods that employ small basis sets (AM1 or $X(\alpha)$), do not give accurate geometries and therefore yield poor results. Hartree-Fock calculations give reasonable geometries, but without correlation corrections, they yield poor energies.

In the remainder of the chapter, the majority of the results will be from calculations at the MP2/6-31+G(d,p) level using geometries optimized at the MP2/6-31+G(d) level. The advantage of this approach is that it includes correlation in the optimizations and is tractable for fairly large systems. The overall accuracy is improved by incorporating additional polarization functions in the final energy calculations. To account for zero point vibrational energies (ZPE), analytical frequencies have been computed at the Hartree-Fock level (ZPE scaled by 0.9135[62]). Most calculations have been completed with the Gaussian series of programs.[63] This overall scheme has proven to be a useful compromise between efficiency and accuracy. The results are reported in the text as transition state energies and these approximate $\Delta H^{\ddagger}$ values at 0 K. The conversion to 298 K (integration of the heat capacities) has only a modest effect on the relative energies (about 1 kcal/mol for most of the reactions) and is not addressed here. In early work, optimizations were completed without correlation corrections (HF/6-31+G(d)). As a result, some values in this chapter may differ from those in the original publications.

4.3 Anti periplanar vs. Synperiplanar Transition States

There is now a large body of computational evidence that indicates that there is a strong preference for antiperiplanar transition states in gas-phase β-eliminations.[43,44,51-55] For example in the reaction of fluoride with ethyl chloride, both the anti and syn transition states are perfectly periplanar (C_s symmetry), but anti is favored by 13.7 kcal/mol. In addition to the large energetic difference, there are significant structural differences between anti and syn transition states. In Figure 3, it can be seen that proton transfer has progressed to a greater extent in the syn transition state. The C_β-H_β distance is 1.43 Å in the anti transition state, but it is stretched to 1.48 Å in the syn transition state. In contrast, leaving group expulsion has progressed to a greater extent in the anti transition state. These results suggest that the anti and syn transition states lie at different points on the variable transition state spectrum and that the syn transition state has more E1cb-character. This effect has been noted in solution.[7] The enhanced E1cb-character of the syn transition state can be understood with the qualitative M.O. theory presented in Section 2.7. In the syn orientation, there is less π-overlap so the interaction between the α and β carbons is limited. Therefore as the β-proton is removed, electron flow to the α-carbon is restricted and cleavage of the C_α-X bond is retarded. As a result, proton transfer must progress to a greater extent before it is beneficial to shift electron density into the σ* orbital of the C_α-X bond. As noted earlier, the reduced interaction between the α and β carbons in the syn orientation biases the system towards asynchronous mechanisms and an E1cb-like transition state is not unexpected.

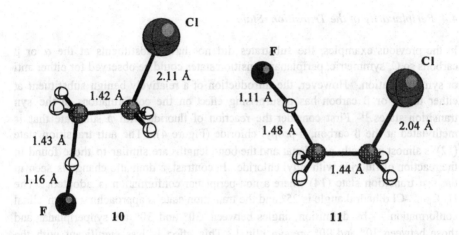

Figure 3. Transition states for the fluoride-induced eliminations of ethyl chloride (MP2/6-31+G(d)).

The anti preference in gas-phase eliminations varies considerably depending on the nature of the reactants and is reduced in systems with late transition states. For example, the β–elimination of fluoride with ethyl fluoride is endothermic (ΔH = ~ +9 kcal/mol) and consequently has late transition states. Evidence for this can be found in the geometries (Figure 2). In both transition states, proton transfer is nearly complete and the C_β-H_β bonds are stretched by ~ 66% in each case. Leaving group expulsion is well underway in the anti transition state and the C_α-F bond is stretched by about 40 %. Once again, the syn transition state appears to have more E1cb-character and leaving group expulsion is lagging behind (the C_α-F bond is stretched by only 20 %). Nonetheless, it is clear that these are much later transition states than those shown in Figure 3 for ethyl chloride. Along with later transition states, the ethyl fluoride system has a much smaller anti preference, only 7.3 kcal/mol. This can be rationalized in the following way. With later transition states, the π-bond is more developed and rehybridization of the α and β carbons is nearly complete; consequently, anti and syn transition states have almost the same torsional strain. Moreover the long bond lengths to the departing groups reduce the polarization of the carbon π-orbitals (see Schemes 3 and 4). All of these effects combine to attenuate the bias towards anti elimination. This is logical because when a system has two, product-like transition states leading to the same product, the energies of these transition states should be similar.

4.4 Periplanarity of the Transition State

In the previous examples, the substrates did not have substituents at the α or β carbons so C_s symmetric, periplanar transition states could be observed for either anti or syn elimination. However, the introduction of a relatively benign substituent at either the α or β carbon has a surprising effect on the conformations of the syn transition states.[51] First consider the reaction of fluoride with a substrate that is methylated at the β carbon, n-propyl chloride (Figure 4). The anti transition state (12) is almost perfectly periplanar and the bond lengths are similar to those found in the reaction of fluoride with ethyl chloride. In contrast, a dramatic change is seen in the syn transition state (14) where a non-periplanar conformation is adopted. The H_β-C_β-C_α-Cl dihedral angle is 25° and the transition state is approaching a syn clinal conformation — by definition, angles between -30° and 30° are synperiplanar and those between 30° and 90° are syn clinal.[5] This effect is less significant with the substituent at the α carbon and a smaller dihedral angle is seen in the syn transition state (15) of the reaction of fluoride with isopropyl chloride (Figure 4). A twist angle of ~ 7° is observed for the C_β-C_α bond. It should be noted that these results differ from those reported previously at the Hartree-Fock level.[51] Without correlation, larger twist angles are observed (> 30°) and the transition states are truly syn clinal. This highlights the importance of correlation in optimizations, but also points to a potential energy surface that is relatively flat with respect to the H_β-C_β-C_α-X dihedral angle (see below).

The appearance of twisted transition states makes one wonder about the importance of periplanarity in syn eliminations. To address this question, the advantages and disadvantages of a synperiplanar transition state must be weighed. The obvious advantage is that it allows for the maximum π-overlap in the transition state, an important stabilizing force and the key to a concerted elimination. The disadvantage is that it requires a completely eclipsed conformation and therefore the system will suffer from torsional strain. In the synperiplanar transition state of the ethyl chloride elimination (Figure 3), considerable E1cb-character was observed so π-bonding must be limited and may be associated with only a minor energetic advantage. The relatively small preference for periplanarity in the syn elimination of fluoride + ethyl chloride can be illustrated with a vibrational frequency analysis of the transition state. The frequency associated with rotation around the C_β-C_α bond is only 33 cm[-1], and therefore periplanarity has a weak restoring force. Given the trivial preference for periplanarity in the unsubstituted system, it is not surprising that a modest increase in torsional strain (methylation) could lead to a twisted transition state.

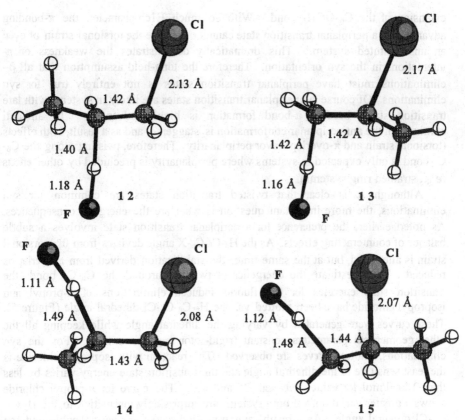

Figure 4. Transition states for the fluoride-induced eliminations of *n*-propyl chloride and isopropyl chloride (MP2/6-31+G(d)).

This result is general and one can expect to see twisted transition states for many syn eliminations. For example, calculations on systems with a variety of substituents at the α or β carbons (-F, -CN, -SiH₃) also exhibit syn clinal transition states with H$_\beta$-C$_\beta$-C$_\alpha$-Cl dihedral angles varying between 20 and 55°. Moreover there is an example of an unsubstituted system giving a syn clinal transition state.[53] In the reaction of hydroxide with ethyl methyl ether, the syn transition state has a dihedral angle (H$_\beta$-C$_\beta$-C$_\alpha$-OCH₃) of 35°. In this system, there is the combination of a very poor leaving group (CH₃O⁻) and a strong base (HO⁻), so a transition state with a high degree of E1cb-character is expected. In fact, the syn transition exhibits only a slight extension of the C_α-OCH_3 bond. With so much E1cb-character, the π-bonding advantage of a periplanar transition state cannot overcome the torsional strain of even an unsubstituted system. This dramatically demonstrates the weakness of π-interactions in the syn orientation. Therefore the long-held assumption that all β-eliminations must have periplanar transition states is not entirely true for syn eliminations. Of course synperiplanar transition states are likely in systems with late transition states because π-bond formation is highly advanced. In an anti elimination, the antiperiplanar conformation is staggered, and as a result, both effects (torsional strain and π-overlap) favor periplanarity. Therefore, twisting along the C_β-C_α bond is only expected in systems where periplanarity is precluded by other effects (*e.g.*, strained ring systems).

Although it is clear that twisted transition states are common for syn eliminations, the more important question is what are the energetic consequences. As noted earlier, the preference for a periplanar transition state involves a subtle balance of counteracting effects. As the H_β-C_β-C_α-X angle deviates from 0°, torsional strain is alleviated, but at the same time, the stabilization derived from π-overlap is reduced. To investigate the energetics of twisting around the C_α-C_β bond, the transition state energies for the fluoride-induced eliminations of *n*-propyl and isopropyl chloride have been plotted vs. the H_β-C_β-C_α-Cl dihedral angle (Figure 5). These curves were generated by varying the dihedral angle while keeping all the other geometrical parameters constant (rigid-rotor approximation). For the syn eliminations, shallow curves are observed. The reaction with isopropyl chloride is the least sensitive to the dihedral angle and the transition state energy varies by less than 2 kcal/mol for values between -25 and 45°. The curve for *n*-propyl chloride shows a greater variation, yet both systems are surprisingly insensitive to the H_β-C_β-C_α-Cl dihedral angle. As a result, syn transition states are very flexible and can accommodate a wide range of orientations with respect to rotation around the C_α-C_β bond. In contrast, the anti eliminations are much more sensitive to the dihedral angle and the transition state energy rises rapidly for deviations greater than 15° from periplanarity.

The plots in Figure 5 are not consistent with a simple torsional strain effect because if eclipsing interactions were the only factor, a double-well potential should be observed with a maximum near 0° (fully eclipsed) and minima for two twisted transition states (+ or - rotation). Instead, each curve has a single minimum and no maximum is seen near 0°.

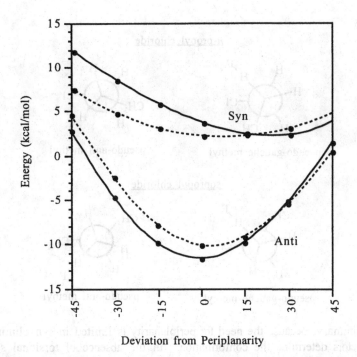

Figure 5. Plots of transition state energy vs. deviation from periplanarity for the E2 reactions of fluoride with *n*-propyl and isopropyl chloride. The upper curves are for syn elimination and the lower for anti elimination. The *n*-propyl system is represented by a solid line and the isopropyl by a dashed line. Calculations derived from rigid-rotor approximation at the MP2/6-31+G(d,p) level.

In the twisted transition state of the *n*-propyl chloride system, the methyl substituent could be either pseudo-gauche or pseudo-anti to the departing chloride, but only the pseudo-gauche transition state is observed (Scheme 5). For isopropyl chloride, the methyl substituent could be either pseudo-gauche or pseudo-anti to the departing proton, but again only the pseudo-gauche transition state is observed. Apparently the methyl substituent provides a stabilizing interaction with the departing group. This is not surprising because it is polarizable and can stabilize the developing charge at the site of the departing group. As a result, the system preferentially rotates in one direction to reduce the strain of an eclipsed (periplanar) conformation. It should be noted that the pseudo-gauche orientation is not universal for twisted transition states and that other substituents can shift the preference to pseudo-anti.

Scheme 5

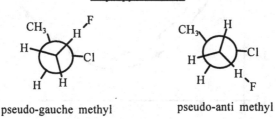

n-propyl chloride

pseudo-gauche methyl   pseudo-anti methyl

isopropyl chloride

pseudo-gauche methyl   pseudo-anti methyl

In summary, because the need for periplanarity is limited in syn eliminations, subtle factors determine the conformation. In the absence of torsional strain, a periplanar transition may result, but otherwise substituents can force the system into a twisted, syn clinal transition state. Present examples indicate that rotation is only favorable in one direction depending on the nature of the substituent.

4.5 Regiochemistry

To investigate the regioselectivity of gas-phase eliminations, the reactions of fluoride with *sec*-butyl chloride and *sec*-butyl fluoride have been examined. In these systems, three elimination products are possible: in order of increasing stability, 1-butene, Z-2-butene, and E-2-butene (Eq. 17).

$$CH_3CHCH_2CH_3 \xrightarrow{F^-} CH_2=CHCH_2CH_3 + \underset{CH_3}{\overset{H}{>}}C=C\underset{CH_3}{\overset{H}{<}} + \underset{H}{\overset{CH_3}{>}}C=C\underset{CH_3}{\overset{H}{<}} \qquad (17)$$

Table 8. Transition State Energies for the Fluoride-Induced Eliminations of *sec*-Butyl Chloride and *sec*-Butyl Fluoride.[a]

Substrate	Product	Transition State	
		Anti	Syn
sec-Butyl Chloride	1-Butene	-9.8	2.4
	Z-2-Butene	-10.5	3.5
	E-2-Butene	-11.5	1.8
sec-Butyl Fluoride	1-Butene	-1.9	4.6
	Z-2-Butene	-1.5	7.5
	E-2-Butene	-2.7	6.4

[a] kcal/mol.

The transition state energies for the anti and syn pathways to these products are listed in Table 8. Starting with *sec*-butyl chloride, it can be seen that Saytzev selectivity is obeyed for anti elimination. The differences are small, but the reactivity order matches the product stability. That is, the lowest barrier is for the formation of E-2-butene and the highest is for the formation of 1-butene. With anti stereochemistry, the span of transition state energies is 1.7 kcal/mol. For comparison, the energy span for the products is 2.8 kcal/mol experimentally.[64] Therefore, about half of the variation in alkene stability is manifested in the transition states. A different pattern is seen for the syn eliminations. Formation of E-2-butene is still the preferred path, but the barrier for forming 1-butene is smaller than that for forming Z-2-butene. The reluctance to form cis alkenes *via* syn elimination has been observed in solution and is the result of steric interactions in the transition state.[65] Although the syn transition state is not periplanar (eclipsed), it does suffer from a gauche interaction between the methyl groups. This is more severe than in the anti transition state because even in a syn clinal arrangement, the CH_3-C_α-C_β-CH_3 dihedral angle is small (~30°).

In the reaction of fluoride with *sec*-butyl fluoride, the pattern is more complicated. With anti stereochemistry, the lowest energy pathway leads to E-2-butene, but 1-butene is favored over Z-2-butene. In addition, the preference for Saytzev elimination (E-2-butene vs. 1-butene) is reduced to 0.8 kcal/mol. This is somewhat surprising because with later transition states (the reaction is endothermic), one might have assumed that alkene stability (*i.e.*, formation of 2-butene) would have played a larger role than in the *sec*-butyl chloride system. Apparently, the E1cb-character of the fluoride/alkyl fluoride system is significant enough to erode the Saytzev selectivity. This assumes that the 1° β-carbon is more acidic than the 2° β-carbon. Although there is little data on the gas-phase acidity of alkanes, indirect evidence seems to justify this assumption.[66] Moreover, calculations on the reaction of NH_2^- with propane show a small preference for the abstraction of a 1° proton. With syn stereochemistry, Saytzev elimination no longer dominates and 1-butene is the preferred product. This is not surprising because π-bonding and consequently alkene stability are less important in the syn orientation. As a result, acidity is the key component in the selectivity. However, the magnitude of the effect is noteworthy. There is a 2.6 kcal/mol shift in the 2-butene/1-butene preference in going from anti to syn elimination.

The results fit well with conclusions derived from condensed phase experiments. First, with a small base (F⁻) and a good leaving group (Cl⁻), the calculations predict Saytzev elimination. Second, the calculations show that the preference for Saytzev elimination is reduced when a poorer leaving group (F⁻) is employed. Finally, these results suggest that the Saytzev preference is greatly diminished when the elimination follows a syn pathway. In solution, syn elimination generally leads to Hofmann selectivity.[67] Taken together, the computational results confirm that the regiochemistry of eliminations is a balance of several factors and that the small, inherent preference for Saytzev elimination can be overcome, particularly in E1cb-like mechanisms.

4.6 Eliminations in Cyclopentyl and Cyclohexyl Ring Systems

Eliminations in ring systems have provided some of the best insight into transition state conformations. For example, Cristol used a dichloronorbornane system to provide evidence that anti eliminations require periplanar transition states.[68] The trans isomer gives a relatively fast syn elimination whereas the cis isomer gives a slow anti elimination. The difference in rates was explained by the fact that periplanarity is possible for the syn elimination of the trans isomer, but not for the anti elimination of the cis isomer (ring constraints).

fast syn elimination   slow anti elimination

In a related study, DePuy showed that cyclopentyl tosylates gave more syn elimination than cyclohexyl tosylates.[69,70] To rationalize this result, DePuy pointed out that the chair conformation of a six membered ring allows for an antiperiplanar, but not a synperiplanar arrangement, and therefore syn elimination should be retarded. In contrast, the envelope conformation of a five membered ring allows for periplanar transition states in both the anti and syn orientations.   As a result, syn elimination should be more competitive in the smaller ring.

To gain further insight into the effect of ring size on eliminations, *ab initio* calculations have been carried out on the reactions of fluoride with cyclopentyl and cyclohexyl chloride.[52] The transition states for the cyclohexyl system are given in Figure 6 and energies are listed in Table 9.   For the syn elimination, there are two possible conformations because the departing chloride could be in either an axial (**17**) or equatorial (**18**) position. The leaving group must be axial for an anti elimination (**16**) or a highly unstable, cyclic trans alkene would be formed.   There is nothing unusual about these transition states in terms of structure or energy, and they are similar to those found for the analogous reaction of fluoride with *sec*-butyl chloride (giving Z-2-butene).   Like the *sec*-butyl system, the cyclohexyl system does not give a synperiplanar transition state and an H_β-C_β-C_α-Cl dihedral angle of ~35° is observed in **17** and **18**. Using the best syn transition state for comparison (**17**), the anti preference of the cyclohexyl system (15.3 kcal/mol) is only slightly greater than that of the *sec*-butyl system (14.0 kcal/mol for forming Z-2-butene). In other words, the cyclohexyl system does not have a strong bias against syn elimination.   As noted earlier, syn eliminations have considerable conformational freedom with respect to the H_β-C_β-C_α-X angle so it is not surprising that the geometric constraints of a six-membered ring would have little effect on the stability of the syn transition state.

Table 9. Transition State Energies for the Fluoride-Induced Elimination Reactions of Cyclohexyl and Cyclopentyl Chloride [a]

	Ring	
Pathway	Cyclohexyl	Cyclopentyl
anti	-9.9	-10.2
syn	5.4[b]	-0.9
	8.1[c]	

[a] kcal/mol. [b] axial chlorine. [c] equatorial chlorine.

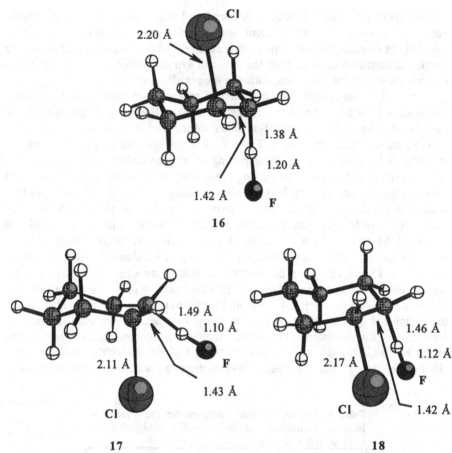

Figure 6. Transition states for the fluoride-induced eliminations of cyclohexyl chloride (MP2/6-31+G(d)).

If syn elimination is not disfavored in the cyclohexyl system, then DePuy's experimental results require that it be preferentially stabilized in the cyclopentyl system. The transition states for the reaction of fluoride with cyclopentyl chloride are given in Figure 7. The anti transition state (19) resembles the one for cyclohexyl chloride (18) in terms of structure and energy. However, the syn transition state (20) adopts an almost perfectly periplanar conformation with an $H_\beta-C_\beta-C_\alpha-Cl$ dihedral angle of ~2°. Moreover, the syn transition state is unusually stable and as a result, the cyclopentyl system has a small anti preference (only 9.3 kcal/mol). In fact, this is the smallest anti preference that we have observed for a β–elimination in a fluoride/alkyl chloride system.

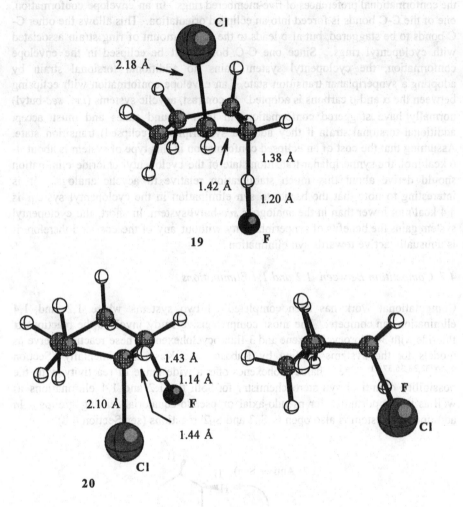

Figure 7. Transition states for the fluoride-induced elimination reactions of cyclopentyl chloride (MP2/6-31+G(d)). Newman projection of the syn transition state is shown on the right.

This result challenges our conclusion that periplanarity is not a powerful stabilizing force in syn eliminations. To explain this unusual observation, one must consider the conformational preferences of five-membered rings. In an envelope conformation, one of the C-C bonds is forced into an eclipsed orientation. This allows the other C-C bonds to be staggered, but also leads to the small amount of ring strain associated with cyclopentyl rings. Since one C-C bond must be eclipsed in the envelope conformation, the cyclopentyl system gains no additional torsional strain by adopting a synperiplanar transition state – an envelope conformation with eclipsing between the α and β carbons is adopted. In contrast, acyclic systems (*i.e.*, *sec*-butyl) normally have staggered conformations in their ground states and must accept additional torsional strain if they adopt a synperiplanar (eclipsed) transition state. Assuming that the cost of an eclipsed conformation in this type of system is about 4-6 kcal/mol, the synperiplanar transition state of the cyclopentyl chloride elimination should derive about this much stabilization relative to acyclic analogs. It is interesting to note that the barrier to syn elimination in the cyclopentyl system is 4.4 kcal/mol lower than in the analogous *sec*-butyl system. In short, the cyclopentyl system gains the benefits of synperiplanarity without any of the cost and therefore it is unusually active towards syn elimination.

4.7 Competition Between 1,2 and 1,4 Eliminations

Computational work has been completed on two systems where 1,2 and 1,4 elimination can compete. The most comprehensive study involves the reactions of fluoride with 3-chlorocyclohexene and 3-fluorocyclohexene. These reactions serve as models for the systems studied by Rabasco and Kass experimentally (Section 3.3).[32,33,36,37,71] The 3-halocyclohexenes offer a wide range of reactivity with the possibility of anti or syn stereochemistry for both the 1,2 and 1,4 eliminations as well as the opportunity for pseudo-axial or pseudo-equatorial leaving groups. In addition, the system is also open to S_N2 and S_N2' reactions (see Section 4.8).

The relative energies of the various transition states are given in Table 10. For the 1,4 eliminations, there are four possible transition states because the leaving group can be pseudo-axial or pseudo-equatorial and anti or syn orientations are possible. Focusing on the results from the chloro system, it can be seen that axial leaving groups are preferred and that the anti preference is surprisingly small.

Table 10. Transition State Energies for the Fluoride-Induced Eliminations of 3-Halocyclohexenes.[a]

		Halogen	
	Stereochemistry[b]	Chlorine	Fluorine
1,4-Eliminations	anti, axial	-14.6	-6.5
	syn, axial	-12.1	-5.3
	anti, equatorial	-7.4	-2.0
	syn, equatorial	-8.7	-1.6
1,2-Eliminations	anti, axial	-13.6	-4.2
	syn, axial	-2.2	4.2
	syn, equatorial	-0.9	4.3

[a] kcal/mol. [b] axial and equatorial refer to halogen.

The best anti transition state (21) is only 2.5 kcal/mol more stable than the best syn (22) transition state (Figure 8). In addition, the anti and syn transition states have similar bond lengths. The same effect is seen for 3-fluorocyclohexene where the anti preference is only 1.2 kcal/mol. These results clearly suggest that 1,4 eliminations have a smaller stereochemical bias than 1,2 eliminations. This is in good accord with the experimental work of Rabasco and Kass which shows only a mild anti preference in 1,4 eliminations initiated by strong bases.[71] The modest selectivity can be rationalized by turning to a simple, qualitative M.O. picture. The interacting orbitals are the same as in the 1,2 elimination, except that the coupling of the proton transfer process to the leaving group expulsion process is through a fully formed π-bond. A cartoon of the orbital picture is given in Scheme 6. Recall that in the 1,2 elimination, it was the relative polarization of the π orbitals at the α and β carbons that led to the preference for anti elimination (Schemes 3 and 4).

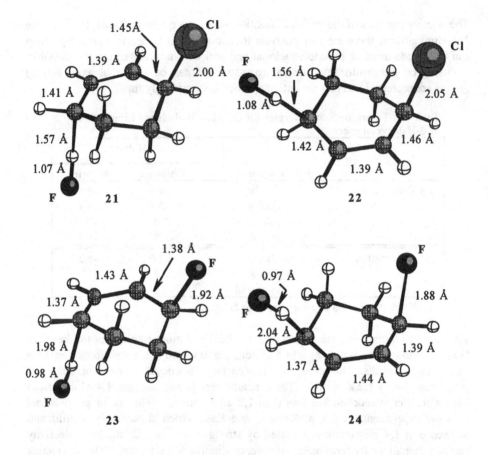

Figure 8. Transition states for the fluoride-induced 1,4 eliminations of 3-chlorocyclohexene and 3-fluorocyclohexene (MP2/6-31+G(d)). In each case, the leaving group is pseudo-axial.

In the 1,4 elimination, the intervening carbons mediate this interaction and the coupling is almost the same in anti and syn arrangements. In other words, the orientation of the departing proton at the δ-carbon has little effect on the polarization of a π orbital on the β–carbon and therefore is of less consequence in determining the strength of the coupling. Because the intrinsic preference for anti stereochemistry is so small in 1,4 eliminations, subtle effects can tip the balance towards syn stereochemistry. The fact that syn elimination is often seen in condensed phase studies[72-74] suggests that solvation phenomena may favor a syn pathway.

Scheme 6

1,2-Eliminations

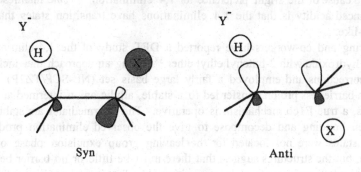

Syn   Anti

1,4-Eliminations

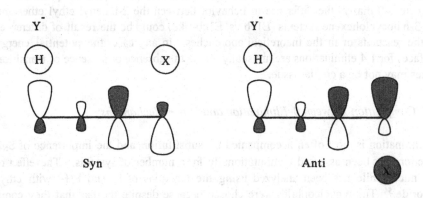

Syn   Anti

The transition states of the 1,2-eliminations are not exceptional and resemble those of saturated analogs; however, they are somewhat more stable because they have a vinyl substituent at the α-carbon (Table 10). For example, the barrier to the anti elimination of 3-chlorocyclohexene is 3.7 kcal/mol lower than that of cyclohexyl chloride. With both of the unsaturated substrates, 1,4 elimination is favored over 1,2 elimination. Using the most stable transition states for comparison, the preference for 1,4 elimination in the chloro system is 1 kcal/mol and in the fluoro system it is 2.3 kcal/mol. This leaves the opportunity for competition between the two mechanisms. Recall that Kass and Rabasco[37] saw a mixture of 1,2 and 1,4 elimination in their gas-phase experiments (Section 3.3). It should be noted that although the 1,2 and 1,4 eliminations give the same product (1,3-cyclohexadiene), the later involves the removal of a much more acidic proton (allylic). In fact, this is probably the cause of the slight preference for 1,4 elimination.[36] One manifestation of the enhanced acidity is that the 1,4 eliminations have transition states that are more E1cb-like.

Nibbering and co-workers have reported a DFT study of the 1,4-elimination reaction of hydroxide with 2-butenyl ethyl ether.[38] Using an approach that included non-local corrections and employed a fairly large basis set (NL-SCF/TZ2P), they found that a barrierless proton transfer led to a stable, allylic anion intermediate. In other words, a true E1cb mechanism is operative. The intermediates are stabilized by hydrogen bonding and decompose to give the observed elimination products. Transition states were not located for the leaving group expulsion phase of the mechanism, but the structures suggest that there must be little or no barrier beyond the inherent endothermicity of the decomposition. As for stereochemistry, the E1cb intermediates associated with the syn and anti pathways have similar energies (anti is favored by ~1 kcal/mol) and therefore only a modest stereochemical preference is expected. Finally, the difference in behavior between the 2-butenyl ethyl ether and the 3-halocyclohexene systems (E1cb vs. E1cb-like) could be the result of differences in the reactants or in the theoretical approaches. In any case, the potential energy surfaces for 1,4 eliminations are unusually flat so the absence or presence of transition states may not be a crucial issue.

4.8 Competition Between β–Elimination and Other Mechanisms

β-elimination is most often accompanied by substitution and the importance of S_N2 reactions has been assessed computationally for a number of systems. The effect of the nucleophile has been analyzed using the reactions of F^- and PH_2^- with ethyl chloride.[54] These nucleophiles were chosen because despite the fact that they come from different rows of the periodic table, they have similar basicities. The gas-phase proton affinities of F^- and PH_2^- are 371.5 and 369.5 kcal/mol, respectively.[75] As a result, their elimination reactions have approximately the same exothermicities and fair comparisons are possible.

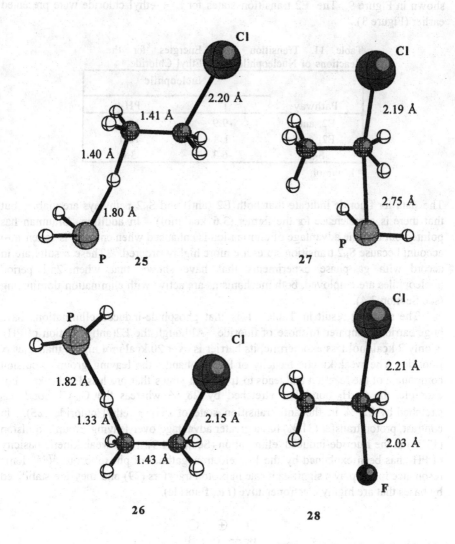

Figure 9. Transition states of the E2 and S_N2 reactions of PH_2^- and F^- with ethyl chloride (MP2/6-31+G(d)).

Transition state energies for these systems are given in Table 11 and structures are shown in Figure 9. The E2 transition states for F^- + ethyl chloride were presented earlier (Figure 3).

Table 11. Transition State Energies for the Reactions of Nucleophiles with Ethyl Chloride [a]

Pathway	Nucleophile	
	F^-	PH_2^-
E2, anti	-9.9	10.2
E2, syn	1.8	18.1
S_N2	-6.3	3.0

[a] kcal/mol.

The data for fluoride indicate that both E2 (anti) and S_N2 pathways are viable, but that there is a preference for the former (3.6 kcal/mol). In addition, Brauman has pointed out that the advantage of elimination is enhanced when entropy is taken into account because S_N2 transition states are more highly ordered.[76] These results are in accord with gas-phase experiments that have shown that when 2nd period nucleophiles are employed, both mechanisms are active with elimination dominating (see Section 3.4).

The startling result in Table 11 is that phosphide-induced eliminations have huge barriers compared to those of fluoride. Although the E2(anti) reaction of PH_2^- is only 2 kcal/mol less exothermic, its barrier is over 20 kcal/mol higher than that of fluoride. The weak kinetic basicity of PH_2^- enhances the leaving group expulsion component of the reaction and leads to transition states that are less E1cb-like. For example, the C_β-H_β bond has stretched by 28 % whereas the C_α-Cl bond has stretched by 22 % in the anti transition state of PH_2^- + ethyl chloride (**25**). In contrast, proton transfer (31 %) has a greater advantage over leaving group expulsion (17 %) in the fluoride-induced elimination (Section 4.3). The weak kinetic basicity of PH_2^- has been explained by the low electronegativity of phosphorous.[47,54] Ionic resonance forms play a significant role in proton transfers (**29**) and they are stabilized by bases that are highly electronegative (*i.e.*, fluoride).

$$\overset{\ominus}{B_1}\text{-}\text{-}\text{-}\overset{\oplus}{H}\text{-}\text{-}\text{-}\overset{\ominus}{B_2}$$

29

With bases such as PH_2^-, covalent bonding plays an important role in the bonding and structures like **29** are destabilized. Moreover, the P-H bond has the wrong polarization for an efficient proton transfer because hydrogen is effectively more electronegative than phosphorous.

$$\overset{\delta+}{R_2P}\!\!-\!\!\overset{\delta-}{H}$$

As a result, the hydrogen is transferred as a proton, but eventually must take on hydride character. This requires extensive electronic reorganization and adds to the barrier.

Going back to the data in Table 11, it is clear that elimination will not play a role for PH_2^- and therefore substitution is the only reasonable pathway. However, the S_N2 barrier is also fairly high and therefore an inefficient substitution is expected. This is consistent with gas-phase experiments where a relatively low rate constant has been measured for the reaction of PH_2^- with CH_3Cl.[77] The results are also in accord with the reactivity pattern found experimentally for sulfur nucleophiles (Section 3.4).[41,42] Finally, it should be noted that 3rd period elements are not intrinsically better for S_N2 reactions. The data in Table 11 show that in the absence of solvation effects, a 2nd period nucleophile is better suited for both S_N2 and E2 reactions than a 3rd period nucleophile of similar basicity. The enhanced S_N2 reactivity of 3rd period nucleophiles in solution must therefore be a solvation effect rather than a manifestation of greater intrinsic reactivity.

Hu and Truhlar[50] have considered the competition between substitution and elimination for the reaction of ClO^- with ethyl chloride. Although the E2 transition state has a higher electronic energy, it is favored over the S_N2 transition state when zero point energies are included. They have completed dynamics calculations (variable transition state theory) on the system and the rates indicate that elimination should be favored over substitution by over an order of magnitude at room temperature. This study provides a solid theoretical foundation for the assumption that dynamics would favor the E2 process (Brauman's entropic effect[76]). Isotope effects have also been calculated (see Section 4.9).

The importance of substrate structure has been evaluated in several instances for the reactions of fluoride with alkyl chlorides.[51,52,54] Of course, it is expected that increasing substitution at the α-carbon will destabilize the crowded S_N2 transition state, but in contrast, have a slight stabilizing effect on the E2 pathway because the alkene product is more highly substituted and the reaction is more exothermic. Table 12 lists transition state energies for the eliminations and substitutions of several fluoride/alkyl chloride reactions.

Table 12. Transition State Energies for the Elimination and Substitution Reactions of Fluoride with Alkyl Chlorides.[a]

Substrate	E2 (anti)	S_N2
Ethyl Chloride	-9.9	-6.2
n-Propyl Chloride	-11.4	-7.0
Isopropyl Chloride	-10.2	-4.4
Cyclopentyl Chloride	-10.0	-5.2
Cyclohexyl Chloride	-9.9	-1.8
3-Chlorocyclohexene	-13.6[b]	-7.0[c]

[a]kcal/mol. [b]Most stable 1,2-elimination. The anti and syn S_N2' transition states have energies of -4.6 and -5.0 kcal/mol, respectively.

Addition of a methyl group at the β-carbon (n-propyl chloride) stabilizes the elimination and substitution transition states, but the effect is greater for the former, so the preference for elimination is enhanced (as compared to ethyl chloride). The advantage of the methyl substituent is probably derived from it acting as a polarizable group and stabilizing the charge in the transition state. In contrast, a methyl group at the α-position (isopropyl chloride) has almost no effect on the elimination, but increases the S_N2 barrier by 1.8 kcal/mol. Evidence of crowding in this S_N2 transition state is found in the C-F and C-Cl bond distances which are significantly longer than those in the 1° systems. The overall order of S_N2 reactivity (n-propyl > ethyl > isopropyl) has also been observed in recent gas-phase experiments by McMahon.[78] Cyclopentyl chloride provides another example of a 2° halide, but the effect of the α-substitution is attenuated because the bipyramidal geometry of the S_N2 transition state reduces eclipsing interactions in the 5-membered ring and some torsional strain is relieved. Moreover, the cyclopentyl system benefits from having substituents at the β-carbon (see above). In contrast, an S_N2 transition state (bipyramidal carbon) adds eclipsing interactions to the chair conformation of a 6-membered ring and the torsional strain is increased.[79,80] As a result, substitution is at a great disadvantage in the cyclohexyl system. This reactivity pattern has also been observed in solution.[81] Both reactions of the unsaturated system, 3-chlorocyclohexene, are significantly stabilized by the α-vinyl substituent. Of course, the elimination is stabilized because the product is now a conjugated diene. As for the substitution, there is gas-phase and condensed phase data that supports allylic activation of S_N2 processes.[21,41,82] The S_N2 reaction gains the most from the α-vinyl substituent, but the elimination still maintains a substantial advantage compared to acyclic systems. Finally, it is interesting to note that the stability of the E2 transition state is surprisingly insensitive to the structure of the substrate. With the exception of the unsaturated system, all the E2 transition states have about the same relative energy. Consequently, it is the stability of the S_N2 transition state that mainly determines the selectivity.

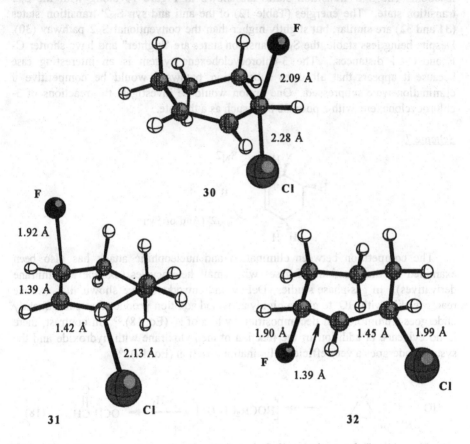

Figure 10. Transition states for the S_N2 and S_N2' reactions of fluoride with 3-chlorocyclohexene (MP2/6-31+G(d)).

With allylic halides, there is the possibility of an S_N2' reaction where the nucleophile attacks a vinylic carbon (see Scheme 7). In early work, Bach looked at this mechanism for simple substrates.[83,84] More recently, we have investigated this process for the reaction of fluoride with 3-chlorocyclohexene. Because fluoride could add to either side of the ring, anti and syn transition states are possible for the S_N2' reaction. The S_N2' transition states are shown in Figure 10 along with the S_N2 transition state. The energies (Table 12) of the anti and syn S_N2' transition states (31 and 32) are similar, but slightly higher than the conventional S_N2 pathway (30). Despite being less stable, the S_N2' transition states are "tighter" and have shorter C-F and C-Cl distances. The 3-chlorocyclohexene system is an interesting case because it appears that all three substitution pathways would be competitive if elimination were suppressed. One option would be investigate the reactions of 3-chlorocyclohexene with a poor E2 base such as a thiolate.

Scheme 7

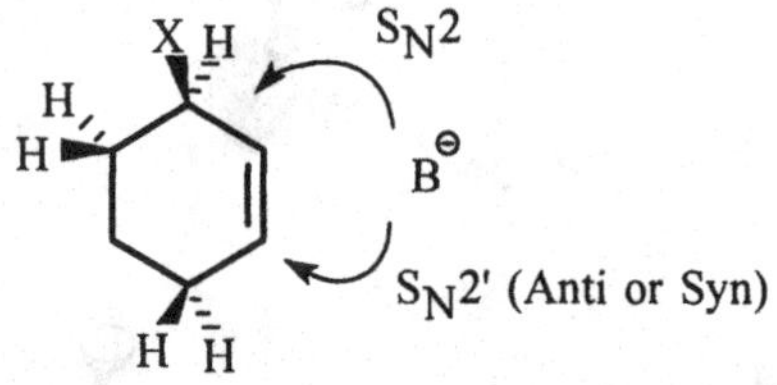

The competition between elimination and nucleophilic attack has also been examined for the reactions of bases with small heterocycles (oxirane and thiirane derivatives). In gas-phase studies, DePuy and coworkers have shown that oxirane reacts readily with HO^- to give a highly activated addition product that subsequently undergoes a unimolecular decomposition by loss of H_2 (Eq. 18).[85] In contrast, there is no evidence of addition in the reaction of methyloxirane with hydroxide and the system undergoes a very efficient elimination reaction (Eq. 19).[86]

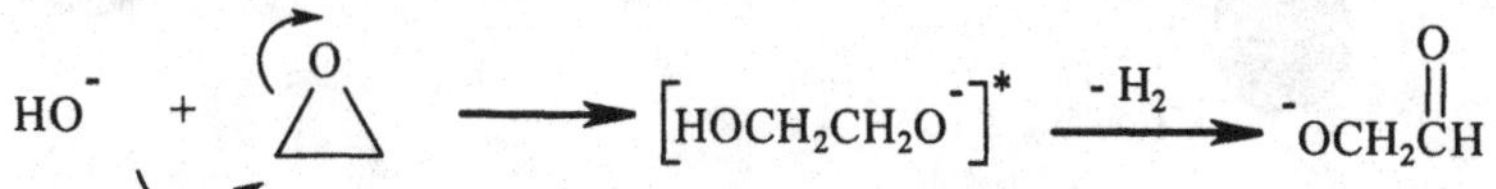

$$HO^- + \triangle\!\!\!O \longrightarrow [HOCH_2CH_2O^-]^* \xrightarrow{\;-H_2\;} {}^-OCH_2\overset{O}{\overset{\|}{C}}H \qquad (18)$$

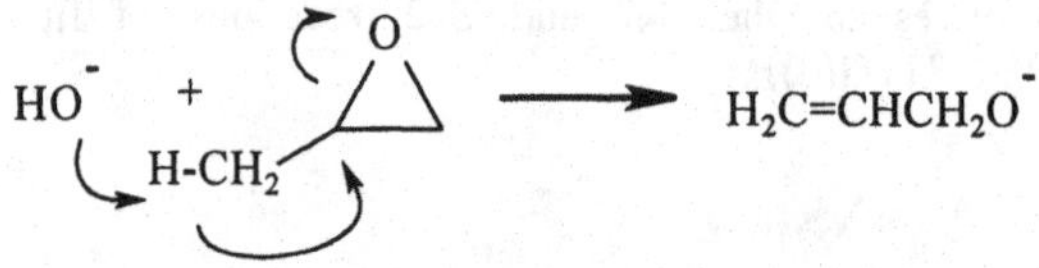

$$HO^- + \triangle\!\!\!O \longrightarrow H_2C=CHCH_2O^- \qquad (19)$$

To investigate the selectivity of these systems, *ab initio* calculations have been applied to the reactions of HO⁻ with methyloxirane and HS⁻ with methylthiirane.[55] The most stable transition states for elimination and addition are shown in Figure 11 and the energies for all the transition states are listed in Table 13. In the reaction of HO⁻ with methyloxirane, the preferred transition state is for anti elimination and it is 2.6 kcal/mol more stable than the best S_N2 transition state (addition at carbon 2). The E2(anti) transition state (33) is E1cb-like and has a significant stretch in the C_β-H_β distance (34 %), but only a modest increase in the cleaving C-O distance (10 %). However, it has less E1cb-character than the transition state of an acyclic analog, HO⁻ + CH₃CH₂OCH₃ (Eq. 20).[53,55]

$$HO⁻ + CH_3CH_2OCH_3 \longrightarrow H_2O + CH_2{=}CH_2 + ⁻OCH_3 \qquad (20)$$

In the E2(anti) transition state of this reaction, the C_β-H_β stretch is 38 % and the C-O stretch is only 5 %. Obviously, ring strain relief in the oxirane system is enhancing the leaving group ability of the alkoxide and therefore pushing the system towards a more synchronous elimination. In terms of transition state energies, the barrier to anti elimination in the methyloxirane system is 7 kcal/mol smaller than in the acyclic analog (Eq. 20); consequently, about 25 % of methyloxirane's ring strain (27 kcal/mol)[87] has been released at the transition state despite the modest increase in the C-O distance. The addition transition state (34) exhibits a greater C-O extension (23 %) and therefore should gain more from ring strain relief. This is clearly the case and the S_N2 barrier for methyloxirane is over 20 kcal/mol below that of the S_N2 reaction of HO⁻ with CH₃OCH₂CH₃ (attack at the CH₂ group).

Table 13. Transition State Energies for the Reactions of HY⁻ with Heterocycles and Acyclic Analogs.[a]

Substrate	Reaction	Y	
		O	S
Y △ CH₃	E2 (anti)	-9.2	13.4
	E2(syn)	-4.8	18.9
	Addition at C2	-6.6	-1.2
	Addition at C1	-4.0	4.4
CH₃CH₂YCH₃	E2 (anti)	-2.2	31.6
	E2 (syn)	2.2	36.0
	S_N2	14.3	25.4

[a] kcal/mol.

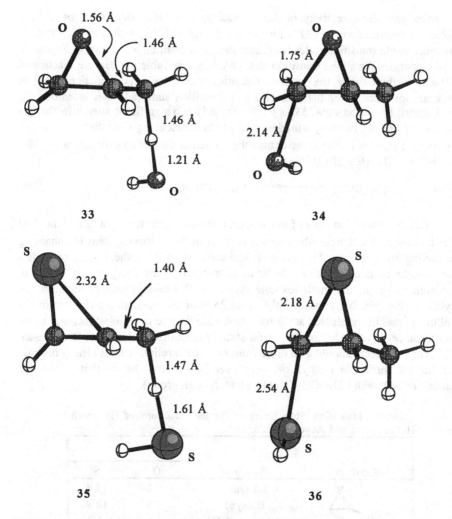

33

34

35

36

Figure 11 Transition states for the elimination and addition reactions of HO⁻ with methyloxirane and HS⁻ with methylthiirane (MP2/6-31+G(d)).

The net result is that the elimination preference is much smaller in the heterocycle (2.6 kcal/mol) than in the simple ether (16.5 kcal/mol). Nonetheless, this energetic advantage along with the entropic bias towards elimination (see above) is sufficient to lead exclusively to the elimination product.

The situation for the HS⁻ + methylthiirane system is completely different. Here elimination faces a large barrier (13.4 kcal/mol) and is unlikely under gas-phase conditions. This provides another example where a 3rd period nucleophile is incapable of inducing elimination. In the E2 (anti) transition state (Figure 11), there is a considerable extension of the breaking C-S bond (27 %) so substantial ring strain relief is expected (**35**). Compared to the analogous reaction of $CH_3CH_2SCH_3$, the elimination barrier is 18.2 kcal/mol smaller suggesting that most of thiirane's ring strain (~20 kcal/mol)[87] has been relieved. Nonetheless, the S_N2 addition has a relatively low barrier (-1.2 kcal/mol) and will clearly dominate (**36**). As with the methyloxirane system, addition at the least substituted carbon is preferred. A remarkable feature of the addition transition state is its high stability compared to the S_N2 reaction of the acyclic analog. The barrier of HS⁻ addition to methylthiirane is 26.6 kcal/mol lower than that of the S_N2 reaction of HS⁻ with $CH_3CH_2SCH_3$ even though the strain energy of thiirane is only ~20 kcal/mol. Despite appearing to release more ring strain, cleavage of the C-S bond in the addition transition state has progressed to a lesser extent (20 %) than in the elimination. The unusual stability of the addition transition state as well as the anomalous ring strain relief have been examined in detail. The full analysis is beyond the scope of this chapter, but the enhanced stability is the result of reduced steric interactions in the bipyramidal transition state.[88] Gas-phase reactions have not been reported for methylthiirane, but it is known that HS⁻ reacts readily with thiirane by an addition mechanism.[89]

Nibbering has presented an interesting variation on a gas-phase elimination.[34,35] When NH_2^- and other amides react with diethyl sulfide, ethyl thiolate is formed in what would appear to be a simple E2 reaction.

$$NH_2^- + CD_3CD_2SCH_2CH_3$$

4% $\rightarrow$ $CD_3CD_2S^- + NH_3 + C_2H_4$ (21)

3% $\rightarrow$ $CH_3CH_2S^- + NH_2D + C_2D_4$ (22)

28% $\rightarrow$ $CD_3CHDS^- + NH_2D + C_2H_4$ (23)

65% $\rightarrow$ $CH_3CHDS^- + NH_3 + C_2D_4$ (24)

However, deuterium labeling studies (Eq. 21-24) indicate that an α',β elimination dominates where deprotonation initially occurs at an α position and then an intramolecular elimination leads to the observed products (Scheme 8). This type of mechanism also has been observed in solution.[90,91]

Scheme 8

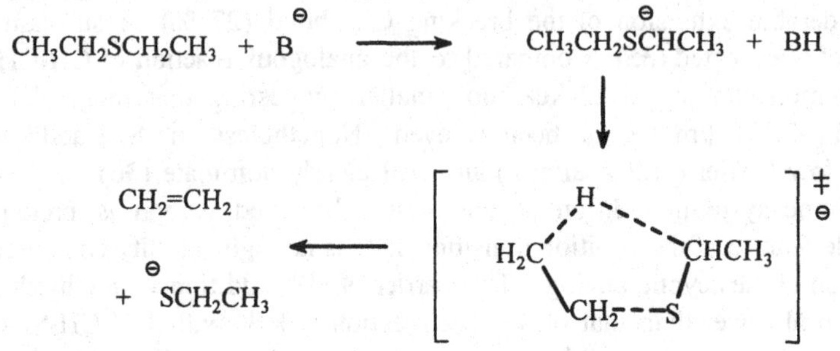

$$CH_3CH_2SCH_2CH_3 \ + \ \overset{\ominus}{B} \ \longrightarrow \ CH_3CH_2\overset{\ominus}{S}CHCH_3 \ + \ BH$$

$$CH_2{=}CH_2$$

$$+ \ \overset{\ominus}{S}CH_2CH_3 \ \longleftarrow \ \left[\begin{array}{c} H \cdots \\ H_2C \diagup \quad \diagdown CHCH_3 \\ \diagdown CH_2 \cdots S \diagup \end{array} \right]^{\ddagger}_{\ominus}$$

An interesting aspect of this reaction is that the α',β pathway is preferred despite requiring a syn conformation for the intramolecular elimination. To examine this unusual preference, *ab initio* calculations have been completed on the reaction of NH_2^- with $CH_3CH_2SCH_3$.[56] An energy profile for the system is shown in Figure 12. From the initially formed ion-dipole complexes, the step with the lowest barrier is abstraction of an α-proton. This is not unexpected because alkyl sulfides are moderately acidic and deprotonation by NH_2^- is exothermic.[31] The E2 barrier is 3.0 kcal/mol larger and apparently does not compete. Once the α-anion ($CH_3CH_2SCH_2^-$) is formed and NH_3 departs, the proton transfer is effectively irreversible under gas-phase conditions. At this point, the only mechanistic option in the gas phase is the α',β elimination pathway. The transition state for the intramolecular elimination is less stable than the E2 (anti) transition state, but this is immaterial because the proton transfer is irreversible and the system cannot reach the E2 pathway. The barrier to intramolecular elimination is below the energy of the separated reactants, so it is viable and can occur with a reasonable rate constant. The computational results indicate that there is nothing unusually favorable about the intramolecular syn elimination and that it is the high acidity of the α-position that causes the selectivity.

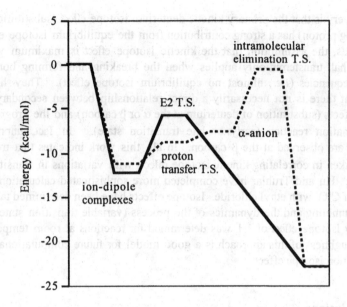

Figure 12. Energy profile for the reaction of NH_2^- with $CH_3CH_2SCH_3$. Solid line is E2 elimination and dashed line is α',β elimination.

4.9 Isotope Effects in E2 Reactions

Jensen and Glad have used *ab initio* methods to predict isotope effects in gas-phase eliminations.[92-95] They have assumed that the isotope effect can be estimated on the basis of changes in the zero point energies (ZPE). Of course deuterium substitution lowers the ZPE because the vibrational frequencies of bonds to deuterium are reduced. By calculating vibrational frequencies and ZPE's with and without deuterium substitution, the effect on the potential energy surface can be quantified. This approach ignores two important factors that play a role in determining kinetic isotope effects. First, proton tunneling is ignored. Second, a dynamics treatment is required to connect transition state energies to relative rates. Both of these factors can play a critical role in determining the effect of deuterium substitution on kinetics; consequently, this method is not reliable for predicting absolute values, but it can highlight trends that could be useful in evaluating experimentally determined kinetic isotope effects. Their results are extensive and cannot be described here in detail. A key discovery is that the primary kinetic deuterium isotope effect (substitution of the transferring proton) has a strong contribution from the equilibrium isotope effect. In other words, the assumption that the kinetic isotope effect is maximum when the proton is half transferred only applies when the breaking and forming bonds have similar frequencies (*i.e.*, almost no equilibrium isotope effect). They have also shown that there is not necessarily a direct relationship between secondary kinetic isotope effects (substitution of deuterium at the α or β carbon) and the progression of the elimination reaction (early or late transition state). In fact, unpredictable variations are observed at the β carbon. In all, this work indicates that much care must be taken in correlating kinetic isotope effects with variations in transition state structures. Hu and Truhlar have completed more sophisticated calculations on the reaction of ClO⁻ with ethyl chloride. Isotope effects have been determined taking into account tunneling and the dynamics of the process (variable transition state theory). A primary isotope effect of 3.1 was determined for reactions at room temperature.[50] The thoroughness of this approach is a good model for future computational studies of elimination isotope effects.

5. Conclusions

Several important conclusions have resulted from these studies:

(1) β-eliminations intrinsically prefer an anti pathway. The anti preference varies with the nature of the reactants, but there is no evidence that syn elimination would be preferred in the gas phase (except possibly in endothermic eliminations). Therefore, the appearance of syn elimination in solution is probably a result of solvation phenomena or ion pairing.

(2)   Although anti eliminations have a strong preference for a periplanar transition state, syn eliminations do not require absolute periplanarity and are often twisted (syn clinal) because the π-bonding component of the transition state is less important. As a result, syn eliminations have exceptional flexibility with respect to the H_β-C_β-C_α-X dihedral angle and a wide range of conformations is possible. For example, syn elimination is easily accommodated in the chair conformation of a six-membered ring.

(3)   The regioselectivity of eliminations involves a subtle balance of effects. Saytzev elimination is preferred in the cases investigated here, but Hofmann elimination is only at a slight disadvantage. The preference for Saytzev elimination is reduced in E1cb-like systems and lost in syn eliminations.

The results suggest that eliminations do not have a strong, intrinsic preference in terms of regioselectivity so a combination of minor effects must be taken into account to explain the observed selectivity.

(4) In simple allylic systems, there appears to be a small preference for 1,4 elimination over 1,2 elimination. In addition, the advantage of anti stereochemistry is greatly diminished in 1,4 eliminations.

(5) For 2nd period nucleophiles, elimination is favored over substitution in all the cases that have been considered. With 3rd period nucleophiles, there is a strong bias against elimination. The stability of E2 transition states appears to be fairly insensitive to the structure of the substrate, but the stability of S_N2 transition states varies significantly depending on the substitution at the α-carbon.

Future computational work is needed in many areas. The effect of base size and strength on regioselectivity needs to be examined. Greater variations in leaving group ability should be investigated. The effect of substrate structure on the preference for 1,4 vs. 1,2 elimination needs further study. Work on these and other topics should provide valuable new insights into the complex mechanisms of β-eliminations.

Acknowledgments

I would like to thank all the collaborators and colleagues who have made this work possible. These include C. H. DePuy, S. R. Kass, V. M. Bierbaum, J. M. Lee, P. Freed, J. J. Rabasco, and G. N. Merrill. The generous financial support of the National Science Foundation and San Francisco State University is also greatly appreciated.

References

1. E. Baciocchi, *Acc. Chem. Res.* **12** (1979).
2. D.V. Banthorpe, *Elimination Reactions* (Elsevier, New York, 1963).
3. R.A. Bartsch, *Acc. Chem. Res.* **8**, 128 (1975).
4. R.A. Bartsch and J. Zavada, *Chem. Rev.* **80**, 454 (1980).
5. W.H. Saunders, Jr. and A.F. Cockerill, *Mechanisms of Elimination Reactions* (John Wiley & Sons, New York, 1973).
6. W.H. Saunders, Jr., *Acc. Chem. Res.* **9**, 19 (1976).

7. J.R. Gandler in *Chemistry of Double-Bonded Functional Groups* , ed. Patai, S. (Wiley, Chichester, U.K., 1989).

8. T.H. Lowry and K.S. Richardson, *Mechanism and Theory in Organic Chemistry* (Harper-Collins, New York, 1987).

9. J.F. Bunnett, *Angew. Chem., Int. Ed. Engl.* **1**, 225 (1962).

10. A. Saytzev, *Ann. Chem.* **179**, 296 (1875).

11. A.W. Hofmann, *Ann. Chem.* **78**, 253 (1851).

12. W.H. Saunders, Jr., *et al., J. Am. Chem. Soc.* **87**, 3401 (1965).

13. H.C. Brown and O.H. Wheeler, *J. Am. Chem. Soc.* **78**, 2199 (1956).

14. H.C. Brown, *et al., J. Am. Chem. Soc.* **78**, 2193 (1956).

15. D.H. Froemsdorf and M.D. Robbins, *J. Am. Chem. Soc.* **89**, 1737 (1967).

16. R.A. Bartsch, *et al., Tetrahedron Lett.* , 2621 (1972).

17. D.J. Cram, *et al., J. Am. Chem. Soc.* **78**, 790 (1956).

18. M. Schlosser, *et al., Helv. Chim Acta* **56**, 1530 (1973).

19. R.P. Thummel and B. Rickborn, *J. Org. Chem.* **37**, 4250 (1972).

20. C. Margot, *et al., Tetrahedron* **46**, 2425 (1990).

21. C.K. Ingold, *Structure and Mechanism in Organic Chemistry* (Cornell University Press, Ithaca, 1969).

22. M.L. Dhar, *et al., J. Chem. Soc.* , 2093 (1948).

23. W.H. Saunders, Jr. and D.H. Edison, *J. Am. Chem. Soc.* **82**, 138 (1960).

24. A.F. Cockerill, *et al., J. Am. Chem. Soc.* **89**, 901 (1967).

25. J.I. Hayami, *et al., Bull. Chem. Soc. Jpn.* **44**, 1628 (1971).

26. W.F. Bayne and E.I. Snyder, *Tetrahedrom Lett.* , 571 (1971).

27. A. Pross and S. Shaik, *Acc. Chem. Res.* **16**, 363 (1983).

28. R.D. Bach, *et al., J. Am. Chem. Soc.* **101**, 2845 (1979).

29. T. Su and M.T. Bowers in *Gas Phase Ion Chemistry* , ed. Bowers, M.T. (Academic Press, New York, 1979).

30. C.H. DePuy and V.M. Bierbaum, *J. Am. Chem. Soc.* **103**, 5034 (1981).

31. S.G. Lias, *et al., J. Phys. Chem. Ref. Data, Suppl. 1* **17**, 1 (1988).

32. J.J. Rabasco, *et al., J. Am. Chem. Soc.* **116**, 3133 (1994).

33. J.J. Rabasco, Ph.D. Thesis (University of Minnesota, 1993).

34. L.J. de Koning and N.M.M. Nibbering, *J. Am. Chem. Soc.* **109**, 1715 (1987).

35. L.J. de Koning and N.M.M. Nibbering, *J. Am. Chem. Soc.* **110**, 2066 (1988).

36. J.J. Rabasco and S.R. Kass, *Tetrahedron Lett.* **34**, 765 (1993).

37. J.J. Rabasco and S.R. Kass, *Tetrahedron Lett.* **32**, 4077 (1991).

38. F.M. Bickelhaupt, *et al., J. Am. Chem. Soc.* **117**, 9889 (1995).

39. M.E. Jones and G.B. Ellison, *J. Am. Chem. Soc.* **111**, 1645 (1989).

40. R.C. Lum and J.J. Grabowski, *J. Am. Chem. Soc.* **110**, 8568 (1988).

41. C.H. DePuy, *et al., J. Am. Chem. Soc.* **112**, 8650 (1990).

42. S. Gronert, et al., J. Am. Chem. Soc. **113**, 4009 (1991).
43. T. Minato and S. Yamabe, J. Am. Chem. Soc. **110**, 4586 (1988).
44. F.M. Bickelhaupt, et al., J. Am. Chem. Soc. **115**, 9160 (1993).
45. S. Gronert, et al., J. Org. Chem. **60**, 488 (1995).
46. G.N. Merrill, et al., J. Phys. Chem. , submitted
47. S. Gronert, J. Am. Chem. Soc. **115**, 10258 (1993).
48. L.A. Curtiss, et al., J. Chem. Phys. **94** (1991).
49. M.J.S. Dewar, et al., J. Am. Chem. Soc. **107**, 3902 (1985).
50. W.-P. Hu and D.G. Truhlar, J. Am. Chem. Soc. **118**, 860 (1996).
51. S. Gronert, J. Am. Chem. Soc. **115**, 652 (1993).
52. S. Gronert, J. Org. Chem. **59**, 7046 (1994).
53. S. Gronert, J. Am. Chem. Soc. **114**, 2349 (1992).
54. S. Gronert, J. Am. Chem. Soc. **113**, 6041 (1991).
55. S. Gronert and J.M. Lee, J. Org. Chem. **60**, 4488 (1995).
56. S. Gronert and P. Freed, J. Org. Chem. **61**, 9430 (1996).
57. A.D. Becke, Phys. Rev. A **38**, 3098 (1988).
58. S.H. Vosko, et al., Can. J. Phys. **58**, 1200 (1980).
59. A.D. Becke, J. Chem. Phys. **98**, 5648 (1993).
60. C. Lee, et al., Phys. Rev. B **37**, 1988 (1988).
61. J.C. Slater, Quantum Theory of Molecules and Solids (McGraw-Hill, New York, 1974).
62. J.A. Pople, et al., Isr. J. Chem. **33**, 345 (1993).
63. M.J. Frisch, et al. (Gaussian, Inc., Pittsburgh, PA, 1995).
64. D.R. Stull, et al., The Chemical Thermodynamics of Organic Compounds (John Wiley, New York, 1969).
65. D.S. Baily, et al., J. Am. Chem. Soc. **92**, 6911 (1970).
66. C.H. DePuy, et al., J. Am. Chem. Soc. **111**, 1968 (1989).
67. J. Sicher, et al., Coll. Czech. Chem. Commun. **36**, 3633 (1971).
68. S.J. Cristol and E.F. Hoegger, J. Am. Chem. Soc. **79**, 3438 (1957).
69. C.H. DePuy, et al., J. Am. Chem. Soc. **87**, 2421 (1965).
70. C.H. DePuy, et al., J. Am. Chem. Soc. **84**, 1314 (1962).
71. J.J. Rabasco and S.R. Kass, J. Org. Chem. **58**, 2633 (1993).
72. R.J. Moss, et al., J. Org. Chem. **50**, 5132 (1985).
73. R.J. Moss and B. Rickborn, J. Org. Chem. **51**, 1992 (1986).
74. D. Tobia and B. Rickborn, J. Org. Chem. **51**, 3849 (1986).
75. J. Berkowitz, et al., J. Phys. Chem. **98**, 2744 (1994).
76. W.N. Olmstead and J.I. Brauman, J. Am. Chem. Soc. **99**, 4219 (1977).
77. D.R. Anderson, et al., J. Am. Chem. Soc. **105**, 4244 (1983).
78. C. Li, et al., J. Am. Chem. Soc. **118**, 9360 (1996).

79. H.C. Brown, *J. Chem. Soc.* , 1248 (1956).
80. E.L. Eliel, *Stereochemistry of Carbon Compounds* (McGraw-Hill, New York, 1962).
81. P.J.C. Fierens and P. Verschelden, *Bull. Soc. Chim. Belg.* **68**, 580 (1952).
82. C.A. Bunton, *Nucleophilic Substitution at a Saturated Carbon Center* (Elsevier, Amsterdam, 1963).
83. R.D. Bach and G.J. Wolber, *J. Am. Chem. Soc.* **107**, 1352 (1985).
84. R.D. Bach, *et al.*, *J. Org. Chem.* **51**, 1030 (1986).
85. V.M. Bierbaum, *et al.*, *J. Am. Chem. Soc.* **98**, 4229 (1976).
86. C.H. DePuy, *et al.*, *J. Am. Chem. Soc.* **104**, 6483 (1982).
87. C.J.M. Stirling, *Tetrahedron* **41**, 1613 (1985).
88. S. Gronert and J.M. Lee, *J. Org. Chem.* **60**, 6731 (1995).
89. V.M. Bierbaum and S. Gronert, , unpublished results
90. G. Wittig and R. Polster, *Ann.* **599**, 13 (1956).
91. W.H. Saunders, Jr. and D. Pavlovic, *Chem. Ind. (London)* , 180 (1962).
92. S.S. Glad and F. Jensen, *J. Am. Chem. Soc.* **116**, 9302 (1994).
93. S.S. Glad and F. Jensen, *J. Phys. Chem.* **100**, 16892 (1996).
94. S.S. Glad and F. Jensen, *J. Org. Chem.* **62**, 253 (1997).
95. P.A. Nielsen, *et al.*, *J. Am. Chem. Soc.* **118**, 10577 (1996).

COMPUTATIONAL ANALYSES OF PROTOTYPE CARBENE STRUCTURES AND REACTIONS

HOLGER F. BETTINGER and PAUL v. R. SCHLEYER

Computer Chemistry Center, Institut für Organische Chemie, Friedrich–Alexander–Universität Erlangen–Nürnberg, Henkestr. 42, D–91054 Erlangen, Germany

PETER R. SCHREINER

Institut für Organische Chemie, Georg–August–Universität Göttingen, Tammannstr. 2, D–37077 Göttingen, Germany

HENRY F. SCHAEFER III

Center for Computational Quantum Chemistry, The University of Georgia, Athens, GA, 30602, USA

The application of *ab initio* and density functional theory (DFT) to selected problems in carbene chemistry is reviewed. A brief historic introduction is followed by an analysis of the theoretical techniques necessary to describe carbene structures and properties adequately. An evaluation of the computational methods {self-consistent field (SCF), nth order Møller-Plesset (MPn), complete active space (CAS), multireference configuration interaction (MR-CISD), coupled cluster (CC), full configuration interaction (FCI), and DFT} as well as basis set dependencies is presented. Most notably, hybrid Hartree–Fock/DFT results (B3LYP, in conjunction with moderate basis sets, e.g., triple-ζ (TZ) plus polarization) are reasonably close in quality to highest level treatments {e.g., CCSD(T)/TZ2P}, at much lower computational costs. The usual problems associated with pure Hartree-Fock based methods, such as spin contamination, are much less pronounced at hybrid Hartree–Fock/DFT levels. The structures, properties, and reactions of simple alkyl (methylene, ethylidene, dimethylcarbene, 2–butylidene, and *tert*.-butylcarbene), cyclic (parent as well as substituted cyclopropylidenes, cyclobutylidene, cyclopentadienylidene, and cyclopenta-1,3-dien-5-yl carbene), unsaturated (vinyl carbene, vinylidene, propynylidene, and propadienylidene), and aromatic carbenes (phenyl- and naphthylcarbene) are presented systematically.

1   Introduction

The history of quantum chemistry is closely intertwined with methylene, which is probably the most thoroughly studied highly reactive molecule. The controversial but nonetheless fruitful interplay between experiment and theory led to the *"unraveling of the structure and energetics of the CH_2 molecule."*[1] The history of research on methylene is worth outlining due to its importance to computational quantum chemistry.[1-4]

In their pioneering 1960 paper, Foster and Boys[5] presented the first *ab initio* computations on methylene. When they submitted their paper, *no* spectroscopic data were available; their study was thus purely predictive. Using an orthogonalized Slater basis set in a configuration interaction (CI) calculation employing 128 configurations, they obtained a 129° angle for the triplet ground state, and a singlet–triplet energy separation (ΔE_{ST}) of 1.06 eV (24.4 kcal mol^{-1}).

However, just at that time Herzberg and co–workers published their ground-breaking results on the flash photolysis of diazomethane.[6,7] They were able to observe methylene, and the spectra obtained suggested a linear structure in a nontotally symmetric electronic state, although there were some inconsistencies in the moments of inertia. An alternative bent structure ($\Theta_e = 140°$) was rejected due to the lack of side bands, which should arise from transitions from higher rotational levels of the ground state. As the $^3\Sigma_g^-$ nature of the ground state of methylene and its structure seemed to be on a solid experimental basis, the results of Foster and Boys were generally regarded as a failure of *ab initio* theory.[8]

Although some semiempirical[9,10] and *ab initio* studies[11,12] gave a bent triplet ground state of methylene, the experimental results were seriously challenged first in 1970 by Bender and Schaefer.[13] Their study, which employed configuration interaction with single and double substitution out of the SCF wave function (CISD) and a double–ζ basis set, was called by Gaspar and Hammond,[14] "*by far the most elaborate calculation carried out to that date on methylene, or indeed almost any molecule.*" Bender and Schaefer predicted a 135° angle and a bond length of 1.096 Å for triplet methylene. This was in sharp contrast to Herzberg's adopted interpretation (180°, 1.03Å) of the CH_2 spectrum, but agreed well with the rejected structure characterized by a 140° angle and a bond length of 1.07Å. The flat $^3B_1$ hyperface obtained by Bender and Schaefer predicted a linearization barrier of 7 kcal mol^{-1}.

Also in 1970, ESR spectroscopy of matrix isolated methylene independently performed by Bernheim *et al.*[15] and by Wassermann *et al.*[16] predicted a bent triplet state with an angle between 128° and 143° with the most probable value being 136°. This was in remarkable agreement with Bender and Schaefer's prediction! Herzberg reinterpreted his older work and graciously concluded in 1971 that the triplet is bent between 128° and 148°.[17]

Further ESR experiments and *ab initio* investigations, summarized by Harrison in his 1974 review,[18] confirmed the bent structure of triplet methylene, although the experimental studies favored different angles. To arrive at a final theoretical answer on the triplet methylene structure, McLaughlin, Bender, and Schaefer[19] performed CISD calculations using a larger basis set than in their 1970 paper.[13] Employing such a DZP basis in a CI calculation was without precedent in 1972. In good agreement with their earlier CISD/DZ result, they obtained an angle of 134.0°, which they estimated to be reliable to within 2°. Their theoretical prediction was confirmed by Bunker and Jensen,[20] who obtained an angle of 133.84 ± 0.05°

experimentally for the triplet ground state. However, the barrier for inversion of the molecule is low, 1940 ± 80 cm^{-1} (5.5 kcal mol^{-1}), and there are only two vibrational levels below that barrier.

Not only the geometry of the ground state, but also the singlet–triplet energy splitting of methylene has been the subject of heated debates between experimentalists and theoreticians. *Ab initio* computations predicted the energy difference between the equilibrium geometries of both states to be $T_e = 11 \pm 2$ kcal mol^{-1},[21,22] in acceptable agreement with several experiments ($T_e = 8.5 \pm 0.9$ kcal mol^{-1}).[23-25] In 1976, however, Lineberger as well as Ellison and co–workers predicted a singlet-triplet energy gap of 19.5 kcal mol^{-1} using laser photodetachment techniques on CH$_2^-$ anion.[26,27] Again, the theoretical results were considered to be incorrect by experimentalists,[1] although several higher level calculations[28-31] and a number of indirect experimental measurements[32-40] consistently gave much lower values for T_e than Lineberger's determination.[26] This considerable discrepancy was resolved by Hayden *et al.*[41] and McKellar *et al.*[42] The latter group using far infrared laser resonance spectroscopy with Zeeman modulation published the most accurate determination of the singlet–triplet splitting of methylene:

$$T_o = 3165 \pm 20 \text{ cm}^{-1} \text{ (9.05} \pm 0.06 \text{ kcal mol}^{-1})$$

Note that McKellar's estimate for the zero–point vibrational energy correction has the wrong sign resulting in $T_e < T_o$. It is now universally agreed that $T_e > T_o$ for the $\tilde{a}\ ^1A_1$ state of methylene.[43] Hence, the McKellar's estimates for T_e are not given.

Sears and Bunker reinterpreted the CH$_2^-$ photoelectron spectrum and showed that it is in accord with the small singlet–triplet gap.[44] Finally Lineberger *et al.* also obtained a revised singlet–triplet gap around 9 kcal mol^{-1} using photodetachment of the CH$_2^-$ anion, and conceded that the earlier result was due to "hot" bands.[45] Fitting McKellar's experimental data using the MORBID Hamiltonian,[46] adding a relativistic correction,[47] and a Born–Oppenheimer diagonal correction[48] gives the current benchmark values for comparison between experiment and theory:

$$T_e = 9.372 \text{ kcal mol}^{-1}$$
$$T_o = 9.155 \text{ kcal mol}^{-1}$$

The success of quantum chemistry is, not limited to methylene. Other carbenes have been studied by electronic structure theory as well. The following contribution will focus on theoretical studies of the structures, properties, and reactions of simple "prototype" carbenes, like unsaturated, cyclic and alkyl-, vinyl-, as well as aryl-substituted carbenes. While experimental carbene chemistry has been reviewed very extensively,[49-60] there are only a limited number of older theoretical overviews.[61,62] Although this article does not attempt to cover the theoretical literature on carbenes exhaustively, it should provide insights and leading references to computational research on carbenes.

2   Electronic Structure and Electronic Structure Computations

The electronic structure of carbenes can be understood following Walsh's approach[63] by following the changes in the nature and the energies of the molecular orbitals (MO's) comprising methylene first at a qualitative level. Starting with the linear $D_{\infty h}$ geometry, it is easy to comprehend what happens to each MO upon bending. The MO's are built from s and p atomic orbitals (AO's). As shown in Scheme 2.1, combinations of the hydrogen $1s$ orbitals yield one σ_g and one σ_u orbital. Symmetry adapted combinations (meaning that only orbitals of the same point group are considered) of the carbon σ_g, σ_u, and π_u orbitals yield two bonding ($2\sigma_g$ and $1\sigma_u$), two degenerate non–bonding ($1\pi_u$), and two anti–bonding ($3\sigma_g$ and $2\sigma_u$) molecular orbitals. Although the $1s$ atomic orbital on carbon is also of σ_g symmetry, it cannot mix effectively (i.e., the overlap is negligibly small) with the σ_g hydrogen AO's since the carbon $1s$ AO is about 250 kcal mol^{-1} below the hydrogen orbitals. As a general consequence, the carbon "core orbital" has very little effect on the chemistry of carbon compounds, and is very often kept "frozen" in MO calculations.

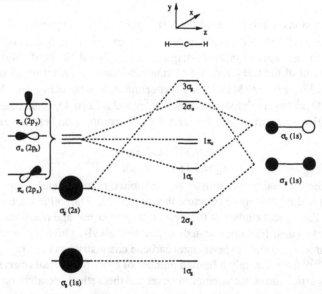

Scheme 2.1 Orbitals of linear methylene through symmetry adapted combination of atomic orbitals .

Occupying the orbitals with the eight valence electrons in energetic order yields the lowest energy configuration $(1\sigma_g)^2(2\sigma_g)^2(1\sigma_u)^2(1\pi_u)^2$. While this procedure is straightforward for the σ orbitals, the energy degeneracy of the $1\pi_u$ orbitals gives rise to three linear electronic states, namely $^1\Sigma_g^+$, $^1\Delta_g$, and $^3\Sigma_g^-$ (Table 2.1).[64] Applying Hund's rule for arranging two electrons (possibly with α and/or β spins) in two degenerate orbitals results in a $^3\Sigma_g^-$ triplet ground state ($\alpha\alpha$ or $\beta\beta$) for linear methylene.

Bending of the linear $D_{\infty h}$ to the bent C_{2v} form lifts the degeneracy of the two non–bonding π_u orbitals (Scheme 2.2). The energy and shape of one of the π_u orbitals does not change upon bending (to become $1b_1$) because the hydrogens lie in the nodal plane. The second π_u MO, which is fully symmetric in C_{2v} symmetry ($3a_1$), is lowered due to orbital mixing with the low-lying carbon $2s$ AO. As the $1\sigma_u$ orbital is *anti*-bonding for the hydrogens, its energy increases with decreasing H–C–H angle (to become the $1b_2$ MO). The carbon $2s$ contribution to the $2\sigma_g$ *decreases* upon bending to $2a_1$. Thus, for bent CH_2 the following energy ordering of the molecular orbitals is expected:

$$1a_1 \; 2a_1 \; 1b_2 \; 3a_1 \; 1b_1$$

Moreover, the $3a_1$ and $1b_1$ molecular orbitals approach each other energetically with increasing H-C-H angle until they become degenerate in linear configurations.

The occupation of the MOs of bent methylene (Fig. 2.2) is straightforward, and a $^3B_1$ state results for the triplet (Table 2.1). Three states arise in the singlet manifold depending on the occupation of the energetically similar $3a_1$ and $1b_1$ molecular orbitals. The state with the doubly occupied $3a_1$ orbital is the lowest lying singlet state of methylene, labeled $\tilde{a}\;^1A_1$, with the $\tilde{b}\;^1B_1$ and $\tilde{c}\;^1A_1$ states being higher lying singlet states.[43]

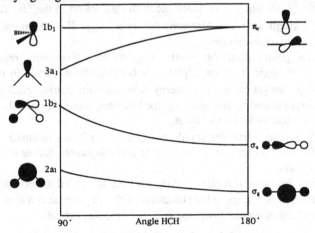

Scheme 2.2 Walsh diagram for methylene.

For the singlets, the occupation of a energetically higher lying molecular orbital does not necessarily result in a higher lying state. Although the sum of the orbital energies increases, the electron–electron repulsion can decrease. If this decrease is more than the orbital energy sum increases, the resulting state is more favorable despite the occupation of higher lying MOs.

Hence, the bent $^3B_1$ state not necessarily is lower in energy than the linear $^3\Sigma_g^-$, although the sum of the orbital energies is lower for the bent geometry. These *a priori* not obvious facts were the reason for the long lasting uncertainty about the geometry of the methylene ground state.

TABLE 2.1: Electronic states and their primary electronic configuration for linear and bent methylene.

linear methylene, $D_{\infty h}$		bent methylene, C_{2v}	
$^3\Sigma_g^-$	$(1\sigma_g)^2(2\sigma_g)^2(1\sigma_u)^2(1\pi_u)^2$	$\tilde{X}\ ^3B_1$	$(1a_1)^2(2a_1)^2(1b_2)^2(3a_1)^1(1b_1)^1\ (\alpha\alpha)$
$^1\Delta_g$	$(1\sigma_g)^2(2\sigma_g)^2(1\sigma_u)^2(1\pi_u)^2$	$\tilde{a}\ ^1A_1$	$(1a_1)^2(2a_1)^2(1b_2)^2(3a_1)^2\ (\alpha\beta)$
		$\tilde{b}\ ^1B_1$	$(1a_1)^2(2a_1)^2(1b_2)^2(3a_1)^1(1b_1)^1\ (\alpha\beta-\beta\alpha)$
$^1\Sigma_g^+$	$(1\sigma_g)^2(2\sigma_g)^2(1\sigma_u)^2(1\pi_u)^2$	$\tilde{c}\ ^1A_1$	$(1a_1)^2(2a_1)^2(1b_2)^2(1b_1)^2\ (\alpha\beta)$

Several conclusions can be drawn from Walsh's[63] correlation diagram for CH_2. The consequences not only are important for methylene, but even more so for understanding of the energies and the geometries of substituted carbenes.

- The H–C–H angle is expected to be the smallest for the $\tilde{a}\ ^1A_1$ state. The electron lone–pair occupies the $3a_1$ orbital which is in–plane with the CH bonding electron pairs. According to VSEPR theory,[64] the repulsion of the lone pair electrons with the CH bonds results in a small H–C–H angle.
- Similarly, the smaller in–plane electron interaction in the $^3B_1$ state (which has only a single in–plane electron) favors a larger H–C–H angle, since the CH–CH repulsion dominates.
- In turn, geometrical constraints, e. g., bulky substituents, resulting in a larger R–C–R angle favor the triplet state because the $3a_1$ and $1b_1$ orbitals are very similar energetically. The energy necessary for pairing electron spins is not overcompensated by the energy gained by occupying the only slightly lower lying $3a_1$ orbital with two electrons.
- The opposite is true for small angles, e. g., carbenes in small rings. The $3a_1$ and $1b_1$ orbitals separate energetically and this favors double occupation of the $3a_1$ orbital.
- Mixing the $2s$ central carbon atom with the π_u orbital lowers the energy of the $3a_1$ orbital. A change in hybridization of the $3a_1$ orbital changes its energy. The more σ character, the lower the energy of the $3a_1$ orbital.

2   Electronic Structure Computations

2.1   General Considerations

An adequate theoretical description of the lowest lying singlet state of methylene is not trivial due to the similar energies of the $3a_1$ and $1b_1$ molecular orbitals. In linear methylene these orbitals are degenerate and their occupation results in a $^1\Delta_g$ state with two degenerate components, $^1A_1$ and $^1B_1$, in terms of the C_{2v} subgroup of $D_{\infty h}$. The electron configuration of the $^1\Delta_g$ state is then:

$$(1a_1)^2(2a_1)^2(1b_2)^2[(3a_1)^2 - c\,(1b_1)^2],\ \text{with}\ c = 1.$$

Thus, for a proper description of the $^1\Delta_g$ state, a wave function is needed which takes into account the two configuration state functions (CSF's), constituting this electronic state. Clearly, a single-reference (single-configuration) Hartree–Fock (HF) wave function is not suited for this problem. Although bending of the molecule lifts the degeneracy to give an $^1A_1$ and a $^1B_1$ state, the HF wave function

$$(1a_1)^2(2a_1)^2(1b_2)^2(3a_1)^2\ (\alpha\beta)$$

is expected to be a poor approximation to the molecular wave function of the $^1A_1$ state, because it does not account for the contribution of the second configuration. This nearly degenerate situation would be better approached with a two–configuration function

$$(1a_1)^2(2a_1)^2(1b_2)^2[(3a_1)^2 - c\,(1b_1)^2],\ \text{with}\ c \neq 1.$$

In contrast, the $^3B_1$ state can be described with one CSF, and the HF wave function is thus expected to be a reasonable approximation to the correct wave function. As a consequence, HF theory treats the triplet and singlet state of methylene in an unbalanced way. The triplet state is described "better" than the singlet state resulting in singlet–triplet energy separations which are too large.

2.2   Evaluation of theoretical methods

The two–configuration character of the singlet ground state of methylene makes the correct computation of the singlet–triplet energy separation (ΔE_{ST}) of carbenes notoriously difficult and time-consuming. High level treatments, which give perfect agreement with the experimental ΔE_{ST}, are feasible for methylene due to its the small size. However, limited computational resources often do not yet allow such high level calculations for larger carbenes. Thus, knowing the performance of practicable quantum chemical methods for computing the ΔE_{ST} of CH_2 allows an evaluation of the quality of theoretical results for larger molecules.

The time–independent Schrödinger equation within the limits of a particular basis set can be solved "exactly" by configuration interaction (CI) methods if the CI space is not truncated.[a] Such full configuration interaction (FCI) studies of methylene employing several basis sets of double–ζ quality calibrate results for ΔE_{ST} as well as for the geometries.[43,66,67] Although the FCI/DZP ΔE_{ST} values are more than 3 kcal mol^{-1} above the firmly established experimental value,[42] the "exact" solution of the Schrödinger equation in the space spanned by the double–ζ basis set, allows for a systematic comparison of the performance of various computational methods (Table 2.2). The results presented were obtained at selected levels of theory employing two different DZP basis sets (one uses the same {DZP},[43] the other {DZP'}[66,67] uses different orbital exponents for the polarization functions for the $^3B_1$ and the $^1A_1$ states) and the popular, somewhat smaller split–valence 6–31G* basis set.

The semiempirical AM1 method is obviously not suited to study the singlet–triplet separation of carbenes; the ΔE_{ST} value is 30 kcal mol^{-1} for methylene. Similarly, HF/DZP theory predicts a ΔE_{ST} of 26.4 kcal mol^{-1}, more than twice as large as the FCI result. The triplet state is too stable relative to the singlet state as a consequence of neglecting the energy contributions from the second configuration. This still holds true for rather simple correlation treatments which are based on the HF reference wave function, like MP2, MP4, and CISD. The ΔE_{ST} values predicted by these methods are higher than FCI by 7.9, 4.5, and 2.8 kcal mol^{-1}, respectively.

The explicit inclusion of the second configuration

$$(1a_1)^2(2a_1)^2(1b_2)^2(1b_1)^2$$

in a two–configuration SCF (TC–SCF) treatment improves the ΔE_{ST} appreciably. Results similar to TC–SCF are obtained with the complete active space SCF (CASSCF) method,[68-71] which not only takes two configurations into account, but all configurations which can be obtained by distributing the active electrons into the active orbitals. Thus, CASSCF can be visualized as a full configuration interaction treatment within a small "active" window comprised of the chemically most important orbitals. However, dynamic correlation effects (as considered in MRCI) are not included.

---

[a] However, the Born-Oppenheimer approximation still is used.[65]

TABLE 2.2 Singlet–triplet energy separations of methylene with various theoretical methods and DZP or 6–31G* basis sets.

Method	Reference	T_e	T_0
AM1	–	–	30
HF/DZP	72	26.4	26.0
(TC)SCF/DZP [a]	72	13.6	13.1
CASSCF(2,2)/6–31G* [b]	73	14.5	14.0
CASPT2-g1(2,2)/6–31G* [b,c]	73	–	15.4
MP2/6-31G* [d]	73	20.9	20.4
MP4SDTQ/6-31G*//MP2/6-31G* [d]	73	17.1	16.7
B3LYP/6-31G*	73	13.7	13.3[g]
CISD/DZP [d]	43	15.5	15.0
(TC)CISD/DZP [a,d]	43	12.6	12.1
CISDTQ/DZP [d]	43	12.7	12.2
CCSD/DZP [d]	43	13.6	13.1
CCSD(T)/DZP [d]	43	12.9	12.4
FCI/DZP [d,e]	43	12.66	12.2
FCI/DZP' [d,f]	66,67	11.97	–
Experiment	42	[h]	9.1
Best corrections to experiment	46-48	9.4	9.2

[a] (TC) designates a two–configuration reference for the singlet state, for the triplet state one configuration is used.

[b] Singlet geometry optimized at CASSCF/6–31G*; triplet at ROHF/6–31G*.

[c] Single point energies were calculated using CASPT2 for both singlet and triplet.

[d] One core orbital frozen.

[e] C(9s5p1d/4s2p1d), H(4s1p/2s1p) with Cartesian d-type polarization functions carbon ($\alpha_d = 0.75$) and p type polarization functions on each hydrogen ($\alpha_p = 0.75$).

[f] C(9s5p1d/4s2p1d), H(4s1p/2s1p) with d-type polarization functions carbon ($\alpha_d = 0.74$ and $\alpha_d = 0.51$ for $^3B_1$ and $^1A_1$ states, respectively) and p type polarization functions on each hydrogen ($\alpha_p = 1.0$).

[g] including the 0.2 kcal mol^{-1} ZPVE correction taken from Ref. 42

[h] 8.6 kcal mol-1; note that the zero–point correction has the wrong sign

Two– and multi–configuration methods generally are improved when dynamic correlation is considered. Dynamic electron correlation can be included at the CASPT2 level by means of a Møller-Plesset second order perturbation (MP2) treatment within the active space configurations. The improvement is considerable, but shows that non-dynamic effects dominate (Table 2.2). Including the two most important reference configurations in a CISD treatment (TC-CISD) gives the FCI ΔE_{ST} value for the methylene.

Density functional theory (DFT) has become a powerful tool in quantum chemistry the past few years.[74-77] The electron density is the basic variable instead of the many-body wave function of traditional quantum chemical methods. The formal analogy with HF theory is revealed in the Kohn-Sham formulation of DFT. The HF exact exchange for a single determinant is replaced by an exchange-correlation functional of the density. This introduces electron correlation into the DFT treatment. The HF/DFT hybrid methods use linear combinations of HF exchange and some exchange functionals. Becke's three parameter functional[78] employing the correlation functional of Lee, Yang, and Parr[79] (B3LYP) has proven to be especially useful in quantum chemical studies.[73,77,80,81] The B3LYP method treats the singlet and triplet states of methylene in a balanced way; the error relative to the FCI result is only 1.1 kcal mol^{-1}. This good agreement suggests that B3LYP can be a valuable "low cost" method in theoretical studies of carbene reactions.

Highly correlated methods, although formally based on the HF configuration, give excellent agreement with FCI. The coupled cluster method including single and double excitations (CCSD) out of the HF reference improves ΔE_{ST} significantly; the error relative to the FCI results is only 0.9 kcal mol^{-1}. Adding connected triple excitations *via* the CCSD(T) method eliminates that small CCSD error and agrees very well with the FCI results. The CISDTQ method, which is also based on a single HF wave function, reproduces the FCI values as well. However, it should be kept in mind that although a CISDTQ treatment of an eight electron system comes close to the FCI calculation, this will not be true for larger systems. Here, the differences between CCSD(T) or CISDTQ and FCI predictions are expected to be more pronounced.

2.3 Effect of basis set

It is apparent from the FCI/DZP $\Delta E_{ST}(CH_2)$ result that basis sets larger than DZP are needed for better accuracy. Unfortunately, even for molecules as small as CH_2, basis sets larger than DZP are beyond current capabilities for FCI treatments.[43] Going from a double-ζ to a triple-ζ quality basis set improves the HF and CASSCF results only marginally (Table 2.3). In contrast, MP2 and CISD give better agreement with experiment, although the error is still 4-8 kcal mol^{-1}. Both CASPT2-g1 and B3LYP deviate by 2.3 kcal mol^{-1} from experiment when Dunning's correlation consistent triple-ζ quality basis set, including f functions on carbon, is used. Again, B3LYP seems to offer a cheap alternative to study the chemistry of carbenes.

The G2 theory[82,83] achieves high accuracy by combining results of several lower level computations to approximate a large high level calculation which is not practical for larger systems.[84] G2 heats of formation at 0K of $^3B_1$ and $^1A_1$ states of methylene deviate from experiment only by -1.0 kcal mol^{-1} and 1.4 kcal mol^{-1}, respectively.[83] Although these errors are small, their opposite sign results in a too small $T_o = 6.7$ kcal mol^{-1} and $T_e = 6.9$ kcal mol^{-1}).

Very large basis sets combined with highly correlated methods can reproduce the experimental singlet-triplet splitting of methylene. Bauschlicher, Langhoff, and Taylor[85] demonstrated that a second–order CI (SOCI) calculation with large Gaussian basis sets including g functions gives $T_o(CH_2) = 9.1$ kcal mol^{-1}.

DZP basis sets are not appropriate for accurate calculations of molecular geometries (Table 2.4). The FCI/DZP bond lengths are too long and the HCH angles are too small. Larger basis sets improve the results and an estimated complete basis set gives excellent agreement with experiment.

TABLE 2.3 Singlet–triplet energy separations of methylene with various theoretical methods and basis sets.

Method	Reference	T_e	T_0
HF/6-311+G**	This work	28.7	28.4
MP2/6-311+G** [a]	This work	17.4	17.0
B3LYP/6-311+G**	This work	12.2	11.8
B3LYP/cc-pVTZ	This work	11.8	11.4
CISD/6-311+G** [a]	This work	14.7	14.3
CISD/cc-pVTZ [a]	This work	13.2	12.8
CASSCF(2,2)/cc-pVTZ [b,c]	73	–	14.3
CASPT2-g1(2,2)/cc-pVTZ [b,d]	73	–	11.4
SOCI/[5s4p3d2f1g/4s3p2d]ANO//SOCI/TZ2P[a,e]	85	9.1	9.0
SOCI/TZ3P(2f,2d)+2diff// (TC)CISD/TZ3P(2f,2d)+2diff [e,f,g]	72	9.4	9.0
MP4SDTQ/9s7p2d1f/5s2p//CMRCI	86	11.5	–
CCSDT-1/9s7p2d1f/5s2p//CMRCI [h]	86	10.1	–
TD-CCSD/6s5p3d2f/4s3p2d ANO [a]	87	9.2	8.9
G2	82,83	6.9	6.7
est. complete basis CMRCI [i]	88	9.2	–
est. complete basis RCCSD(T) [i]	88	9.5	–
Best corrections to experiment	46-48	9.4	9.2

[a] One core orbital frozen.

[b] Singlet geometry optimized at CASSCF/6–31G*; triplet at ROHF/6–31G*.

[c] Single point energies were calculated using CASSCF for singlet, ROHF for triplet.

[d] Single point energies were calculated using CASPT2 for both singlet and triplet.

[e] SOCI includes all configurations having no more than two electrons in external orbitals.

[f] (TC) designates a two–configuration reference for the singlet state, for the triplet state one configuration is used.

[g] One frozen core and one deleted virtual orbital.

[h] The CCSDT-1 model adds to CCSD some linear contributions from triple excitations (see Refs. 89-91)

[i] Based on all–electron contracted multireference configuration interaction (CMRCI) or RCCSD(T) calculations employing the cc–pVXZ basis sets up to X=5 (quintuple-ζ).

TABLE 2.4 Geometric parameters of $\tilde{a}\,{}^1A_1$ and $\tilde{X}\,{}^3B_1$ $(r_e/\text{Å}, \Theta_e/°)$

Method	Reference	$\tilde{a}\,{}^1A_1$		$\tilde{X}\,{}^3B_1$	
		r_e	Θ_e	r_e	Θ_e
SCF/DZP	72	1.101	103.7	1.076	129.4
(TC)SCF/DZP [a]	72	1.104	102.7	–	–
(TC)CISD/DZP [a,b]	72	1.117	101.5	1.085	131.8
CISDTQ/DZP [b]	43	1.120	101.4	1.088	132.2
CCSD/DZP [b]	43	1.119	101.2	1.087	132.1
CCSD(T)/DZP [b]	43	1.120	101.3	1.088	132.2
FCI/DZP [b,c]	43	1.120	101.4	1.088	132.2
FCI/DZP' [b,d]	67	1.120	101.8	1.084	132.5
B3LYP/6–311+G**	This work	1.114	101.5	1.080	135.4
B3LYP/TZ2P	This work	1.109	101.9	1.077	135.2
(TC)CISD/TZ3P(2f,2d)+2diff [a,e]	72	1.105	102.3	1.075	133.0
est. complete basis CMRCI [f]	88	1.106	102.2	1.075	133.7
est. complete basis RCCSD(T) [f]	88	1.106	102.2	1.075	133.8
Experiment (vis. abs.) [g]	92	1.11	102.4	1.078	~136
Experiment (IR) [g]	93	1.107	102.4	–	–

[a] (TC) designates a two–configuration reference for the singlet state, for the triplet state one configuration is used.

[b] One core orbital frozen.

[c] C(9s5p1d/4s2p1d), H(4s1p/2s1p) with Cartesian d-type polarization functions carbon $(\alpha_d = 0.75)$ and p type polarization functions on each hydrogen $(\alpha_p = 0.75)$.

[d] C(9s5p1d/4s2p1d), H(4s1p/2s1p) with d-type polarization functions carbon $(\alpha_d = 0.74$ and $\alpha_d = 0.51$ for ${}^3B_1$ and ${}^1A_1$ states, respectively) and p type polarization functions on each hydrogen $(\alpha_p = 1.0)$.

[e] One frozen core and one deleted virtual orbital.

[f] Based on all–electron contracted multireference configuration interaction (CMRCI) or RCCSD(T) calculations employing the cc–pVXZ basis sets up to X=5 (quintuple-ζ).

[g] Zero point geometry (r_0,Θ_0)

3 Energies and Structures of Prototype Carbenes

3.1 General Considerations

For substituted carbenes R_1CR_2, which in the general are less symmetric, the two important molecular orbitals usually are named σ and p.[49,94] The MO of σ symmetry is "in–plane" with the C–R and C–R' bonds, whereas the p orbital is perpendicular to the plane of the molecule. The resulting electronic states are labeled $^1(\sigma^2)$, $^3(\sigma p)$, $^1(\sigma p)$, $^1(p^2)$.

Scheme 3.1 The σ and p orbitals of a carbene.

Substituents can stabilize carbenes either *via* π–type interaction with the *p* carbene orbital or *via* inductive (electronegativity) effects through the σ electron skeleton of the molecule. Luke *et al.* provided a systematic study of substituent effects on the stabilization of the singlet and triplet states.[95,96] The effects of the substituent X (X = Li, BeH, BH$_2$, CH$_3$, NH$_2$, OH, F) on the stabilities of the singlet and triplet carbenes XCH were evaluated using the following equations:

$$XCH \text{ (singlet)} + CH_4 \longrightarrow H_3CX + CH_2 \text{ (singlet)} \qquad (3.1)$$
$$XCH \text{ (triplet)} + CH_4 \longrightarrow H_3CX + CH_2 \text{ (triplet)} \qquad (3.2)$$

According to the MP4SDTQ/6–31G*//HF/3–21G results, singlet states are stabilized both by π donating and π accepting substituents (in different orientations), the only exception being BeH which stabilizes the triplet state more than the singlet. Triplet states also are stabilized by substitution, but the extend differs for a given substituent X. The σ donating and π accepting Li and BeH favor triplet states, while all other first row substituents stabilize the singlet state preferentially. However, the σ contribution to the stabilization of carbenes is small, because both the C–X bonding orbitals and the σ orbital have "sp^2" character. The small rehybridization associated with partially ionic bonding increases or decreases the σ character at the carbene center due to electronegative or electropositive substituents, respectively.

3.2   Alkylcarbenes

The singlet state of methylcarbene (**1**) is found to be asymmetric (i. e. C_1 point group) at almost all levels of theory.[97-99] The structure (see Scheme 3.2) is characterized by an almost parallel alignment of one CH bond with the axis of the vacant carbene *p* orbital; the corresponding HCC angle is 95°-96° at CCSD/TZ2P(f,d) and CISD/DZP.[98] This CH bond is considerably longer (1.112 Å) than the two other CH bonds of the methyl group (1.096 Å, 1.091 Å).[98] However, the ground state of methylcarbene is a C_s symmetric triplet (^{3}A"). [100] As the methyl group donates π-density to the p orbital, the out-of-plane CH bonds are slightly longer than the CH bond which lies in the nodal plane of the p orbital. One methyl group lowers the singlet–triplet gap by 6-7 kcal mol^{-1} resulting in $\Delta E_{ST}(CH_3CH)$ = 2-5 kcal mol^{-1}.[100-104]

Hyperconjugation was confirmed by Sulzbach *et al.* to be the most important effect determining the singlet–triplet separation of simple alkyl carbenes.[104] Using equations (3.1) and (3.2) the authors find that the singlet states of alkylcarbenes are stabilized about twice as much than the corresponding triplet states (see Tables 2.5 and 2.6). Singlet methylcarbene clearly shows the geometric distortions associated with hyperconjugative interaction between a methyl CH bond and the carbene p orbital. The same features, a rather long β–CH bond and a small HCC angle, are found in hyperconjugatively stabilized carbocations.[105] In the triplet carbene, which is electronically related to a radical, a less favorable two center three electron (2c-3e) interaction is involved. Thus, the stabilizing effect is smaller for the triplet state.

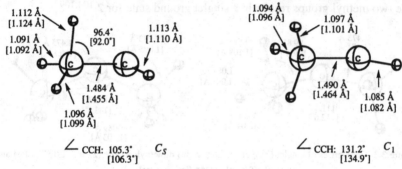

Scheme 3.2 Geometry of singlet (left) and triplet (right) methylcarbene (1) optimized at the CISD/DZP and B3LYP/6–311+G** [in brackets] levels of theory.

The importance of hyperconjugative effects on ΔE_{ST} was rejected by Khodabandeh and Carter (KC) based on various GVB approaches and Mulliken population analyses for methylcarbene.[103] The authors concluded that it is not the singlet state stabilization by the methyl group, but that rather the triplet state destabilization, which results in a smaller singlet–triplet separation with respect to methylene. Their analyses show that the hybridizations of the σ orbitals are almost identical in the singlet states of methylene and of methylcarbene. For the triplet states, however, Mulliken population analyses indicate a decrease in p character when going from methylene to methylcarbene. This minor change in hybridization (7.2% less p character in triplet methylcarbene vs. methylene) was assumed by KC to cause a methylcarbene triplet state destabilization of ca. 4 kcal mol^{-1} resulting in a T_o = 5.1 kcal mol^{-1} at their highest level of theory. However, the Mulliken population analysis is based on a defined but arbitrary partitioning of the molecular orbitals in the Hilbert space, and it is thus not associated with a quantum mechanical observable.[106] Population analyses do not correspond to physical reality, but are an interpretation of it. In contrast, Natural Population Analysis (NPA) by Reed and Weinhold *et al.*[107,108] at B3LYP/6-311+G** gives almost the same hybridization for [3]1 and triplet methylene, but an increase in *p* character in the σ orbital of [1]1 relative to methylene. Applying KC's arguments to the NPA results would suggest a *destabilization* of the methylcarbene singlet state! Analyses in terms of σ effects using results obtained by population analyses is thus not suited to discuss stabilization effects of substituents in carbenes. Rather, isodesmic equations (3.1) and (3.2) should be employed to evaluate the stabilization energies of substituents. This data contradicts KC's conclusions.

The singlet state of dimethylcarbene (2) has C_2 symmetry (just as the isoelectronic 2-propyl cation),[105] whereas the triplet favors C_{2v} symmetry, with staggered methyl conformations.[109] The preferred stabilization of the singlet state by the two methyl groups results in a singlet ground state for **2**.[104,109]

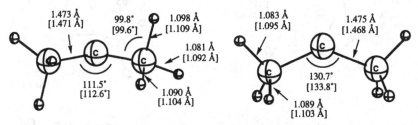

Scheme 3.3 Geometries of singlet (left) and triplet (right) dimethylcarbene (**2**) at CISD/TZ2P+f and B3LYP/6–311+G** [in brackets]

The high level computations (CCSD(T)/TZ2P+f//CISD/TZ2P+f + ZPVE) of Richards *et al.* predict ΔE_{ST} = -1.4 kcal mol[-1]. Sulzbach *et al.*[104] also found the singlet state to be the ground state of **2** (ΔE_{ST} = -2.3 kcal mol[-1]) when employing the method of Hehre, Radom, Schleyer, and Pople at the B3LYP/TZ2P level of theory.[96] This method evaluates the stabilizing effect of the substituents on the singlet ($Stab_S$) and triplet ($Stab_T$) using equations (3.1) and (3.2) and the experimental singlet-triplet splitting of methylene to predict the singlet-triplet splitting of a substituted carbene R_1CR_2.

$$\Delta E_{ST}(R_1CR_2) = \Delta E_{ST}(CH_2) - (Stab_S - Stab_T) \qquad (3.3)$$

An equivalent procedure for correcting the singlet–triplet energy separation obtained at a given level is to apply for the error in the ΔE_{ST} of methylene at the same level. Although this ΔE_{ST} correction seems plausible, it is likely that errors are not negligible. Such a correction gives ΔE_{ST} = 2.3 kcal mol[-1] for **1**, and ΔE_{ST}

= -2.5 kcal mol^{-1} for **2** at B3LYP/TZ2P.[104] Alkyl group stabilization (σ and π) is quite different from hydrogen σ stabilization, and one cannot necessarily expect that the "errors" in computing ΔE_{ST} of methylene to be transferable to larger substituted carbenes.

The *tert*-butyl group has an even more pronounced effect than the methyl group:[104] it stabilizes the singlet by 17.8 kcal mol^{-1} and the triplet by 7.6 kcal mol^{-1} at B3LYP/TZ2P according to equations (3.1) and (3.2). Both Armstrong *et al.*[110] and Sulzbach *et al.*[104] obtain asymmetric structures for *tert*-butylcarbene (**3**) with one very small CCC angle around 80° (see Scheme 3.4). This unusual structure derives from hyperconjugative stabilization of the carbene center by the adjacent CC σ bond, which is elongated considerably. In *tert*-butylfluorocarbene (**4**) the stabilizing effect of the F atom reduces the CC hyperconjugation resulting in a larger CCC angle (100.9°) and a shorter donating CC bond (1.564 Å).[110]

A second *tert*-butyl group does not increase the singlet–triplet gap nearly as much as the first. The steric repulsion between the two *tert*-butyl groups enlarges the central CCC bond angle in singlet di–*tert*–butylcarbene (**5**) to 125°;[104] this increases the *p* character of the carbene σ orbital. Higher *p* character, however, favors the triplet state.

TABLE 3.1 Stabilization (in kcal/mol) of the *tert*-butyl groups in *tert*-butyl– and di–*tert*–butylcarbene according to equations (3.1) and (3.2) at B3LYP/TZ2P.

Carbene	Stab$_S$	Stab$_T$	Stab$_S$ – Stab$_T$
tert-butyl carbene	17.8	7.6	10.2
di–*tert*-butyl carbene	26.6	20.1	6.5

Thus, the second *tert*-butyl group stabilizes the singlet only by an additional 8.8 kcal mol^{-1}, which is about half of the stabilization provided by the first *tert*-butyl group.[104] In contrast, the second *tert*-butyl group stabilizes the triplet state better (by an additional 12.5 kcal mol^{-1}) than the first (7.6 kcal mol^{-1}).[104] Even though the singlet state is stabilized more overall than the triplet state, **5** still has a triplet ground state (ΔE_{ST} = 3 kcal mol^{-1}), due to the geometric distortion which is favorable for the triplet state.[104] Di–*tert*–butylcarbene was the first directly observed alkyl-substituted carbene; its stable ESR signal suggests a triplet ground state.[111]

A similar effect, but in the opposite direction, operates in cyclopropylidene (see below). The small bond angle at the carbene center of ca. 60° decreases the *p* contribution to the carbene σ orbital and favors the singlet state; in the triplet this angle is 69°. The singlet state is stabilized according to equations (3.1) to (3.3) by 25.2 kcal mol^{-1}, whereas the triplet state is destabilized by -0.5 kcal mol^{-1} at B3LYP/6-311+G**.

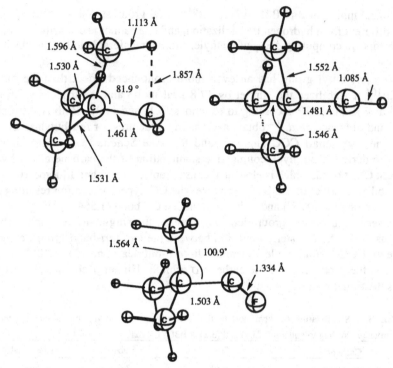

Scheme 3.4 Geometries of singlet (top left) and triplet (top right) *tert*-butylcarbene (**3**), and singlet
tert-butylfluorocarbene (**4**) optimized at B3LYP/6–311+G**.

The relatively large angle at the triplet carbene center is due, as with the triplet CH_2
itself, to the presence of only a single electron in the in–plane hybrid orbital. Thus
a singlet ground state with ΔE_{ST} of -11.1 kcal mol^{-1} and -14.9 kcal mol^{-1} is
obtained for cyclopropylidene at the MRCISD/TZ2P and CCSD(T)/TZ2P levels of
theory, respectively, using B3LYP/TZP geometries and zero point corrections.[81]
The B3LYP/6–311+G** ΔE_{ST} is 14.2 kcal mol^{-1}.

3.3   Unsaturated Carbenes

In his 1975 review, Hartzler recognized four classes of unsaturated carbenes.[112] The
first group constitutes the methylene carbenes, vinylidene $H_2C=C$: being the
prototype. The second class are vinylcarbenes and related α,β unsaturated carbenes
as e. g. aryl carbenes and cyclopentadienylidenes. Propadienylidene, $H_2C=C=C$:, is

the vinylidene carbene prototype. Propargylenes comprise the fourth class; propynylidene is the simplest example.

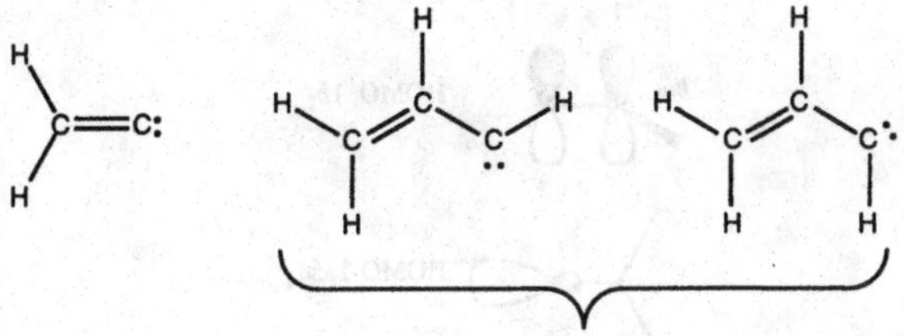

Vinylidene   *trans*– (left) and *cis*–vinylcarbene (right)

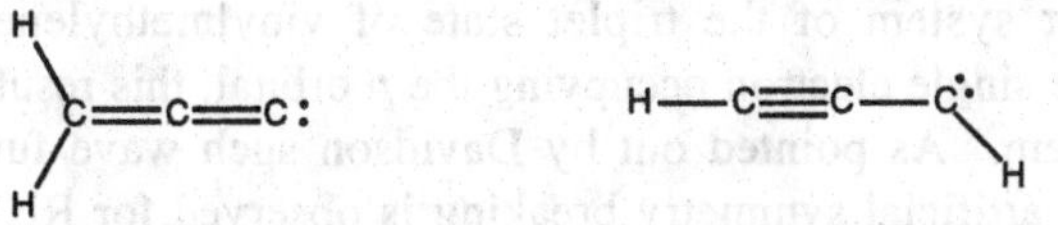

Propadienylidene   Propynylidene

Scheme 3.5 Types of unsaturated carbenes .

Vinylidene (**7**) has a $\tilde{X}\,^1A_1$ singlet ground state; the singlet state is stabilized by 36.9 kcal mol^{-1} whereas the triplet state is destabilized by 23.0 kcal mol^{-1} relative to methylene according to equations (3.1) to (3.3) at B3LYP/6-311+G**. The highest occupied MO (HOMO) is the π–bonding $1b_1$. The lowest unoccupied MO (LUMO) is the carbene p orbital ($2b_2$), which is in plane with the second–highest occupied MO (HOMO-1) of a_1 symmetry, the carbene σ orbital (see Scheme 3.6).

The lowest excited state of **7**, $\tilde{a}\,^3B_2$ is obtained by excitation of one electron from the HOMO-1 to the LUMO. Promotion of one electron from the HOMO to the LUMO results in the second-lowest excited state of **7**, $\tilde{b}\,^3A_2$. The high level calculations (CCSD(T)/TZ(2df,2pd)//CISD/TZ(2df,2pd) of Vacek *et al.*[113] give T_e (–45.6 kcal mol^{-1}) and T_o (–46.6 kcal mol^{-1}) values which are in very good agreement with the experimental data (T_e = –46.8, T_o = –47.6 kcal mol^{-1}) of Ervin *et al.*[114] The B3LYP/6–311+G** values (see Table 2.6) are T_e = -47.5, T_o = -48.1 kcal mol^{-1}.

Even better results (T_o = –47.4 kcal mol^{-1}) were obtained by Stanton and Gauss[115] using the equation-of-motion coupled–cluster method (EOM–CCSD) in conjunction with a large atomic natural orbital (ANO) basis set. The $\tilde{b}\,^3A_2$ state is 63.2 kcal mol^{-1} (experiment: 63.5 kcal mol^{-1}), and the $\tilde{A}\,^1A_2$ is predicted to be 71.7 kcal mol^{-1} above the singlet ground state at the EOM–CCSD level of theory.

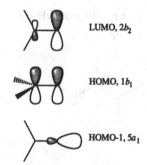

LUMO, $2b_2$

HOMO, $1b_1$

HOMO-1, $5a_1$

Scheme 3.6 Frontier orbitals of vinylidene in C_{2v} symmetry.

The adjacent π system of the triplet state of vinylmethylene (**8**) allows delocalization of the single electron occupying the p orbital; this results in an allyl radical–like π system. As pointed out by Davidson such wave functions pose special problems as artificial symmetry breaking is observed for RHF with small basis sets.[62] The singlet state of vinyl carbene also profits from the stabilizing effect of the adjacent double bond. Whereas the $^1(\sigma p)$ state of vinyl carbene gives rise to an allyl radical type π system similar to the $^3(\sigma p)$ state, the analogous interaction is less favorable in the $^1\sigma^2$ state.

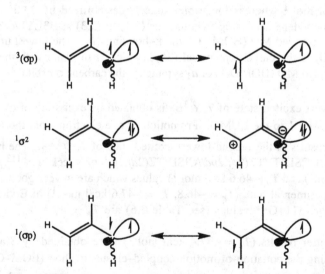

Scheme 3.7 Resonance structures for electronic states vinylmethylene (**8**).

Using MRCI/DZP Yoshimine *et al.* found that the two possible C_s minima (*cis* and *trans*) of the vinylmethylene triplet ground state are isoenergetic and have allyl radical–like π structures.[116] The barrier for interconversion of the *cis* and *trans* isomers is 5.7 kcal mol^{-1}. The most stable singlet species are asymmetric $^1(\sigma p)$ states which also have π systems similar to an allyl radical. The *cis* and *trans* C_1 isomers are found to be isoenergetic. A singlet–triplet energy splitting of 12 kcal mol^{-1} was obtained by Yoshimine *et al.*[116] The conventional C_s symmetric $^1A'$ $^1\sigma^2$ carbenes, both the *cis* and *trans* isomers, do not correspond to minima, but are 2.1 kcal mol^{-1} (for the *trans* conformation) and 1.3 kcal mol^{-1} (for the *cis* conformation) higher lying transition structures for the interconversion of the asymmetric C_1 ^1A minima (see Scheme 3.8). It should be noted that ESR experiments[117,118] agree with the triplet nature of the vinylmethylene ground state, but suggest a singlet–triplet energy separation of less than 1 kcal mol^{-1}.[117]

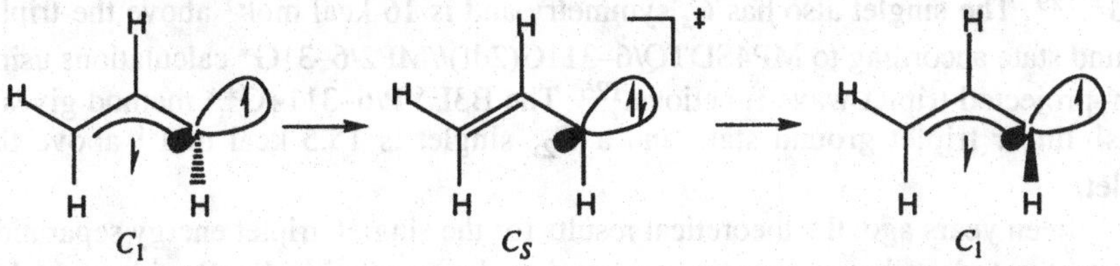

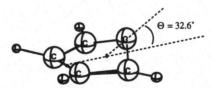

Scheme 3.8 Allyl type asymmetric structures are minima for singlet *cis*–**8**, the closed shell carbene corresponds to a transition structure. Note that the same is true for the isoenergetic *trans* conformer.

Cyclopentadienylidene (**9**) is found to have a C_{2v} symmetric $^3B_1$ ground state.[119,120] The lowest singlet state is an open–shell $^1A_2$, (6.3 kcal mol^{-1} at CISD+Q/TZ2P//CISD/DZP) followed by a $^3A_2$ state (6.8 kcal mol^{-1}).[120] The geometries of the $^1A_2$ and $^3A_2$ states are essentially the same and are characterized by short C–C: bonds (1.386 Å at CISD/DZP). The closed–shell singlet cyclopentadienylidene has a non-planar C_s symmetric structure with an folding angle (see Scheme 3.9) of 23.2° at TCSCF/3–21G and 32.6° at B3LYP/6–311+G**, it and is 12 and 10 kcal mol^{-1} above the ground state at CISD/6–31G*//TCSCF/3–21G[119] and B3LYP/6–311+G**, respectively.

$\Theta = 32.6°$

Scheme 3.9 Definition of folding angle Θ in the $^1A'$ state of cyclopentadienylidene (**9**).

Propadienylidene (**10**) was only recently identified by IR spectroscopy as a secondary product of cyclopropenylidene (**11**) irradiation in a matrix.[121] Before the detection of **10**,[122] its ground state was correctly predicted to be the C_{2v} symmetric singlet.[123-125] The triplet is calculated to be 30 – 40 kcal mol[-1] higher in energy using MP4SDTQ and large basis sets with MP2 and MP3/6–31G* geometries.[125,126] Adiabatic excitation energies for the higher singlet states of **10** were reported by Stanton *et al.* very recently.[127]

The C_{2v} symmetric singlet is the ground state of **11**. The C_1 triplet state is 52 and 49.6 kcal mol[-1] higher in energy at MP4SDTQ/6–311G(2df)//MP2/6–31G* and B3LYP/6–311+G**, respectively.

Propargylene (**12**) also has been identified experimentally, and a quasi–linear structure was suggested for the triplet ground state based on ESR measurements.[128] The IR spectrum obtained for matrix isolated [3]**12** agrees with the one calculated for the C_s structure (see Scheme 3.5) using a non harmonic approximation at UMP2/6–31G*.[129] The singlet also has C_s symmetry and is 16 kcal mol[-1] above the triplet ground state according to MP4SDTQ/6–311G(2df)//MP2/6–31G* calculations using spin–projected triplet wave functions.[126] The B3LYP/6–311+G** method gives a quasi–linear triplet ground state and a C_{2v} singlet is 15.5 kcal mol[-1] above the triplet.

Fifteen years ago, the theoretical results for the singlet–triplet energy separation of unsaturated carbenes were considered to be unreliable by Davidson in his review.[62] Today electronic structure theory is capable, at least for small prototype carbenes, to give singlet–triplet energy separations approaching experimental accuracy, as this very short summary of theoretical results documents.

As the singlet–triplet energy separations of prototype carbenes discussed above were obtained at various levels of theory, we provide data at a uniform level in Table 3.2. Due to its good performance for methylene, B3LYP/6-311+G** seems well suited to study the singlet–triplet energy splitting of larger carbenes as well as the stabilization of substituents vs. methylene.

TABLE 3.2 Singlet-triplet energy separation of simple carbenes and stabilization energies vs. methylene at B3LYP/6-311+G**.

Carbene	$T_e{}^a$	$T_0{}^a$	Stab$_S$	Stab$_T$
methylene	12.2 (9.4) [b]	11.8 (9.1) [c]	–	–
methylcarbene	4.9	4.7	15.8	8.6
dimethylcarbene	-1.4	-1.4	26.6	15.1
1-propylidene	5.0	5.6	14.3	8.1
tert-butyl carbene	2.0	2.4 [d]	17.0	7.7
cyclopropylcarbene	-5.8	-5.4	29.2	11.3
cyclopropylidene	-13.3	-14.2	25.2	–0.5
cyclopropenylidene	-51.2	-49.6	72.6	11.2
cyclobutylidene	-5.7	-5.8	28.9	11.3
cyclopentylidene	-8.0	-9.1 [d]	33.2	13.1
cyclopentadienylidene	9.8	9.1 [d]	25.3	22.6
vinylidene	-47.5 (-46.8) [e]	-48.1 (47.6) [e]	36.9	–23.0
propadienylidene	-30.6	-28.4	34.9	–5.2

[a] Experimental value in parentheses, when available.

[b] Expt. Refs. 42,46-48

[c] Expt. Ref. 42

[d] ZPVE correction at B3LYP/6-31G*.

[e] Expt. Ref. 114

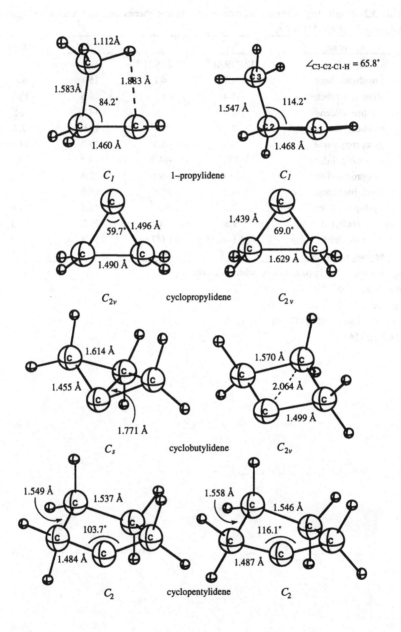

C_1 1–propylidene C_1

C_{2v} cyclopropylidene C_{2v}

C_s cyclobutylidene C_{2v}

C_2 cyclopentylidene C_2

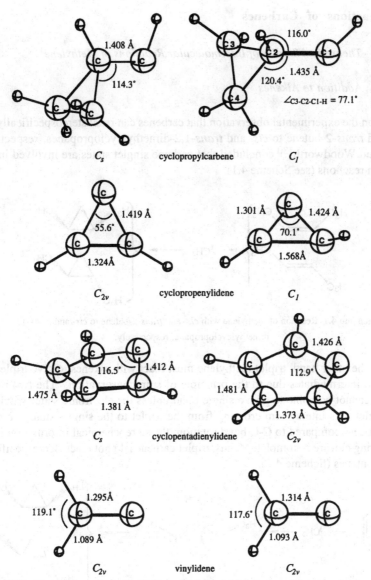

C_s cyclopropylcarbene C_1

C_{2v} cyclopropenylidene C_1

C_s cyclopentadienylidene C_{2v}

C_{2v} vinylidene C_{2v}

Scheme 3.10 Geometries of singlet (left) and triplet (right) 1-propylidene, cyclopropylidene, cyclobutylidene, cyclopentylidene, cyclopropylcarbene, cyclopropenylidene, cyclopentadienylidene, and vinylidene at B3LYP/6–311+G**.

4 Reactions of Carbenes

4.1 Theoretical Studies of Intermolecular Reactions of Methylene

4.1.1 Addition to Alkenes

Based on the experimental observation that carbenes can react stereospecifically with *cis*- and *trans*-2-butene to *cis*- and *trans*-1,2-dimethylcyclopropanes, respectively, Skell and Woodworth[130] concluded that carbene singlet states are involved in such addition reactions (see Scheme 4.1).

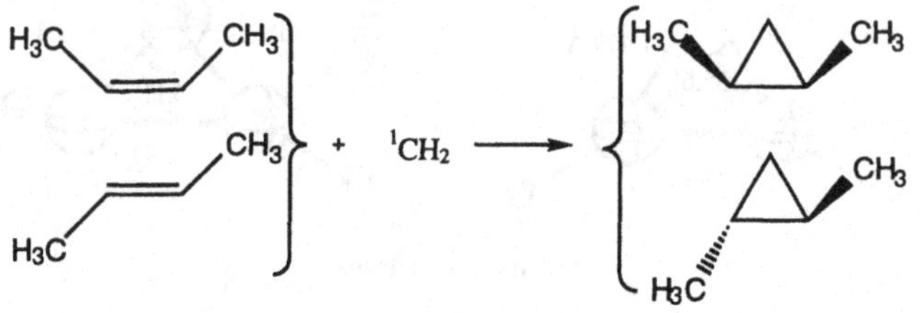

Scheme 4.1 Reaction of methylene with *cis*- and *trans*-2-butene to *cis*- and *trans*-1,2-dimethylcyclopropane, respectively.

On the other hand, triplet methylene must react with alkenes to give triplet 1,3-biradical intermediates due to the principle of spin conservation. The two radical centers cannot combine to give a single bond unless the electronic state switches to the singlet. As intersystem crossing from the triplet to the singlet state is a rather slow process compared to C-C bond rotation, the stereochemical information is lost before ring closure is complete. Thus, triplet carbenes do not react stereospecifically with 2-butenes (Scheme 4.2).

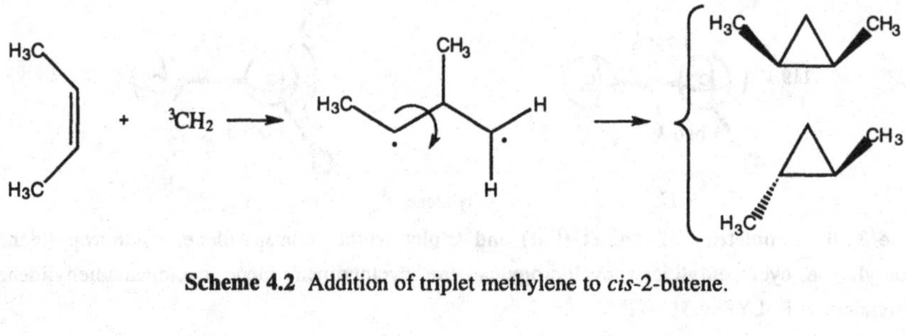

Scheme 4.2 Addition of triplet methylene to *cis*-2-butene.

The similarly electron deficient character of carbenes and carbenium ions led Skell and Garner to propose an initial π approach (Scheme 4.3) for the addition of carbenes to alkenes.[131] The C_{2v} symmetric σ trajectory (Scheme 4.3), which is a least motion process (the relative orientation of the educts is the same as in the product), was shown to be symmetry forbidden for a σ^2 carbene by Hoffmann.[132]

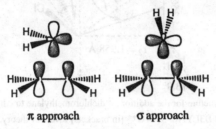

π approach σ approach

Scheme 4.3 The π and σ approach of methylene on ethylene.

Using extended Hückel theory (EHT), Hoffmann found a reaction path corresponding to a π approach. This result was confirmed by Zurawski and Kutzelnigg who calculated the reaction path at SCF and with the connected electron pair approximation (CEPA–2) using a double–ζ quality basis set including d polarization functions.[133] The energy decreased monotonously; the reaction path revealed no barrier. Two stages can be distinguished during the concerted addition of singlet methylene. During the initial π approach, the carbene acts as an electrophile, accepting electron density from the π system. Thus, a HOMO(ethylene) - LUMO(carbene) interaction dominantes at first. At the end of the electrophilic stage, the p carbene orbital is partially populated, and the system was described by Zurawski and Kutzelnigg[133] as the "half-formed" cyclopropane, first suggested by Skell and Garner.[131]

The second stage comprises a nucleophilic interaction between the σ carbene orbital and the π^* of ethylene. The HOMO(carbene) - LUMO(ethylene) interaction becomes important. The carbene hydrogens pivot around the carbene carbon and approach the cyclopropane configuration. Zurawski and Kutzelnigg pointed out that the initial electrophilic interaction is crucial as it more than compensates for the repulsive potential associated with the approaching educts; thus the reaction has no barrier.[133]

The addition of the increasingly nucleophilic carbenes CCl_2, CF_2, CFOH, $C(OH)_2$ involve barriers (8, 27, 37, and 45 kcal mol^{-1}, respectively) at HF/4-31G.[134] The approach of the substituted carbenes is essentially the same as that of methylene. Thus, an initial electrophilic π approach is followed by a nucleophilic phase in which the carbene HOMO interacts with the ethylene LUMO.

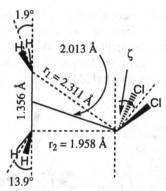

Scheme 4.4 Transition structure for the addition of dichloromethylene to ethylene at the HF/4–31G and B3LYP/6–311+G** [in brackets] levels of theory.

However, there are subtle differences in the HF/4-31G transition structures obtained by Rondan *et al.* for the four substituted carbenes reflecting the electrophilicity of the involved carbene.[134] The ratio of newly forming bond lengths, r_1/r_2, the tilt angle ζ of the CXY plane with respect to the original ethylene plane, and the distortion away from planarity of the ethylene CH_2 groups indicate a clear trend of decreasing electrophilicity along the series CCl_2, CF_2, HOCF, and $C(OH)_2$ (Scheme 4.4 and Table 4.1).

TABLE 4.1 Characteristic data of the TS's for addition of varoius carbenes to ethylene at HF/4–31G.

Carbene	E_a/kcal mol^{-1}	tilt angle ζ/degree	r_1/r_2
CCl_2	8	36	1.18
CF_2	27	43	1.23
CF(OH)	37	48	1.32
$C(OH)_2$	45	58	1.30

For an electrophilic carbene only a slight asymmetry in the newly forming bond lengths (r_1/r_2), a small tilt angle, and a small ethylene distortion are expected. Especially the large ζ and the pronounced distortion of ethylene suggest a predominantly nucleophilic interaction in the TS for addition of $C(OH)_2$. Another indicator of electrophilic or nucleophilic character in the addition TS is the degree of charge transfer. Whereas in the CCl_2 TS 0.29 electrons have been transferred from ethylene to the carbene according to Natural Population Analyses, a *reverse* charge transfer of 0.06 electrons from $C(OH)_2$ to ethylene is obtained in the TS. Rondan *et al.* showed that variations in the transition state structures can be explained using a frontier molecular orbital treatment.[134] In the "earlier" CCl_2-ethylene TS, the

HOMO(CCl_2) - LUMO(ethylene) overlap integral associated with nucleophilic character is considerably smaller than the HOMO(ethylene) - LUMO(CCl_2) overlap integral associated with electrophilic character. The opposite is found for $C(OH)_2$, the nucleophilic character of the carbene predominates in the TS.

4.1.2   Insertion Into Single Bonds

Alkanes were considered for a long time to be an almost inert class of organic compounds. Thus, the insertion of carbenes into CH bonds of alkanes (Scheme 4.5) is an unusual reaction in organic chemistry.

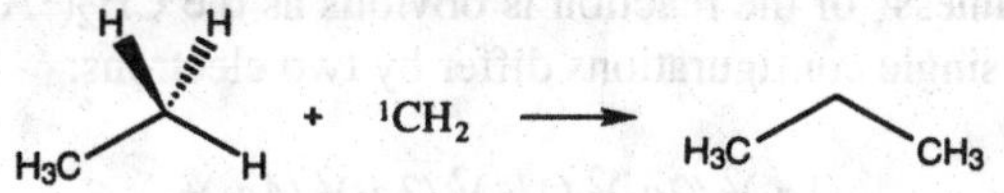

Scheme 4.5 Reaction of singlet methylene with ethane to propane.

Activation of alkane CH bonds with transition metals or electrophiles is of current interest to chemists.[135,136] Although the reaction of carbenes with alkanes, known since the 1950's, is almost useless synthetically due to the lack of selectivity,[137-140] the CH insertion mechanism is very interesting. The reaction of dihydrogen (H_2), a rather simple alkane model, with methylene has been studied in detail theoretically.

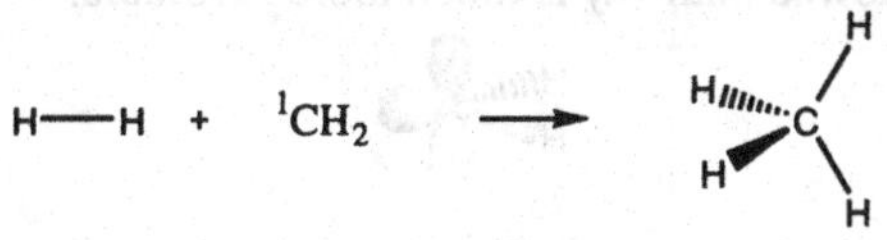

Scheme 4.6 Reaction of singlet methylene with dihydrogen.

Triplet methylene, which behaves like a radical due to the two unpaired electron spins, abstracts a hydrogen atom from H_2 to give a methyl radical and a hydrogen atom. Baskin's *et al.* CISD/DZ study found a barrier of 10-15 kcal mol^{-1}.[141] Triplet methylene attacks the H_2 molecule in a linear fashion as shown in Scheme 4.7.

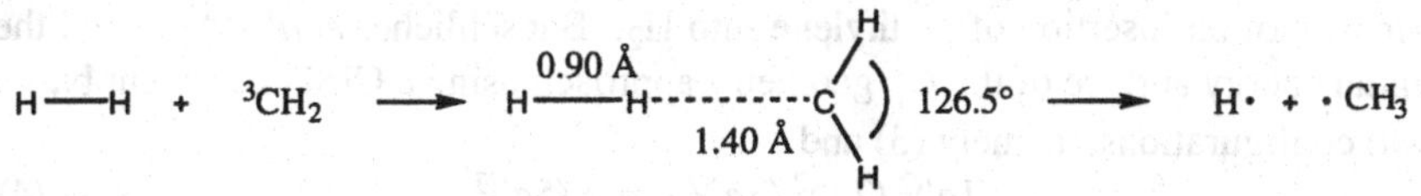

Scheme 4.7 Reaction of triplet methylene with dihydrogen. Transition structure geometry at CISD/DZ.

Zurawski and Kutzelnigg pointed out the relationship between the H_2 insertion reaction and the singlet methylene double bond addition.[133] The least motion pathway, maintaining C_{2v} symmetry, is a Woodward-Hoffmann forbidden pericyclic reaction involving four electrons.

Scheme 4.8 The σ approach of methylene to hydrogen keeping C_{2v} symmetry.

The "forbiddenness" of the reaction is obvious as the $CH_2(^1A_1) + H_2$ and CH_4 (in C_{2v} symmetry) single configurations differ by two electrons:

$$CH_2(^1A_1) + H_2: \qquad (1a_1)^2\,(2a_1)^2\,(1b_2)^2\,(3a_1)^2\,(4a_1)^2 \qquad\qquad (1)$$
$$CH_4\ (C_{2v}\ \text{symmetry}): \qquad (1a_1)^2\,(2a_1)^2\,(1b_2)^2\,(3a_1)^2\,(1b_1)^2 \qquad\qquad (2)$$

Using a three-reference CISD approach with a DZ basis set, Bauschlicher *et al.* predicted a barrier of 27 kcal mol^{-1} for the least motion reaction.[142] As the barrier was not obtained by full optimization of all internal transistion state coordinates, the reported value is certainly a few kcal mol^{-1} too high. Nonetheless, the results nicely demonstrate the qualitative correctness of the Woodward - Hoffmann rules,[143] since the symmetry-allowed pathway is much more favorable.

Scheme 4.9 The π approach of methylene to dihydrogen.

For the non-least motion approach in C_s symmetry (see Scheme 4.9) the configurations become:

$$(1a')^2\,(2a')^2\,(1a'')^2\,(3a')^2\,(4a')^2 \qquad\qquad (3)$$

In principle, a single-configuration HF wave function should suffice to describe the non-least motion insertion of methylene into H_2. Bauschlicher *et al.*[144] studied the potential energy surface of the C_s geometry approach using a CISD treatment based on two configurations, namely (3) and

$$(1a')^2\,(2a')^2\,(3a')^2\,(4a')^2\,(5a')^2 \qquad\qquad (4)$$

A total of around 600 points were calculated on the PES applying constrained symmetries. Following the spirit of Bauschlicher's research, we adopted the following coordinate system:

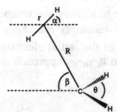

Scheme 4.10 Coordinate system for the non-least motion approach of methylene to dihydrogen.

In this coordinate system the least motion approach corresponds to $(\alpha, \beta) = (90°, 0°)$. For $R > 3$ bohr (ca. 1.59 Å) an attractive reaction channel was discovered characterized by angles $(0°, 90°)$. When approaching each other, the relative orientation of the molecules changes, but the potential is still attractive. Thus, Bauschlicher *et al.* concluded that the insertion of methylene into H_2 proceeds without a barrier if a non-least motion approach is adopted.[144] By analogy to the addition of methylene to alkenes, the two-stage mechanism suggested by Skell and Garner[131] was confirmed theoretically by Kollmar,[145] who studied the insertion PES with semiempirical methods. Kollmar's surface is characterized by a small but sizable barrier of 5 kcal mol^{-1} for the non-least motion insertion, and a relatively small barrier of 13 kcal mol^{-1} for the "forbidden" reaction.[145] Using the correlated CEPA method and basis sets of double as well as triple-ζ quality, the barrier for the non-least motion insertion vanishes,[146,147] in agreement with Bauschlicher's CI results.[144]

Going from the hydrogen molecule to the alkane insertion, several more approaches are available to the carbene.[148] A comparative study of the possible insertion reactions was presented by Bach *et al.*[148] The following discussion requires an understanding of the group orbitals of the -CH_2- fragment, whose construction is described by Jorgensen and Salem.[149] The bonds to the two hydrogens in the methylene group are comprised of one orbital of σ and one of π symmetry (see Scheme 4.11). Replacement of one hydrogen by a methyl group to give ethane gives rise to σ_{CHCH3} and π_{CHCH3} group orbitals. Note that these are both in–plane.

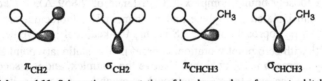

Scheme 4.11 Schematic representation of in–plane carbene fragment orbitals.

The approach of the carbene to alkanes was characterized by Bach *et al.* as σ_{CH2} or π_{CH2} depending on whether the empty *p* carbene orbital attacks the σ or the π orbital of the methylene group in the initial electrophilic interaction (see Scheme 4.12). For ethane, a σ_{CHCH3} and π_{CHCH3} approach also is possible.

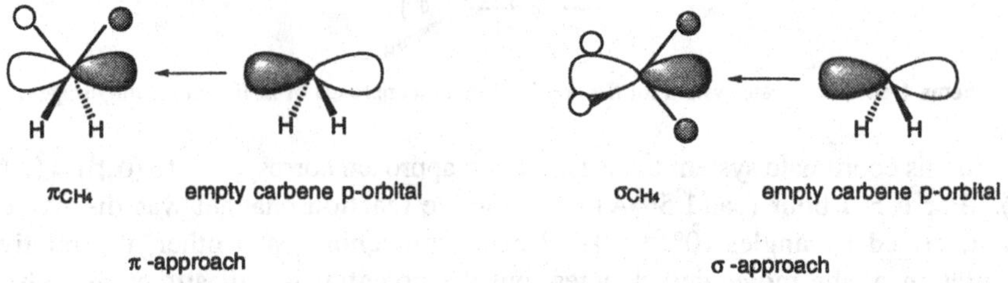

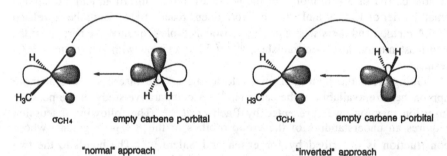

Scheme 4.12 The insertion of a methylene into methane CH-bonds via π- and σ- approaches.

Yet another approach of methylene to methane and ethane was considered by Gano's and Gordon's groups,[150-153] later termed "inverted" approach by Bach *et al.*: a hydrogen migrates from the hydrocarbon to the smaller lobe of the σ carbene orbital as sketched in Scheme 4.13.

Scheme 4.13 Methylene insertion into ethane *via* the "normal" and "inverted" σ-approaches.

The insertion of methylene into methane was found to be a barrierless process, both by Gano *et al.* and by Bach *et al.*[148] At the QCISD/6-31G* level of theory there exists a loosely bound complex (CC distance of 2.859 Å), 1.7 kcal mol^{-1} more stable than isolated methylene and methane. The molecular arrangement resembles the σ_{CH2} approach, but the TS is only 0.4 kcal mol^{-1} above the complex (1.1 kcal mol^{-1} with zero point vibrational energy). The stationary point associated with the π_{CH2} approach is 0.7 kcal mol^{-1} above the complex, and is a second order stationary point. The inverted π_{CH2} approach is much less favorable, its TS is 4.5 kcal mol^{-1} higher than the complex. No σ_{CH2} TS could be found at higher levels (QCISD/6-311G* and QCISD(T)/6-31G*) and at MP2/6-31G*. A negative activation barrier of 1.8 kcal mol^{-1} is obtained with single point calculations at QCISD(T)/6-311+G(2df,p)//QCISD/6-31G*.[a] However, Bach et al. concluded that the insertion of methylene into methane occurs without a barrier via the σ_{CH2} approach.[148]

Very similar results were obtained for the insertion of methylene into ethane. No TS could be found by Bach et al. for the σ_{CH2} approach at MP2/6-31G*, and reaction coordinate following showed the insertion to occur without a barrier.[148] Only a very small barrier of 0.5 kcal mol^{-1} relative to the educts (MP4SDTQ/6-31G*//MP2/6-31G*) is obtained for an inverted π_{CH2} reaction. The other possible approaches (π_{CH2} and inverted σ_{CH2}) involve second order stationary points at HF/6-31G*.

The results of Bach et al. suggest "that singlet methylene attacks simple hydrocarbons in a σ_{CH2} fashion, that the hydrogen will migrate to the larger lobe of the carbene σ orbital, and that the insertion will take place with little or no activation barrier."[148]

4.2   Theoretical Studies of Intramolecular Reactions of Alkylcarbenes

4.2.1   1,2-Hydrogen migrations

Carbenes with α–H–atoms readily undergo [1,2]-H shifts to alkenes. Experimental studies agree that [1,2]-H shifts occur on the singlet potential energy surface.[57,61] Matrix experiments suggest that intersystem crossing of singlet methylcarbene to the lower triplet state can not compete with hydrogen migration.[58,154] The rearrangement is so facile that the existence of [1]1 as a true intermediate in a potential energy minimum with a finite lifetime was questioned for a long time. Photolyses of matrix isolated diazoethane (13) and nascent 3-methyldiazirine (14) at 8 K did not yield singlet or triplet 1 as an observable product.[154]

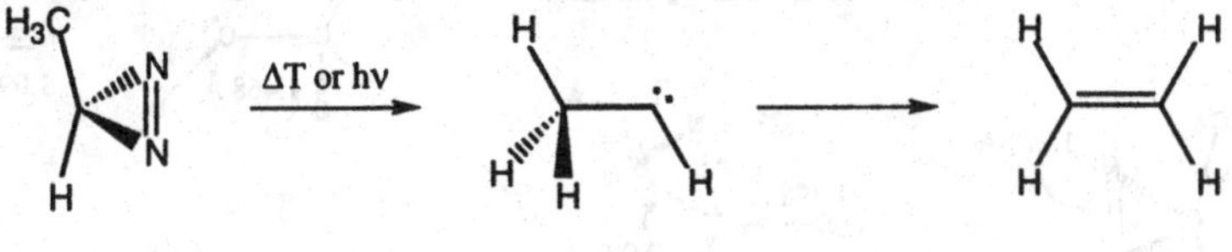

14   **1**

Scheme 4.14 Thermolysis or photolysis of 3–methyldiazirine (14) yields ethylene.

---

[a] A negative activation barrier indicates that the TS might not exist at the level employed for the single point calculation.

However, trapping of singlet ethylidene by CO in a CO-doped Ar matrix competes with the facile [1,2]-H shift.[154] Laser flash photolysis (LFP) of 14 in the presence of pyridine does not give the pyridinium ylide.[155] The pyridinium ylide probe method was previously successfully employed to intercept carbenes, e.g. cyclopropyl– and dicyclopropylcarbene,[156] adamantylidene,[157] homocubanylidene,[158] and dimethylcarbene.[159,160]

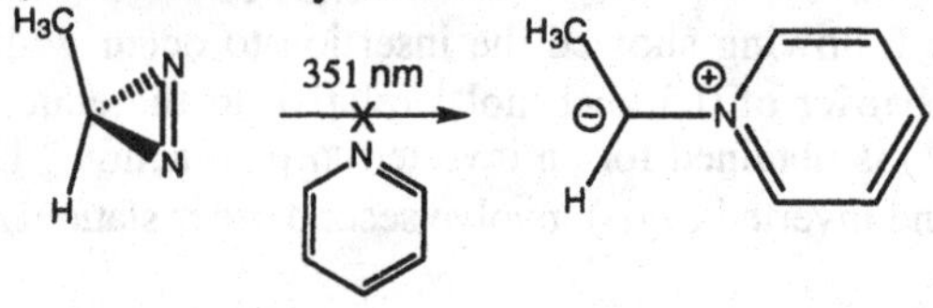

Scheme 4.15 Failure of the pyridinium ylide probe for trapping of methylcarbene.

Perdeuterated ethylidene-d_4, however, has been captured by pyridine, and a life time of 500 ps in pentane at ambient temperatures was deduced (assuming that the second order rate constant of the reaction of 1-d_4 with pyridine is 10^9 M^{-1}s^{-1}).[155]

The photochemical generation of carbenes by decomposition of alkyldiazirines yields (in contrast to pyrolysis) a complex mixture of stable products.[155] Both diazirine and carbene excited states have been discussed as product forming species.[155] There is indeed experimental evidence for hydrogen migration in the dimethyldiazirene excited (n-σ*) state.[159] This state also is assumed to lead to nitrogen extrusion.

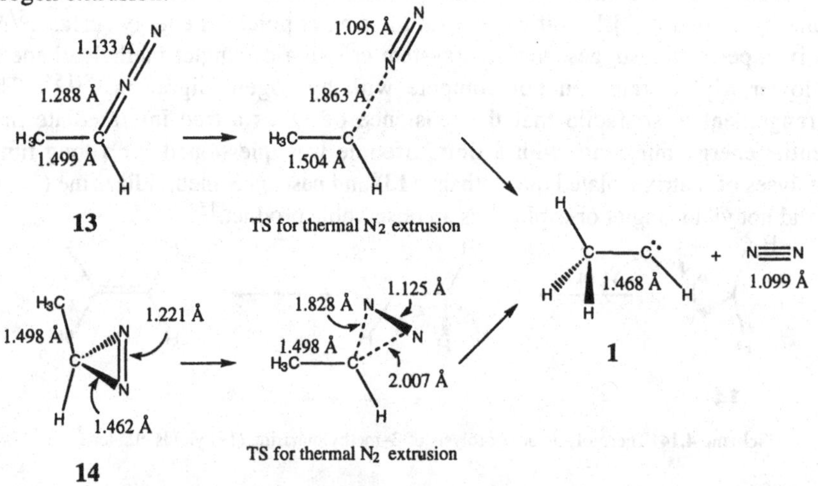

Scheme 4.16 Diazoethane (13) and methyldiazirine (14) transition structures (at CISD/TZP) for thermal nitrogen extrusion to give 1.

A recent high level *ab initio* study by Miller *et al.* suggests that the thermal N_2 extrusions of both **13** and **14** are stepwise processes, thus involving ^{1}**1** as an intermediate (Scheme 4.16).[161] Similarly, decomposition of 2,2-dimethyl-1-diazopropane does not show any tendency to bypass free *tert*-butylcarbene according to MP2/6–31G* + ZPVE results of Armstrong *et al.*[110] The barrier for thermal N_2 extrusion is 26.9 kcal mol^{-1} from **13** and 35.3 kcal mol^{-1} from **14** at CCSD(T)/TZP//CISD/TZP. The barriers for the reverse reaction from **1** and nitrogen to **13** and **14** are 3.3 kcal mol^{-1} and 11.7 kcal mol^{-1}, respectively. Miller *et al.* argued that, because of the high barrier for extrusion of N_2 from **14**, the chance of excess kinetic energy being present to drive the reaction from **1** to ethylene is great.[161] Thus, if possible after all, it should be easier to trap the carbene if **13** instead of **14** is used as the precursor .

The [1,2]-H shift in ^{1}**1** to yield ethylene was studied extensively using many body perturbation theory. In 1980, Nobes *et al.* predicted a barrier of 2.1 kcal mol^{-1} at the MP3/6–31G**//HF/4–31G level of theory, a result with was consistent with the elusive nature of ethylidene.[162] The MP4SDQ/6–31G**//HF/6–31G* level gives a similar barrier of 2.6 kcal mol^{-1}. Corrections to the MP4SDQ/6–31G** energy for (a) triple substitution correlation and (b) a further split in the valence *s,p*-basis functions yielded a classical barrier of -1.1 kcal mol^{-1}.[97,102] Thus, it was concluded that ^{1}**1** is not a true local minimum and that it does not exist as an intermediate species.[97,102]

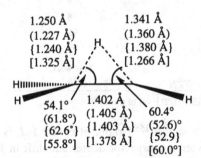

Scheme 4.17 TS for [1,2]-H shift of ^{1}**1** to ethene at the CCSD/TZ2P(f,d), the MP2/6–31G* (in parentheses), the B3LYP/6–311+G** {in braces} and the HF/6–31G* [in brackets] levels of theory.

The basic assumption in employing single point calculations is that the geometry obtained at a lower level is essentially converged. If a high level calculation predicts significantly different structures compared to a low level computation, then the prediction of a high level single point energy evaluation on a low level structure is not reliable. Indeed, re-optimization of the TS for [1,2]-H shift at MP2/6–31G* led to a substantial change in geometry.[99] The inclusion of electron correlation in the MP2 treatment produced an earlier TS. The distances of the migrating hydrogen from the migration origin and terminus are reversed from 1.325 Å and 1.266 Å at RHF to 1.227 Å and 1.360 Å at MP2 (see Scheme 4.17), reflecting the shift of the TS along the reaction coordinate towards ethylidene.

Single point calculations at MP4SDTQ/6-311G**//MP2/6-31G* predict a barrier of 0.6 kcal mol[-1].[99] The existence of [1]1 as a true local minimum was shown beyond any doubt by the high level calculations of Ma and Schaefer (Table 4.2).[98] Using the CISD, CCSD, and CCSD(T) methods with large basis sets up to TZ2P(f,d) for geometry optimization, the authors predict a barrier of 1.2 kcal mol[-1] for the [1,2]-H migration in [1]1 at their highest level of theory (CCSD(T)/TZ2P(f,d)//CCSD(T)/TZ2P+d + ZPVE). We obtain the same barrier at B3LYP/6–311+G**.

TABLE 4.2: Calculated Activation Barriers (in kcal mol[-1]) for the [1,2]-H Shift in Methylcarbene to Ethene.

level of theory	reference	activation energy E_a
AM1	99	14.9
HF/DZP	98	11.3
MP2/6-31G*	99	1.8
MP2/6-311G**//MP2/6-31G*	99	0.1
CISD/DZP [a]	98	4.3
CISD/TZ2P(f,d) [a]	98	2.9
CCSD/DZP [a]	98	3.9
CCSD/TZ2P(f,d) [a]	98	2.5
MP4SDTQ/6-311G**//MP2/6-31G*	99	0.6
B3LYP/6-311+G**	This work	1.2
CCSD(T)/TZ2P(f,d)//CCSD(T)/TZ2P	98	1.2

[a] ZPVE determined at the CISD/DZP level of theory

The experimentally observed[155] isotope effect for 1-d_4 is also evident in the *ab initio* results. The activation energy for deuterium shift in 1-d_4 is 0.3 kcal mol[-1] higher. Including thermodynamic corrections at 298 K and 1 atm, a free energy barrier $\Delta G^{\ddagger} = 1.9$ kcal mol[-1] was predicted for the rearrangement of 1-d_4 to ethylene-d_4.[98] This result is in good agreement with the experimental estimate of 2.3 kcal mol[-1] as an upper limit for the barrier of this reaction.[155]

The hydrogen atom which is best aligned with the vacant carbene *p* orbital migrates in ethylidene. Accordingly, the high *exo* selectivity (*exo* : *endo* = 138 : 1) in the thermal rearrangement of brex-5-ylidene (**15**) was ascribed to the preferential alignment of σ_{CH} and the vacant carbene *p* orbital in the ground state.[163] On the other hand, cyclohexylidene (**16**) derivatives are known to rearrange with a much lower selectivity,[164-166] a striking result in view of the very favorable alignment of the axial σ_{CH} and the vacant p orbital.[167]

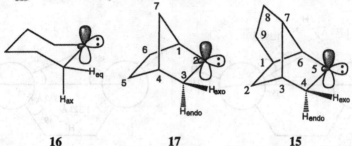

Scheme 4.18 The axial CH bond in cyclohexylidene (**16**) is, in contrast to the equatorial CH bond, aligned favorably with the vacant carbene p orbital.

Evanseck and Houk determined the transition structures for the [1,2] migration of axial and equatorial hydrogen in **16** at RHF/6-31G*.[167] The double Newman projection given below shows the obvious similarity of both TS's (Scheme 4.19).

Both structures resemble half-chair cyclohexane very closely, although the equatorial TS is slightly more flattened (torsional angles of 8° and 14° vs. 19° and 21°). Evanseck and Houk found the two TS's to be almost degenerate energetically (ΔE_a = 0.1-0.2 kcal mol^{-1} at RHF/6-31G* and MP2/6-31G*//RHF/6-31G*).

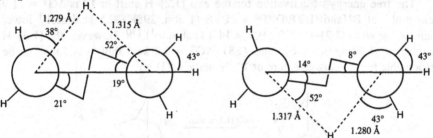

Scheme 4.19 Double Newman projections of the transition structures for [1,2]-H shift of axial (left) and equatorial (right) hydrogen in **16** at HF/6-31G*.

Thus, the authors concluded that the alignment of axial and equatorial hydrogens in **16** is not important for the selectivity, because "*in the transition structures the molecule distorts to nearly the same geometry regardless of which hydrogen migrates.*"[167]

Severe torsional interactions in the *endo* transition structure for the rigid norborn–2–ylidene (**17**) were suggested to be responsible for the high *exo* selectivity (see double Newman projection in Scheme 4.20).[168] The partial CH bonds of the migrating hydrogen are eclipsed more (8° and 9° torsional angles with the norbornane CC bonds) in the *endo*-TS (vs. 35° and 26° in the *exo*-TS).

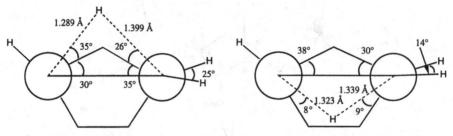

Scheme 4.20 Double Newman projections of the transitions structures for [1,2]-H shift of *exo* (left) and *endo* (right) hydrogen in norborn–2–ylidene (**17**) at HF/6-31G*.

Furthermore, the torsional angle between the *exo*–hydrogen at C3 and the bridgehead hydrogen at C4 is more nearly eclipsed in the *endo*-TS (14°) than in the *exo*-TS (25°). These unfavorable interactions in the *endo* transition structures were said to account for the 1.8 kcal mol⁻¹ higher energy, resulting in the observed selectivity in the norbornane system. Similar non–bonding interactions (as in the norborn–2–ylidene transition structures) disfavor the *endo* transition structure for [1,2]-H shift in **15**.

The free energy of activation for the *exo* [1,2]–H shift in **17** is $\Delta G^{\ddagger} = 11.9$ kcal mol⁻¹ at BHandHLYP/DZP[a] + ZPVE (1 atm, 298K), 2.5 kcal mol⁻¹ lower than for the *endo* [1,2]–H shift ($\Delta G^{\ddagger} = 14.4$ kcal mol⁻¹).[170] However, the [1,3]–H migration yielding nortricyclene (**18**) ($\Delta G^{\ddagger} = 5.2$ kcal mol⁻¹) is found to be responsible for the elusive nature of **17** (Scheme 4.21).[170]

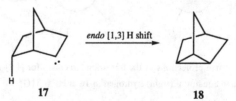

Scheme 4.21 The [1,3] H shift in norborn-2-ylidene yields nortricyclene.

Substitution at the migration terminus enlarges the H migration barrier.[59] For example, the experimental activation parameters for methylchlorocarbene are

---

[a] In the BHandH functional the exact term for the HF exchange integral is replaced by a mixture of the HF exchange and Becke's[169] "half and half" density functional.

$E_a = 4.9$ kcal mol^{-1} and $\Delta S^{\ddagger} = -16.7$ eu.[171] Surprisingly, the MP4SDTQ/6-311G**//MP2/6-31G* result, $\Delta E^{\ddagger} + \Delta ZPVE = 11.5$ kcal mol^{-1} and $\Delta S^{\ddagger} = -3.1$ eu, is in stark disagreement with the experiment.[99] Experimental results suggest that the reaction proceeds classically at high temperatures, but quantum mechanical tunneling becomes important at low temperatures.[a],[174],[175]

4.2.2 Vinylidene to Acetylene Rearrangement

In the 1980s, the direct experimental observation of the $\tilde{X}^1A_1$ ground state of **7** in the UV photoelectron spectra of the $\tilde{X}^2B_2$ H$_2$CC$^-$, $\tilde{X}^2B_2$ D$_2$CC$^-$, and $\tilde{X}^2A_1$ HDCC$^-$ anions ended the long lasting uncertainty about the existence of **7** as a bound intermediate in chemical reactions.[114],[176] When rotational broadening was included in the simulation of line shapes, an estimated lifetime of 0.04 - 0.20 ps resulted for **7**. The experimental enthalpy of isomerization of acetylene to **7** is around 44 kcal mol^{-1},[114],[176-179] the lower limit for the isomerization barrier is 1.3 kcal mol^{-1}. In the singlet ground state of **7**, the hydrogens are perfectly aligned with the empty in-plane carbene p orbital. Thus, a facile [1,2]-H shift with a low barrier, similar to **1**, is expected.

Early theoretical studies of Dykstra and Schaefer[180] (self-consistent electron-pair method, SCEP/DZP) as well as Pople *et al.*[181] (MP4SDQ/6-31G*//HF/6-31G*) gave similar classical barriers of 8.6 and 8.1 kcal mol^{-1}, respectively. Improvements in electron correlation and basis sets lowered the barrier slightly.[182-186] Krishnan *et al.* obtained a classical barrier of 2.2 kcal mol^{-1} at MP4SDTQ/6-311G**//MP2/6-31G* which was lowered to 0.9 kcal mol^{-1} by inclusion of the HF/6-31G* ZPVE correction.[187] The authors speculated that further improvement of the basis set and the electron correlation treatment might eliminate the activation barrier for rearrangement entirely.

---

[a] However, including tunneling corrections *via* the CD-SCSAG method[172] results in a better agreement between calculated and experimental entropies of activation ($\Delta S^{\ddagger} = -14.9$ eu).[173] Although the corrected activation energy is 7.7 kcal mol^{-1} and thus still too high, the agreement is satisfactorily and the disprecpancy was ascribed to the remaining theoretical error and to solvation effects on experimental rates.[173]

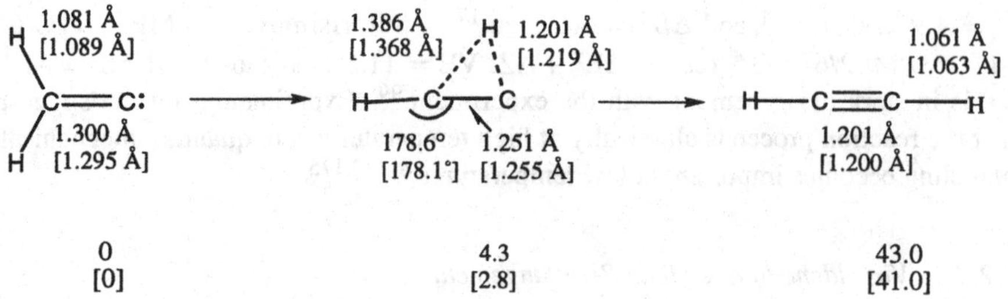

H 1.081 Å
   [1.089 Å]

C═══C:
   1.300 Å
H [1.295 Å]

⟶

H 1.386 Å H 1.201 Å
   [1.368 Å] ⋰ ⋱ [1.219 Å]

H—C⋰⋱C
   178.6° 1.251 Å
   [178.1°] [1.255 Å]

⟶

   1.061 Å
   [1.063 Å]

H—C≡══C——H
   1.201 Å
   [1.200 Å]

0
[0]

4.3
[2.8]

43.0
[41.0]

Scheme 4.22 Isomerization of vinylidene to acetylene and the TS at CCSD/TZ+2P and B3LYP/6–311+G** [in brackets]. Relative energies are given in kcal mol^{-1}.

To arrive at a final answer whether or not [1]7 exists as a minimum on the PES, Gallo *et al.* performed high level computations.[188] The classical barriers obtained at CISD/TZ+2P, CISD/QZ+3P, CCSD/TZ+2P are 6.1 - 7.1 kcal mol^{-1}. Zero point and Davidson[189,190] corrections (for the CISD calculations) lower the barrier to 4-5 kcal mol^{-1}. The CCSD/TZ+2P+f, CCSD/QZ+3P, and CCSDT-1/TZ+2P single point energies improve the classical barriers further to 4.6 - 6.1 kcal mol^{-1}. Estimating the effect of triple substitutions and *f* functions on the geometries leads to the authors' "best" classical barrier of 3.0 kcal mol^{-1}. Gallo *et al.* conclude that the classical barrier for isomerization of [1]7 will *not* vanish upon further refinement of theory.[188] This assessment was confirmed by Petersson *et al.* who estimated a classical barrier of 2.2 ± 0.5 kcal mol^{-1} using the complete basis set-quadratic configuration interaction atomic pair natural orbital CBS-QCI/[6s6s3d2f,4s,2p,1d] APNO model chemistry.[191] A slightly higher classical barrier, 2.85 kcal mol^{-1}, results form CCSD(T), CCSDT, and CCSD(TQ) calculations employing very large basis sets up to cc-pVQZ.[192] Correcting for zero-point vibrational energy gives an activation energy of 1.5 kcal mol^{-1},[192] in excellent agreement with the experimental lower limit of 1.3 kcal mol^{-1}.[114] We compute a classical barrier of 2.8 kcal mol^{-1} at B3LYP/6–311+G**.

Petersson *et al.*[191] also discussed the geometry of the TS, which is almost half-way between vinylidene and acetylene although the reaction is strongly exothermic and the barrier is low. This apparent violation of the Hammond postulate[193] was first noted by Dykstra and Schaefer.[180] Petersson *et al.*[191] suggested that the vinylidene isomerization be separated into two processes: (i) hydrogen migration and (ii) π bond formation. In process (i) one CH bond is broken and a new CH bond is created. Hence, Petersson *et al.* assumed this process to be close to thermoneutral,^{a} with a TS half-way between educt and product, as computed for the [1,2]-H shift TS.

---

^{a} However, it should be noted that the dissociation energy of a *sp*-hybridized CH bond in acetylene (131 kcal mol^{-1}) is much greater than that of a *sp^2*-hybridized CH bond in ethylene (110 kcal mol^{-1}).

The reaction is driven by process (ii), the highly exothermic conversion of the vinylidene lone-pair to a π bond in acetylene. Following the orbital energy of the lone-pair, Petersson *et al.* found that the formation of the new π bond is well underway in the TS. This lowers the barrier for [1,2]-H shift, which would be significant if no favorable process were compensating. The authors concluded that the overall isomerization of vinylidene to acetylene as the sum of two separate processes violates the Hammond postulate, but that *"each of these processes individually satisfies the Hammond postulate."*[191]

4.2.3   Competing Pathways in Intramolecular Reactions of Alkylcarbenes

Although migration of hydrogen is in general the most facile [1,2]-shift, other groups also migrate to the carbene center under certain circumstances. As early as 1961 Phillip and Keating established the ordering of [1,2] migration rates as being $H > C_6H_5 > CH_3$.[57,61] Sulzbach *et al.* selected 2-butylidene (19) to study four distinct rearrangement possibilities of an unstrained acyclic carbene at CCSD(T)/DZP + ZPVE including BHandHLYP/DZP thermodynamic corrections (1 atm, 298 K).[170]

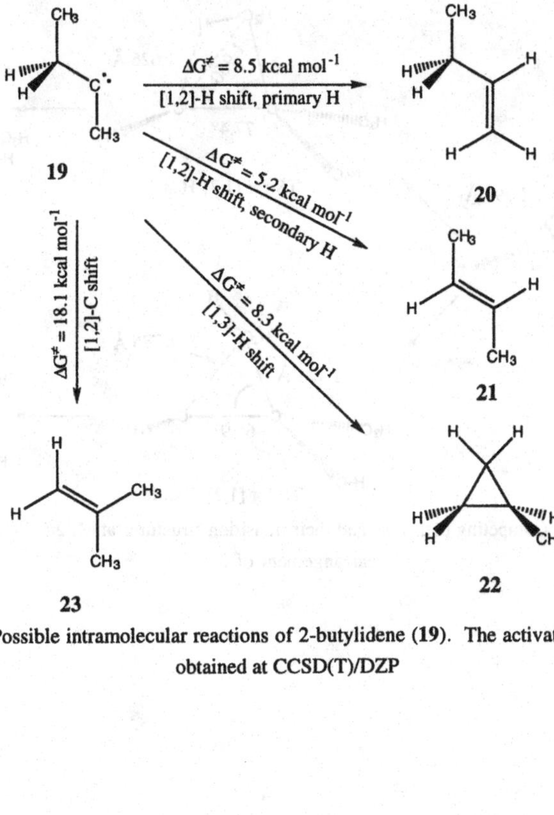

Scheme 4.23 Possible intramolecular reactions of 2-butylidene (19). The activation energies were obtained at CCSD(T)/DZP

The energetically most favorable process, the [1,2]-shift of a secondary H to give *trans*-2-butene (**21**), has a barrier of $\Delta G^{\ddagger} = 5.2$ kcal mol^{-1}. The rearrangement to give *cis*-2-butene (not shown) is almost as facile ($\Delta G^{\ddagger} = 5.9$ kcal mol^{-1}). Hyperconjugation of the methylene H is stronger than that of the methyl H in the lowest energy conformation of **19**, as shown by the different bond lengths and tilt angles of the interacting CH bonds (Scheme 4.24).

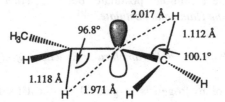

Scheme 4.24 The lowest energy conformation of 2-butylidene (**19**) at CCSD(T)/DZP.

The geometry of the TS for migration of the methylene H is more educt–like than the TS for terminal H migration. Thus, a higher barrier ($\Delta G^{\ddagger} = 8.5$ kcal mol^{-1}) is obtained for the migration of a primary H to give 1-butene (**20**).

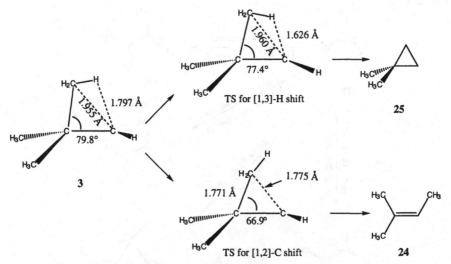

Scheme 4.25 Competing pathways and their transition structures at MP2/6-31G* for the rearrangement of **3**.

The [1,3]-H shift to give methylcyclopropane (22) has a similar barrier ($\Delta G^{\ddagger}$ = 8.3 kcal mol[-1]). The [1,2]-C shift to 2-methylpropene (23) is much less favorable ($\Delta G^{\ddagger}$ = 18.1 kcal mol[-1]). Similarly, the [1,2]-C shift to give 2-methyl-2-butene (24) from 3 was found by Armstrong et al. to be less favorable than the alternative [1,3]-H shift to 1,1-dimethylcyclopropane (25, see Scheme 4.25)) (3.1 vs. 0.1 kcal mol[-1] at QCISD(T)/6-31G*//MP2/6-31G* + ZPVE).[110] The [1,2]-C shift barrier in 3, where the migrating methyl group is tilted towards the carbene center, is considerably smaller than in 19.

In the [1]3 ground state the migrating H is eclipsed and close to the carbene center. The transition structure for [1,3]-H shift is very similar to the educt, resulting in a very low migration barrier (Scheme 4.25). A larger geometric distortion, requiring more energy, is necessary to reach the TS for [1,2] methyl migration yielding 24 (Scheme 4.25). However, photolysis of 2,2-dimethyl-1-diazopropane gives the cyclopropane derivative and the alkene in almost equal amounts.[194] As Armstrong et al.[110] found no evidence for a thermal loss of N_2 bypassing 3, the photolysis of 2,2-dimethyl-1-diazopropane must involve other species than the "free" carbene. Indeed it could be shown that "free" 3, generated by C atom deoxygenation of 2,2–dimethylpropanal, yields 25 exclusively at low temperatures and a mixture of 24 and 25 at higher temperatures.[110,194]

Cyclobutylidene (26), in contrast, gives a 85 : 15 mixture of the [1,2]-C migration and [1,2]-H migration products, methylenecyclopropane (27) and cyclobutene (28), respectively.

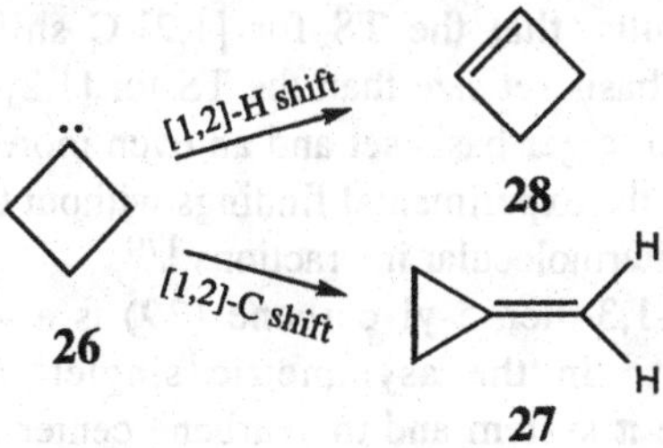

Scheme 4.26 Alternative rearrangement pathways of cyclobutylidene.

Sulzbach et al. studied these reactions at the CCSD(T)/DZP level of theory[170] and located a C_s symmetric "nonclassical" cyclobutylidene minimum structure for [1]26 (Scheme 4.27). Carbon-4 is pentacoordinated with distances to the three other carbons of 1.61 Å (to C1 and C3) and 1.84 Å (to the carbene center C2). The isoelectronic cyclobutyl cation is well studied and has a similar "nonclassical" structure, which can be viewed as a complex of methylene and the allyl cation.[195-201]

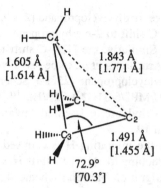

Scheme 4.27 Structure of the "nonclassical" cyclobutylidene at CCSD(T)/DZP.[170] The B3LYP/6–311+G** parameters are given in brackets.

The nonclassical [1]26 structure also facilitates formation of the transition structure for [1,2]-C shift, in which the C4-C2 bond is shortened to 1.635 Å and the C4-C3 bond is elongated to 2.009 Å at CCSD(T)/DZP. Thus, $\Delta G^{\ddagger}$ for [1,2]-C migration in **26** is about 8 kcal mol^{-1} lower than in **19**, and the [1,2]-H shift barrier is higher by about 4 kcal mol^{-1}. However, the higher free energies of activation for [1,2]-C than for [1,2]-H shift in **26**, $\Delta G^{\ddagger}$ = 10.5 and 9.2 kcal mol^{-1}, respectively, disagree with the experimental observations of a greater amount of C–shifted product. Sulzbach *et al.* found, by comparing HF/TZ2P(f,d+), MP2/DZP, and CCSD(T)/DZP results, that the TS for [1,2]-C shift is more sensitive to electron correlation and basis set size than the TS for [1,2]-H migration.[170] Thus, the authors assume that a larger basis set and an even more advanced method than CCSD(T) will reproduce the experimental findings without the need for inclusion of solvent effects or other intermolecular interactions.[170]

Singlet cyclopenta-1,3-dien-5-yl carbene (**29**) is a very interesting species (Scheme 4.28).[202-204] In the asymmetric singlet minimum electrophilic interactions between the π system and the carbene center are shown by the NBO analysis (B3LYP/DZP).[204]

Both the C1-C5 σ bond and the C1–C2 π bond donate electron density into the carbene *p* orbital which is occupied by 0.29 electrons.[204] In accord, the C1–C5 bond is longer than the C5–C4 bond and the C1–C5–C6 angle is small, as typical for molecules stabilized by CC hyperconjugation.

Another asymmetric conformation of [1]**29** exists at the HF/3–21G[202] and CASSCF(6,6)/3–21G[203] levels with the H at C6 pointing "outwards",[202] away from the five membered ring. This outward conformation is 2 kcal mol^{-1} less stable than the inward conformer,[203] shown in Scheme 4.26. At the B3LYP/DZP and

MP2/DZP levels, however, no minimum corresponding to the outward conformer could not be found; optimization resulted in benzvalene (**30**).[204]

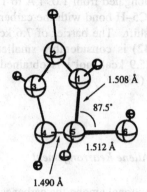

1.508 Å

87.5°

1.512 Å

1.490 Å

Scheme 4.28 Structure of singlet cyclopenta–1,3-dien–5–yl carbene (**29**) at B3LYP/DZP.

Experimentally, the intramolecular addition of the carbene to the butadiene π system proceeds very easily and exclusively in the 1,4 mode.[205,206] A CCSD(T)/DZP//B3LYP/DZP + ZPVE study[204] identified the intramolecular 1,4 addition to give **30** as the most facile intramolecular reaction of **29**. The low 2.6 kcal mol^{-1} barrier is due to the pronounced electrophilic interaction of the carbene p orbital with the π system.[204]

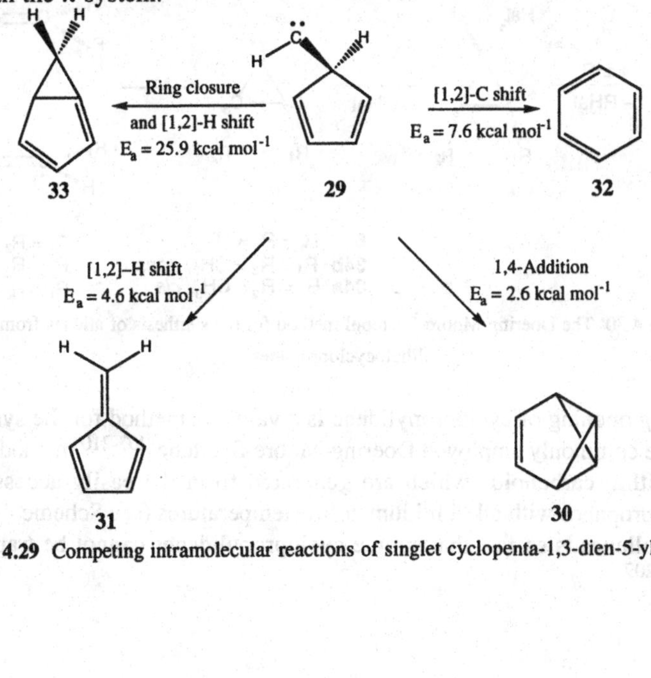

Ring closure
and [1,2]-H shift
$E_a = 25.9$ kcal mol^{-1}

[1,2]-C shift
$E_a = 7.6$ kcal mol^{-1}

33   **29**   **32**

[1,2]–H shift
$E_a = 4.6$ kcal mol^{-1}

1,4-Addition
$E_a = 2.6$ kcal mol^{-1}

31   **30**

Scheme 4.29 Competing intramolecular reactions of singlet cyclopenta-1,3-dien-5-yl carbene.

The [1,2]–H shift to fulvene (31) has a higher barrier (4.6 kcal mol^{-1}). The TS lies very early along the reaction coordinate as the C5–H bond (involved in the H migration) is only slightly elongated from 1.094 Å to 1.116 Å. A rotation around the C6–C5 bond aligns the C5–H bond with the carbene p orbital, i.e. in the best orientation for the [1,2]–H shift. The barrier of 7.6 kcal mol^{-1} computed for the [1,2]–C shift to benzene (32) is considerably smaller than in 2–butylidene35. Finally, a high barrier of 25.9 kcal mol^{-1} is obtained for the rearrangement to bicyclo[3.1.0]hexa-1,3-diene (33).

4.2.4 Other Rearrangements

4.2.4.1 Cyclopropylidene to Allene Rearrangement

Cyclopropylidene (6) is exceptional among the carbenes discussed up to now. Due to the small C-C1-C angle $\phi = 60°$ the singlet state is preferred over the triplet state by 11.1 and 14.9 kcal mol^{-1} at MR-CISD/TZ2P and CCSD(T)/TZ2P, respectively, using B3LYP/TZP geometries,[81] and 14.2 kcal mol^{-1} at B3LYP/6-311+G**. In contrast to typical alkyl carbenes, C2-C3 bond cleavage (see Scheme 4.30) rather than [1,2]-H shift reactions dominate the chemistry of cyclopropylidenes.

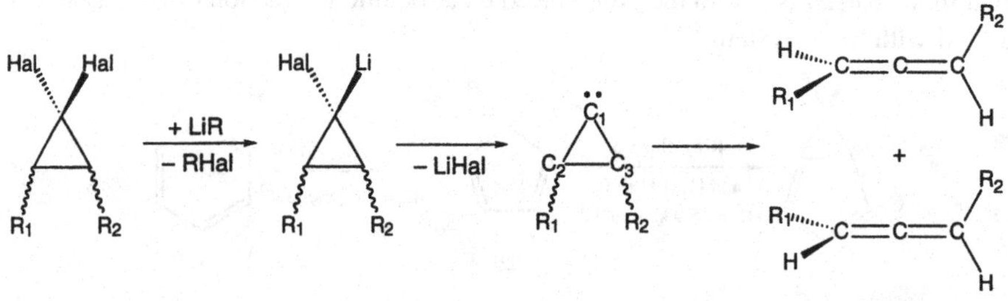

6	$R_1 = R_2 = H$	$R_1 = R_2 = H$
34b	$R_1 = R_2 = CH_3$, *trans*	$R_1 = R_2 = CH_3$, *R*
34a	$R_1 = R_2 = CH_3$, *cis*	$R_1 = R_2 = CH_3$, *S*

Scheme 4.30 The Doering-Moore-Skatebøl method for the synthesis of allenes from 1,1-dihalocyclopropanes.

The ring opening of cyclopropylidene is a valuable method for the synthesis of allenes. The commonly employed Doering-Moore-Skattebøl[207-210] method proceeds *via* α-halolithio carbenoids which are generated from the easily accessible 1,1-dihalocyclopropanes with alkyl lithium at low temperatures (see Scheme 4.30). The reaction to allenes is so fast that simple cyclopropylidenes cannot be trapped with alkenes.[208,209]

The ring opening is very intriguing theoretically. The overall mode of rotation of the methylene groups must be conrotatory in going from cyclopropylidene to allene. In contrast, the analogy to the isoelectronic cyclopropyl cation suggests a disrotatory motion,[211-214] at least in the initial stages of the reaction. The C_s symmetry element perpendicular to the cyclopropylidene ring, which is lost during the reaction, transforms the *two enantiomeric allenes* into each other, but transfers the *one educt* into itself (Scheme 4.31). As this C_s symmetry element is lost during the ring opening, the reaction path must split, i.e., bifurcate, somewhere between the educt and the product.

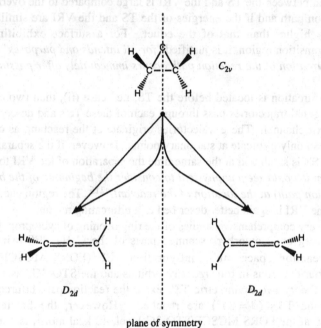

plane of symmetry

Scheme 4.31 During ring opening of cyclopropylidene a C_s symmetry element is lost forcing a bifurcating of the reaction path along the reaction coordinate.

Bifurcation exactly at the TS was shown by Valtazanos and Ruedenberg to be a highly unlikely "*numerical accident.*"[215] A characteristic point for bifurcations is the "valley-ridge-inflection" (VRI) point, which is not a stationary point.[215] At the VRI the bottom of a valley turns into a ridge. The Hessian matrix has a zero eigenvalue and the corresponding eigenvector is perpendicular to the gradient at the VRI. At the VRI, a valley on the PES splits into two valleys. As orthogonal trajectories can only bifurcate at stationary points,[215] an orthogonal trajectory, used as a model for the reaction path, will not split at the VRI. Rather, *bifurcating regions* result when the bifurcation does not coincide with a stationary point.[215]

Valtazanos and Ruedenberg[215] studied the following situations using model surfaces: (i) TS before the bifurcation and (ii) TS after the bifurcation. For case (i) there exists one C_s symmetric transition state. After the TS the VRI point is reached and the reaction path splits into the two exit channels. *"Neither of the two exit channels contains any orthogonal trajectory that connects with the VRI point or the TS or the entrance channel."*[215] Thus, calculation of the IRC starting at the TS will not lead to the reaction products. A *bifurcating transition region* is present, if the separation between the TS and the VRI is large compared to the overall lengths of the reaction path and if the energies of the TS and the VRI are similar and are considerably higher than that of the educt. For a surface exhibiting such a bifurcating transition region, it is justified *"for all intends and purposes"* to assume that *"the bifurcation of the reaction path occurs immediately after passing through the saddle."*[215]

If the bifurcation is located before the TS, i.e., case (ii), than two chiral TS's exist. Orthogonal trajectories pass through each of these TS's and descend into the respective exit channel. These trajectories originate at the reactant, as orthogonal trajectories can only bifurcate at stationary points. However, if the separation of the VRI to the TS's is small and at the same time the separation of the VRI to the educt is large, *"then it would seem unphysical to consider the beginning of the bifurcation (of the reaction path) at the location of the reactant."*[215] The region encompassing the TS and the VRI is again better described as a bifurcating region.

In their very comprehensive studies of the ring opening of cyclopropylidene,[216-219] Ruedenberg and coworkers scanned parts of the C_3H_4 PES using the full optimized reaction space multiconfiguration SCF (FORS MCSCF) method employing four electrons in four *reactive* orbitals and the STO-3G basis set.[216] At this level of theory two asymmetric TS's exist; the reaction path bifurcates at $\phi \approx 80°$ before the TS's ($\phi = 84°$) are reached. However, the barrier for the isomerization at the FORS MCSCF/STO-3G level, 40 kcal mol^{-1}, is considerably larger than at CISD+Q/DZP//RHF/DZP[220,221] (10 kcal mol^{-1}) and that suggested by the elusive character of cyclopropylidene experimentally.[222]

Improving the basis set and enlarging the active space lowers Ruedenberg's barrier considerably; it is 7.5 kcal mol^{-1} at FORS MCSCF(8,8)/DZd + ZPVE.[218] Besides the activation barrier, the geometry of the TS changes to C_s symmetry at the higher MCSCF level. Thus, the location of the bifurcation, which is slightly *before* the TS's when using a minimal basis set, changed to slightly *after* the TS when non-minimal basis sets or larger active spaces are employed. The reaction starts with a disrotatory motion of the methylene groups until the TS is reached.[218]

The reaction path emanating from the TS, which lies on a extremely flat plateau, is still C_s symmetric and the rotation mode of the methylene groups still is disrotatory. Shortly after the TS, the reaction path bifurcates between the VRI and a conical intersection of the $^1A'$ cyclopropylidene ground state PES and a $^1A''$ excited state PES. The motion of the methylene groups switches to a conrotatory mode, and the C_s symmetry path is abandoned. One of the methylene groups keeps its sense of rotation and moves further to the CCC plane, whereas the other methylene group reverses its rotation and moves away from the CCC plane. As the sense of rotation of either of the two groups can reverse with the same probability, two enantiomeric reaction paths are possible. The methylene groups can rotate freely in a cogwheel-like fashion, i.e., conrotatory, leading to perpendicular allene on the descending reaction paths towards the products .

A somewhat different description of the ring opening of cyclopropylidene was obtained by Bettinger *et al.*[81] Two asymmetric TS's were obtained using the B3LYP/TZP, the CISD/TZP, and the CCSD(T)/6-31G* methods, which take dynamical correlation into account. Hence, the bifurcation must occur before the TS's are reached, similar to Ruedenberg's results at the FORS MCSCF/STO-3G level. A CASSCF(4,4)/TZP optimization by Bettinger *et al.* yielded a C_s symmetric TS in agreement with FORS MCSCF/DZd calculations.[218] To ensure that potential diradical character at the TS geometry does not demand a multireference treatment, Bettinger *et al.*[81] inspected the CISD wave function and the T_1 diagnostics[223-225] of the CCSD wave function. Both criteria suggest that a single reference is a valid approximation for the correlation treatment, as the CI coefficients are 0.94, 0.08, and 0.04 and the value of T_1 is below 0.02.

The intrinsic reaction coordinate (IRC)[226-228] calculated at B3LYP/6-31G* (Scheme 4.32) reveals the complex modes of rotation during the ring opening of cyclopropylidene.[81] The reaction starts with a synchronous disrotatory motion of the methylene groups keeping C_s symmetry. At $\phi \approx 80°$ the reaction path bifurcates: although the overall sense of rotation is still disrotatory, one of two methylene groups rotates faster than the other thereby breaking C_s symmetry. In the proximity of the TS, one methylene group *reverses* its sense of rotation so that the overall motion becomes *conrotatory*. In this region of the PES, internal variables characterizing the rotation of the methylene groups change faster than the ring opening angle ϕ. On descending the reaction path further, the motion remains conrotatory until the allene product is reached.

The activation energy for ring opening is 3-5 kcal mol^{-1} according to CCSD(T)/TZ2P and MR-CISD/TZ2P single point calculations on CASSCF(4,4)/TZP, CISD/TZP, and B3LYP/TZP geometries.[81] The reaction enthalpy was determined to be -68 to -69 kcal mol^{-1} at these levels of theory.[81]

138

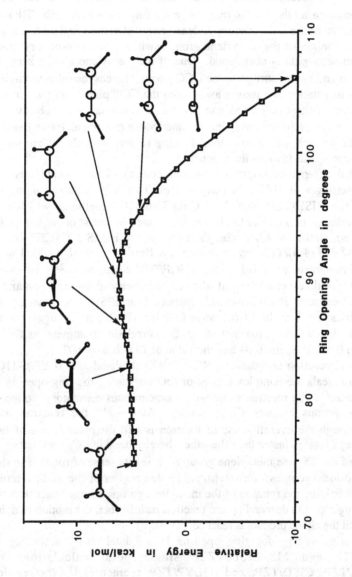

Scheme 4.32 Intrinsic reaction coordinate for the ring opening of cyclopropylidene at B3LYP/6-31G*.

Substitution of cyclopropylidene with 2,3-dimethyl groups has a considerable influence on the ring opening barrier and the shape of the PES.[229] The TS for the ring opening of the *cis* isomer (**34a**) is C_s symmetric and lies very early on the reaction coordinate; ϕ is enlarged by 4.3° and the breaking C2-C3 bond is elongated by 0.067 Å. As one C_s symmetric TS is obtained, the reaction path bifurcation is expected after the TS is passed.

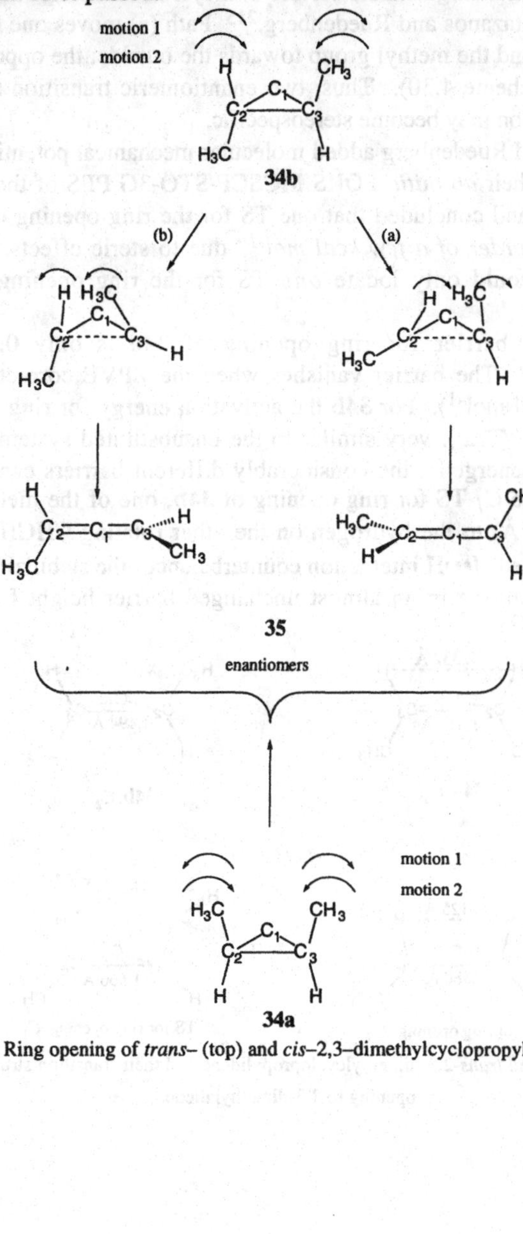

Scheme 4.33 Ring opening of *trans*– (top) and *cis*–2,3–dimethylcyclopropylidene (bottom).

The reaction starts disrotatorily for C_2 *trans*-2,3-dimethylcyclopropylidene (**34b**), but since it has no C_s plane, symmetry is lost upon rotation (see Scheme 4.33). Like the unsubstituted cyclopropylidene, one of the rotating groups keeps its sense of rotation whereas the other reverses during the isomerization to 1,3-dimethylallene (**35**). While the two geometric possibilities arising from this "switch-over" are indistinguishable for the **6**, they can be differentiated for **34b**, as pointed out by Valtazanos and Ruedenberg.[215] Path (a) moves one H nearer to the inside of the ring and the methyl group towards the outside; the opposite is true for motion (b) (see Scheme 4.30). Thus, two enantiomeric transition structures may exist, and the reaction may become stereospecific.

Valtazanos and Ruedenberg added molecular mechanical potentials for the two methyl groups to their *ab initio* FORS MCSCF/STO-3G PES of the unsubstituted cyclopropylidene and concluded that one TS for the ring opening of **34b** is more favorable in "*the order of a few kcal mol⁻¹*" due to steric effects.[219] However, Bettinger *et al.* could only locate *one* TS for the ring opening of the *trans* compound.[229]

The classical barrier for ring opening of **34a** is only 0.5 kcal mol⁻¹ (B3LYP/TZP).[229] The barrier vanishes when the ZPVE correction is applied (ΔZPVE = 0.5 kcal mol⁻¹). For **34b** the activation energy for ring opening is 4.2 kcal mol⁻¹ (B3LYP/TZP), very similar to the unsubstituted system. As **34a** and **34b** are almost isoenergetic, the considerably different barriers cannot be ground state effects. In the C_1 TS for ring opening of **34b**, one of the methyl hydrogens comes close (2.07 Å) to the hydrogen on the other rotating CHCH₃ group. This repulsive non-bonded H•••H interaction counterbalances the stabilizing effect of the methyl groups, resulting in an almost unchanged barrier height from the parent cyclopropylidene.[229]

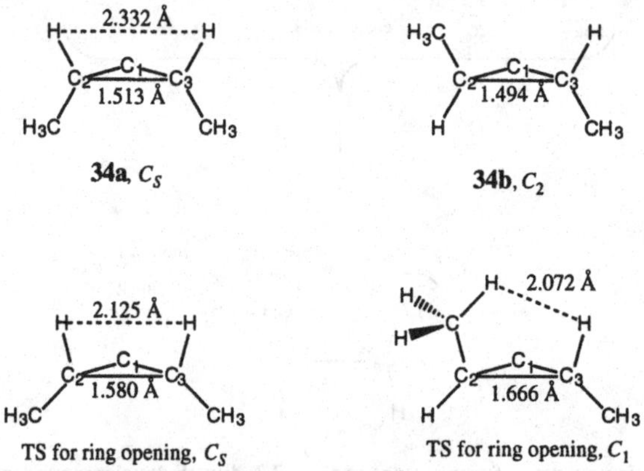

Scheme 4.34 *Cis* and *trans*-2,3-dimethylcyclopropylidene and their transition structures for ring opening to 1,3-dimethylallene.

4.2.4.2 Cycloalkyne to Cycloalkylidenecarbene Rearrangements

Attempts to trap cyclobutyne (**36**), which is highly stained (106 kcal mol^{-1} at MP4SDTQ/6-31G*//MP2/6-31G* + ZPVE[230] according to Scheme 4.35) did not unequivocally show the intermediacy of a free cyclobutyne in rearrangement reactions to cyclopropylidenecarbene (**37**).[231]

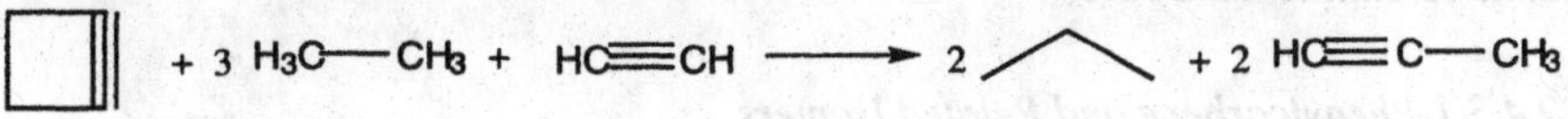

$$\square\!\!\!| + 3\ H_3C\!-\!CH_3 + HC\!\equiv\!CH \longrightarrow 2\ \wedge + 2\ HC\!\equiv\!C\!-\!CH_3$$

Scheme 4.35 Homodesmotic equation for the evaluation of strain in cyclobutyne.

$$\square\!\!\!| \longrightarrow \triangleright\!\!=\!\!C:$$

36   **37**

Scheme 4.36 Rearrangement of cyclobutyne to cyclopropylidenecarbene.

Cyclobutyne is a local C_{2v} minimum on the C_4H_4 PES according to results of Schaefer's group as well as Johnson and Daoust.[230,232-234] However, the potential well is very shallow. The rearrangement to **37** involves a barrier of only 0.4 kcal mol^{-1} using second order CI single point evaluations (SOCI/6-31G*//MCSCF(4,4)/6-31G* + ZPVE, MP4SDTQ/6-31G*//MP2/6-31G* + ZPVE even gave a small negative activation barrier).[230] The transition structure for the rearrangement (Scheme 4.37) is very similar to the educt as expected for an exothermic reaction ($\Delta H \cong 23$ kcal mol^{-1}) with a very low barrier.

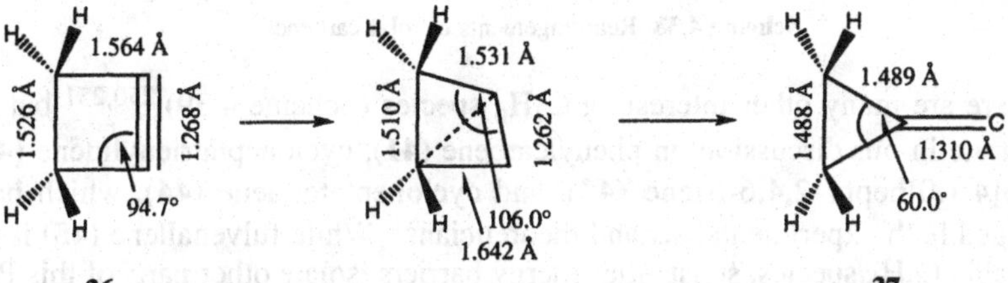

Scheme 4.37 Geometries of cyclobutyne (**36**) and the TS for rearrangement to cyclopropylidenecarbene (**37**) at MCSCF(4,4)/6-31G*.

Due to the smaller ring strain of cyclopentyne (74 kcal mol^{-1}) and cyclohexyne (41 kcal mol^{-1}), these cyclic acetylenes are observable experimentally.[235] The inherent exothermicity of the carbene to alkyne rearrangement (a new CC bond is formed!) more than compensates for the ring strain. Hence, the cycloalkylidenecarbenes isomerize to cycloalkynes. The barrier for rearrangement of cyclopentylidenecarbene is considerably larger than for cyclobutylidenecarbene (10 kcal mol^{-1} vs. 4 kcal mol^{-1} at MP4SDTQ/6-31G*//MCSCF(4,4)/3-21G + ZPVE, respectively), and the corresponding transition structures resemble the carbene geometries.[230]

4.2.4.3 Aromatic Carbenes

4.2.4.3.1 Phenylcarbene and Related Isomers

Aromatic carbenes undergo fascinating rearrangements.[54,236-241] The cascade of rearrangements of p-tolylcarbene (38) affords a spectacular example (Scheme 4.38). The final products, benzocyclobutene (39) and styrene (40), arise after a series of ring expansion-ring contraction steps which interconvert the *para*, *meta*, *ortho*, and *ipso* carbenes.[242-249]

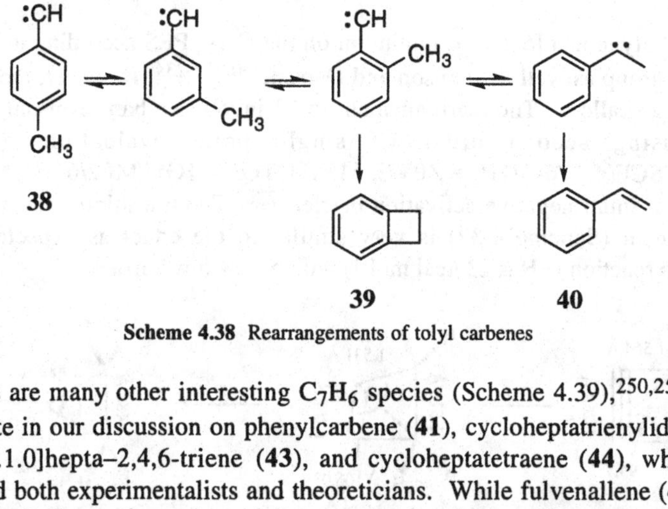

Scheme 4.38 Rearrangements of tolyl carbenes

There are many other interesting C$_7$H$_6$ species (Scheme 4.39),[250,251] but we concentrate in our discussion on phenylcarbene (41), cycloheptatrienylidene (42), bicyclo[4.1.0]hepta–2,4,6-triene (43), and cycloheptatetraene (44), which have challenged both experimentalists and theoreticians. While fulvenallene (45) is the most stable C$_7$H$_6$ species, substantial energy barriers isolate other parts of this PES which can be studied in detail.[251]

Spectroscopic (ESR, UV/VIS, and IR[252-258] as well as fluorescence[259] techniques) low temperature matrix isolation studies suggest that the simplest arylcarbene, phenylcarbene, has a triplet (^{3}A") ground state. However, the singlet-triplet separation has *not* been measured precisely. In most cases, 41 displays singlet reactivity,[255] suggesting a small singlet-triplet energy separation (ΔE_{ST}), probably less than 5 kcal mol^{-1}.[254] The angle at the carbene carbon was estimated

estimated empirically for the triplet (based on the zero-field splitting parameters D/E) to be around 155°.[260]

The first stable photochemical rearrangement product of **41**, cycloheptatetraene (**44**), was characterized spectroscopically.[247] Pyrolytic routes[261-264] also led to **44**, which appears to be the most stable isomer in this region of the C_7H_6 potential surface. Trapping studies[265-272] as well as related experiments on halophenylcarbenes[252] as well as tolylcarbenes[247,273] support the involvement of **44** in phenylcarbene rearrangements.

The proposed intermediacy of **43** in the rearrangement of **41** to **44** is consistent with all available experimental data, but the bicyclic system has so far eluded detection both by direct and indirect means. Benzannelated derivatives of **43** have been trapped chemically (see below)[274-278] and observed directly[279,280] suggesting that **43** should be an energy minimum on the C_7H_6 singlet PES as well.

Cycloheptatrienylidene (**42**) is certainly the most enigmatic species in this series; its role in phenylcarbene rearrangements remained unsolved for a long time. As a special type of a vinylic carbene, the triplet state should be stabilized as found for other π-conjugated carbenes (see above). However, the cyclic conjugated six-π-electron arrangement of the singlet tropylium framework should be aromatic.[281-284] Two different ESR spectra were reported, but these must arise from different species.[285,286] This discrepancy has not been resolved experimentally to date.

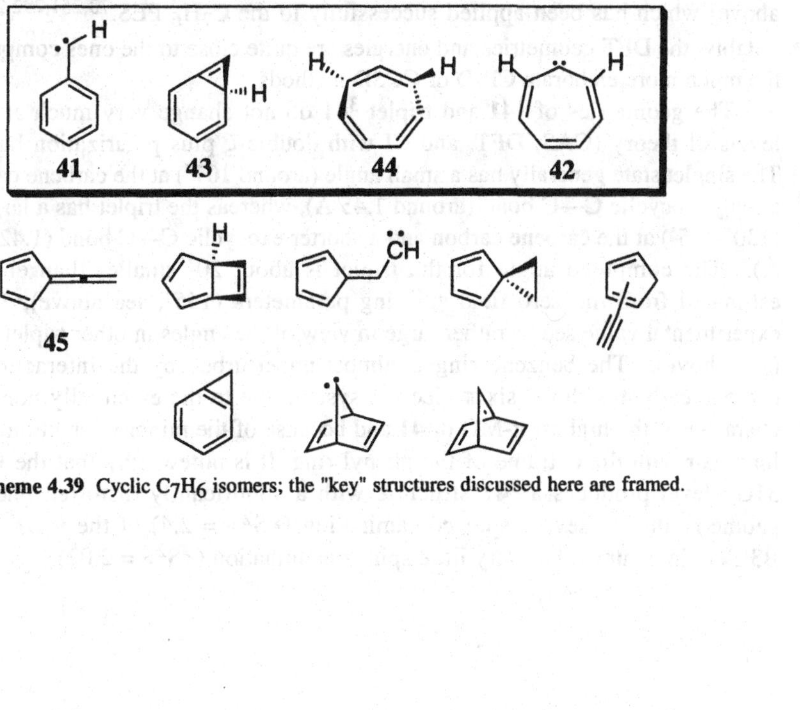

Scheme 4.39 Cyclic C_7H_6 isomers; the "key" structures discussed here are framed.

All theoretical studies (semiempirical[281,282,287-290] or *ab initio*[283,284]) on the ring expansion of **41** found **44** to be considerably more stable than **42**. Investigations on **42** were complicated by the usual problems associated with describing closed vs. open shell character of the singlet state (see Ch. 2), so that reliable estimates for ΔE_{ST} of **42** could not be made for a long time. While *ab initio* methods (at low levels by today's standards) implied that both the $^1A_1$ singlet and the $^3B_1$ as well as the $^3A_2$ triplet states are minima,[283,284] early semiempirical studies predicted that $^1A_1$–**42** should be a transition state for interconverting two enantiomeric **44**.[289,291] The role of the respective open-shell singlets ($^1B_1$ and $^1A_2$) derived from the two triplet states ($^3B_1$ and $^3A_2$) was neglected for a long time and was addressed only very recently.[73,251,292] The most recent and most elaborate studies at DFT and coupled cluster [CCSD(T)] levels by three independent groups identify the $^1A_1$–**42** singlet as a transition structure (see below).[73,251,293]

As discussed above, the open-shell species and some of the transition states require high level treatments. Unrestricted Hartree-Fock theory (UHF) is *a priori* unsuitable; UMPn methods suffer from severe spin contamination, leading to geometries quite different from those obtained at DFT or CISD levels.[73,251,292] Multiconfigurational treatments without (CASSCF)[294] and with inclusion of dynamic electron correlation (CASPT2[295,296] and two-determinant CCSD[297,298]) give excellent results, but are very time-consuming. The best compromise is offered by density functional theory [for a detailed discussion on the applicability of DFT methods (particularly the B3LYP hybrid functional) to carbenes see Ch. 2 above] which has been applied successfully to the C_7H_6 PES.[73,251,292,299] Most notably, the DFT geometries and energies are quite close to the ones computed with the much more elaborate CISD or CCSD methods.

The geometries of ^{1}**41** and triplet ^{3}**41** do not change very much at different levels of theory (CAS, DFT, and CI with double-ζ plus polarization basis sets). The singlet state generally has a small angle (around 107°) at the carbene carbon and a long exocyclic C—C bond (around 1.45 Å), whereas the triplet has a larger angle (130°-135°) at the carbene carbon and a shorter exocyclic C—C bond (1.42 Å - 1.44 Å). The computed angle for the triplet is about 20° smaller than empirically estimated from the zero-field splitting parameters (155°, see above),[260] but the experimental value seems rather large in view of the angles in other triplet carbenes (see above). The benzene ring is almost unperturbed by the interaction of the carbene carbon with the six π-electron system due to the essentially non-bonding character of the highest π-MO of **41** and because of the minor interaction of the σ-lone pair with the σ-frame of the phenyl ring. It is noteworthy that the UMP2/6-31G* level produces a ^{3}**41** structure with a significantly distorted phenyl ring geometry due to severe spin contamination ($<S^2> = 2.4$) of the wave function. B3LYP, in contrast, has very little spin contamination ($<S^2> = 2.02$).

Apparently, ^{1}**41**, with a vacant p-orbital, is stabilized more by phenyl conjugation than the triplet, with a half-filled p-orbital, and the ΔE_{ST} of **41** is smaller than that of methylene (Scheme 4.40). Although the ΔE_{ST} of **41** is not known precisely, the B3LYP/6-311+G*//B3LYP/6-31G*+ZPVE (5.0 kcal mol^{-1})[73] and CCSD(T)/cc-pVTZ (estimated) //BLYP/6-31G* + ZPVE (2.7 kcal mol^{-1})[292] results lie within the experimental estimates ($\Delta E_{ST} \leq 5$ kcal mol^{-1}).[254]

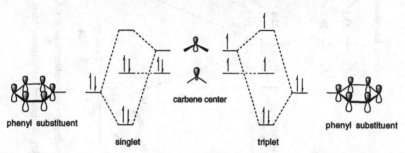

Scheme 4.40 Interaction of a carbene center with a phenyl substituent: the singlet is stabilized more than the triplet.

It is noteworthy that both the G2(MP2,SVP) and G2(RMP2,SVP)/QCI levels[82,300-303] of theory predict a ΔE_{ST} of only 1.4 kcal mol^{-1}, significantly less than at DFT or coupled-cluster levels. In view of the highly accurate CCSD(T)/cc-pVTZ results,[292] one must conclude that the MP2 geometry is bad (compared to the B3LYP/6-31G* geometry)[73] due to spin contamination and that the 6-31G* basis set is too small for accurate QCISD(T) energies[251] when dealing with carbenes.

It was suggested that ΔE_{ST} of **41** computed at a particular level might be correlated by the error found for ΔE_{ST} of methylene at the same level (*vide supra*).[73,251] However, it is not clear that there should be a constant error in computing the ΔE_{ST}'s for such different species. As described above, the phenyl group certainly interacts in a different manner with a carbene carbon than an alkyl group or even a hydrogen. These corrections may thus not be justified.

The allenic C_2-symmetric chiral **44** is a relatively stable species, despite the small angle (around 145° at various optimization levels) at the allenic carbon; the other bonds clearly alternate, supporting the allene character of this molecule.[73,251,281,282,284,288-290,292] The computed IR spectrum (at BLYP/6-31G*) agrees very well with earlier experimental assignments.[252] This agreement not only demonstrates the quality of the computational results, but also rules out the presence of **43** (no IR features agree). However, one must note that allenic structures are "over-stabilized" at BLYP.[304]

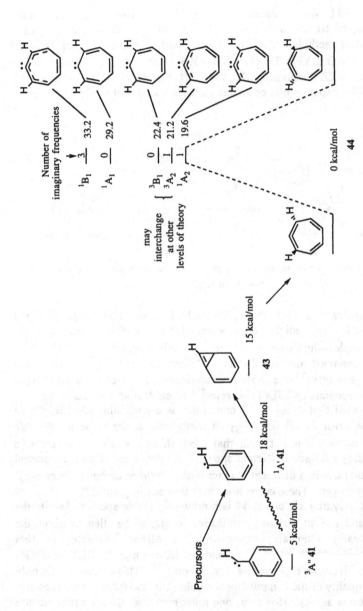

Scheme 4.41 A summary of a part of the phenylcarbene rearrangement surface leading to cycloheptatetraene at CASPT2N/cc-pVDZ//CASSCF(8,8)/6-31G* + ZPVE. Note that this level overestimates the stability of triplet phenylcarbene. More reliable estimates give a ΔE_{ST} of **41** of 3-5 kcal mol^{-1}, in agreement with experimental estimates.

The highly strained **43** formed by vicinal CH insertion of singlet phenylcarbene is slightly lower in energy (around 2.5 kcal mol^{-1} at various levels of theory) than ^{1}**41**.[73,251,292] However, since the formation of **43** from the latter carbene requires an activation energy of at least 13 kcal mol^{-1}, the bicyclic system is unlikely to be observed under these reaction conditions. In addition, its barrier for ring opening to **44** is only 1.0 - 1.5 kcal mol^{-1} (Scheme 4.41). Thus, it would also be difficult to characterize this species *via* attempted generation by other routes, and **43** does not play a significant role in phenylcarbene rearrangements.

The nature and the role of **42** in phenylcarbene rearrangements has been discussed quite controversial. Triplet **42** can related to the cycloheptatrienyl radical by cleaving one of the CH bonds homolytically.[290] As found for the Jahn-Teller distorted radical, strong vibronic coupling leads to two C_{2v} ^{3}**42** states, $^3B_1$ and $^3A_2$.[284] A metastable triplet state ($^3B_1$ or $^3A_2$) has been observed by EPR,[45] but the ground state has not been established experimentally. Despite a considerable number of experimental[261-265,285,286,291,305-310] investigations, the role of **42** in C_7H_6 chemistry has only been identified computationally very recently.[73,251,292] Assigning the correct spin state is complicated by the several low-lying configurations. In the most extensive study,[73] five different spin states for planar, C_{2v}-symmetric **42** were considered: three singlets ($^1A_1$, $^1A_2$, and $^1B_1$) and two triplets ($^3B_1$ and $^3A_2$). The following discussion (energies were determined at CASPT2N(8,8)/cc-pVDZ//CASSCF(8,8)/6-31G* + ZPVE) refers to the cycloheptatrienylidene spin states depicted in Scheme 4.42. Note that the "active space" for the "complete active space calculations (CAS)" was comprised of the seven p orbitals in the ring and the sp^2 hybrid orbital.

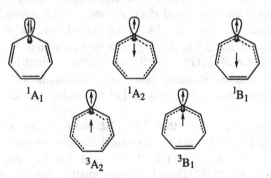

Scheme 4.42 Spin states in **42**.

Triplet States. The $^3A_2$ state can be pictured as a heptatrienyl radical, whereas the $^3B_1$ state resembles a pentadienyl radical, whose terminal atoms interact only weakly with the connecting C-C double bond. In the cycloheptatrienyl radical the corresponding pair of spin states may interconvert *via* pseudorotation (along the Jahn-Teller potential energy well);[311] one form is a minimum while the other is a transition structure. In ^{3}42 the σ-electron cannot interchange with the π-system at SCF levels of theory (UHF and ROHF) and one finds both states to be minima. Inclusion of dynamic electron correlation increases the energy gap between the two triplet states and the $^3B_1$ state is a saddle point at MP2 and BLYP.[292] This situation is reversed at the CASSCF level, where $^3B_1$, now a minimum, is only 2 - 3 kcal mol^{-1} below the $^3A_2$ state (transition structure).[73] Following the imaginary mode of $^3A_2$-42 leads, *via* intermediate geometries of C_s symmetry, to the $^3B_1$ state, quite similar to the situation found in the cycloheptatrienyl radical.

At CASPT2N the electronic energies of the $^3A_2$ and $^3B_1$ states are within 0.2 - 0.4 kcal mol^{-1}, depending on the basis set. Since the topology of the PES for the two states obviously changes on going from CASSCF to CASPT2N, it is inappropriate to apply the CASSCF ΔZPVE correction to the CASPT2N energies, and it is not at all clear which triplet state actually is lower in energy. The change in the relative energies of $^3A_2$ and $^3B_1$ in going from CASSCF to CASPT2N can be attributed to the more delocalized $^3A_2$ vs. $^3B_1$ wave function. Inclusion of dynamic correlation is more important for the more delocalized wave functions, resulting in selective stabilization of the $^3A_2$ state.[312]

However, it was noted that CASPT2N overestimates the effect of dynamic electron correlation and selectively stabilizes $^3A_2$.[42,313-315] Hence, it was concluded that $^3B_1$-42 is most likely the metastable species observed by EPR.[73,251,285,292,316]

Singlet States. In contrast to the very small energy difference between the two triplet states of **42**, the stabilities of three lowest lying singlet states are quite different. At the CASPT2N/6-311G(2d,p) level, the open-shell $^1A_2$ state is 9.3 kcal mol^{-1} lower in energy than the closed-shell $^1A_1$ state, and it is also 20.1 kcal mol^{-1} below the open-shell $^1B_1$ state. The $^1A_2$ is the ground state and it is also slightly lower in energy (by 1.5 and 3.2 kcal mol^{-1}) than the two triplets.

The CASSCF(8,8)/6-31G* vibrational analysis identifies $^1A_2$-**42** as a transition state for the enantiomerization (racemization) of **44**. The activation barrier, 20.5 kcal mol^{-1}, is approximately half of the rotation barrier of allene.[81,317-324]

Some studies suggested the $^1A_1$-**42** state to be a minimum (at CASSCF(8,8)/6-31G*, the lowest a_2 vibrational mode at 59 cm^{-1} leads to **44**) on the PES,[283] while others (at B3LYP/6-31G*) found it to be a transition state for enantiomerization of **44**.[288,289] Thus, it is quite possible that $^1A_1$-**42** is, like $^1A_2$-**42**, a transition state for enantiomerization of cycloheptatetraene. At the equilibrium geometry of $^1A_2$-**42**, $^1A_1$-**42** is an excited state and *vice versa*. Thus, the equilibrium geometries of $^1A_2$-**42** and $^1A_1$-**42** both lie on the lowest singlet potential surface. An a_2 distortion (C-2 and C-7 move out-of-plane in opposite directions) of **42** from planarity allows $^1A_2$ and $^1A_1$ to mix and connects the equilibrium geometries of these two states of **42** to the geometry of **44**.

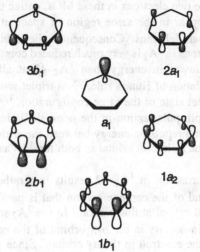

Scheme 4.43 π-MO's for cycloheptatrienylidene (tropylidene) as discussed in the text. The relative energies only are shown qualitatively.

At some intermediate geometries of **42** (between $^1A_2$ and $^1A_1$) these two states must have the same energy, leading to an "*accidental*" Jahn-Teller effect, so that energy lowering caused by mixing of $^1A_2$ and $^1A_1$ is linear, rather than quadratic along the a_2 coordinate. Thus, $^1A_1$, is the transition state for enantiomerization of **44**.

Contrary to early *ab initio* results at HF/STO-3G,[284] the $^1A_1$–**42** state is delocalized with modest bond length alternation ranging from 1.372 to 1.443 Å [CASSCF(8,8)/6-31G*]; DFT and CISD geometries with double-ζ basis are similar.[73] The $^1A_2$ state on the other hand is more localized with single (1.470 Å) and double bonds (1.361 Å) closer to those found in alternating systems.

The $^1B_1$–**42** state is much higher in energy than any of the singlet states due to destabilization by strong Coulomb repulsion caused by occupation of both the a_1 and the $3b_1$ orbitals (with large coefficients on the carbene carbon, see Scheme 4.43). Since these two MOs have atoms in common, the occupying electrons may simultaneously appear in regions where both of these orbitals have electron density. As a consequence, the wave function for the $^1B_1$ state contains high-energy ionic terms.[325-327] This is not the case for the $^3B_1$ state, since the Pauli principle prevents the parallel-spin electrons in the a_1 and $3b_1$ MOs to appear in the same region of space simultaneously.

Electron repulsion in the $^1A_2$ state is negligibly small compared to $^1B_1$, since the two singly occupied MOs in $^1A_2$ (a_1 and $2a_2$) are disjoint (they do not have atoms in common). The two electrons in these MOs, unlike those in the a_1 and $3b_1$ MOs in $^1B_1$, do not appear in the same region of space at the same time, even though they have anti-parallel spins. Consequently, the Coulomb repulsion energy between these two electrons in $^1A_2$ is very much reduced compared to that in $^1B_1$.

Since $^1A_2$-**42** is lower in energy than $^3A_2$-**42** at all levels of theory, it constitutes a formal violation of Hund's rule,[328] as triplet states normally are lower in energy than the singlet state of the same configuration.[329,330] This violation is due to differences in spin polarization in the open-shell singlet and triplet states which effect the Coulomb repulsion energy between the π electrons and the single electron that occupies the a_1 hybrid orbital in both the $^1A_2$ and $^3A_2$ states (Scheme 4.41).[325-327]

Dynamic spin polarization in $^1A_2$-**42** results in a rather large electron spin density in the p-π orbital of the carbene carbon that is parallel to the spin of the electron in the a_1 hybrid orbital at this carbon. In the $^3A_2$ state, spin polarization results in additional spin density in the p-π orbital of the carbenic carbon that is anti-parallel to that of the electron in the a_1 orbital. Since electrons of the same spin in the a_1 and $2b_1$ orbitals are excluded by the Pauli principle from appearing simultaneously in the same region of space, but those of opposite spin are not, spin polarization leads to a lower Coulomb repulsion energy in $^1A_2$ than in $^3A_2$. This is what causes $^1A_2$ to lie below $^3A_2$ at multi-configuration levels of theory (CASSCF and CASPT2).[73]

In summary, it seems unlikely that the $^3A_2$–**42** state has been observed by EPR (*vide supra*). Since $^1A_2$ lies lower in energy than $^3A_2$ and both states have nearly the same equilibrium geometry, intersystem crossing of $^3A_2$ to the lower energy $^1A_2$ state would be likely to depopulate $^3A_2$ so rapidly that $^3A_2$ should not be observable by EPR.

In contrast, $^3B_1$-**42** is much lower in energy than $^1B_1$-**42**. Although the $^1A_2$ state lies slightly below the equilibrium geometry of $^3B_1$ in energy, the difference between the singly occupied π MOs causes these two states to have very different geometries. The CASSCF/cc-pVDZ and CASPT2N/cc-pVDZ energies for the $^1A_2$ state computed at the CASSCF/6-31G* equilibrium geometry of $^3B_1$, are 20.0 kcal mol^{-1} and 14.0 kcal mol^{-1} below those of the corresponding energies at the $^1A_2$ equilibrium geometry, respectively.[73] Thus, at its equilibrium geometry, the $^3B_1$ state lies energetically below any singlet state. This suggests that $^3B_1$ is the cycloheptatrienylidene triplet state detected by EPR. Since $^3A_2$ lies above $^1A_2$, pseudorotation of $^3B_1$ to $^3A_2$, followed by intersystem crossing and vibrational relaxation, would provide a low energy pathway for the lowest triplet state of **42** to reach the equilibrium geometry of **44** (Scheme 4.41).

4.2.3.4.2 Naphthylcarbene and Related Isomers

The naphtylcarbene (**46**) isomers and their rearrangements are expected to be similar to their phenylcarbene analogs, but the experimental facts concerning the naphthylcarbene PES are quite puzzling. Only a relatively small portion of the naphthylcarbene PES had been examined experimentally[202,203,205,206,237,246,247,253,274,275,279,280,331-333] and theoretically (with semiempirical methods)[287] in detail until very recently, when a computational density functional study presented a comprehensive account of all relevant isomers on the $C_{11}H_8$ PES.[80]

One of the more recent experimental studies found "...*no evidence for isomerization of 2-naphthylcarbene to 1-naphthylcarbene...*" at 10 K in matrix isolation.[279] Thus, in marked contrast to phenylcarbene rearrangements (*vide supra*), the interconversion of 1- and 2-naphthylcarbene (1-**46** and 2-**46**) is *not* facile. As for **41**, the rearrangements are suggested to take place in the singlet manifold,[202,205] but ^{1}**46** has not been characterized experimentally. In contrast, the triplet state was generated in various ways and is well characterized.[206,274,275,305,332,333]

The singlet naphthylcarbene rearrangement PES as it is currently known is best summarized in Scheme 4.44. Note that only the "framed" structures have been suggested (simple frame) and have been characterized experimentally to some extent (shadowed frames).

2-Naphthylcarbene undergoes reversible ring closure to 2,3-benzobicyclo[4.1.0]hepta-2,4,6-triene (**47**), but there is no evidence for rearrangement into 1-**46** in solution.[258,263,285,286] This conversion, however, is suggested to take place in the gas phase at high temperatures (375 °C).[203] In contrast to the analogous species on the phenylcarbene PES, **47** has been well characterized in $C_{11}H_8$ solution chemistry.[246,247,279] 1-Naphthylcarbene could not be observed as an intermediate in the 2-**46** rearrangement, as further rearrangements lead to the final common product, cyclobuta[*de*]naphthalene (**48**), at this temperature.

The following discussion refers to the B3LYP/6-311+G*//B3LYP/6-31G* + ZPVE level of theory, unless noted otherwise (Scheme 4.44). Semiempirical methods, used in the early studies of parts of the $C_{11}H_8$ PES, do not give satisfactory results for carbenes and allenes. Not only is the methylene ΔE_{ST} error large at AM1 (30 kcal mol^{-1}), but also carbenoid are favored over allenic structures at this and other semiempirical levels.[80,287]

The ΔE_{ST}'s of 1-**46** and 2-**46** are around 5 kcal mol^{-1}, practically the same as for **41** (*vide supra*), due to very similar MO interactions (Scheme 4.40). The assumption that the singlet state (in which the rearrangements take place) is energetically close to the triplet state thus is justified.[54,236,240,334]

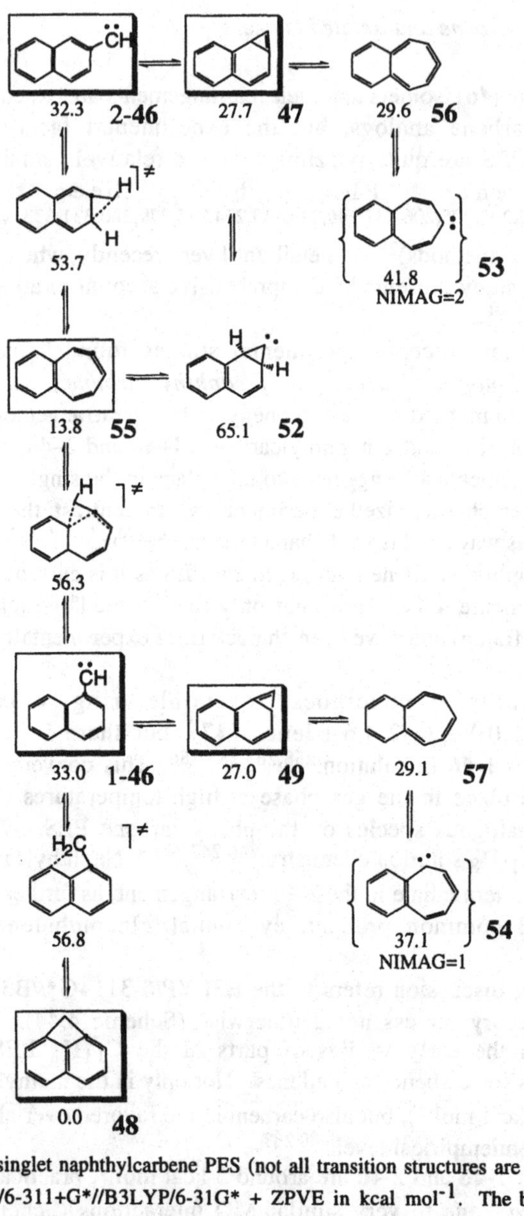

152

Scheme 4.44 The singlet naphthylcarbene PES (not all transition structures are shown).[80] Relative energies at B3LYP/6-311+G*//B3LYP/6-31G* + ZPVE in kcal mol^{-1}. The black dot indicates a carbon label moving from 1-naphthylcarbene to the final product, cyclobuta[*de*]naphthalene.

The structures of the *syn* and *anti* forms of singlet 1-**46** are quite similar, but the *syn* isomers are about 1 kcal mol⁻¹ less stable due to their greater steric repulsion (Scheme 4.45); the carbene carbon angle in the *syn* form is about 2° larger than in the *anti* form. Nevertheless, despite the larger angles in the corresponding triplets (around 135°; singlets are around 107°), the differences in energies of the *syn* and *anti* conformers of triplet 1-**46** also are around 1 kcal mol⁻¹.

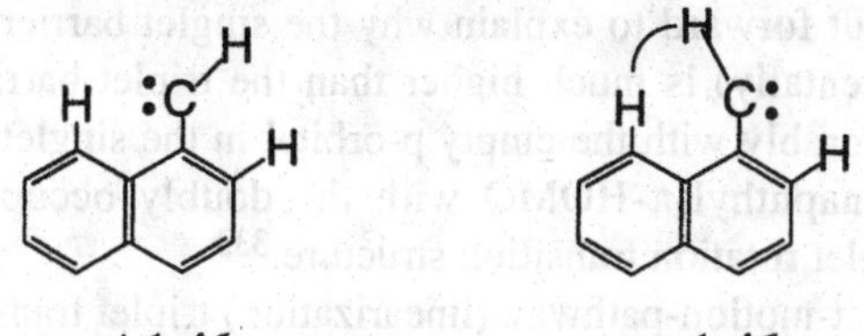

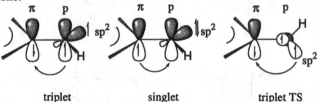

anti-1-**46**   *syn*-1-**46**

Scheme 4.45 Unfavorable H--H repulsion in *syn*-1-**46**.

Like the geometries of **46**, the stabilization energies (ΔH_R, singlet: 25.9 kcal mol⁻¹, triplet: 19.4 kcal mol⁻¹; Scheme 4.46) of 2-**46** are comparable to those of **41**.

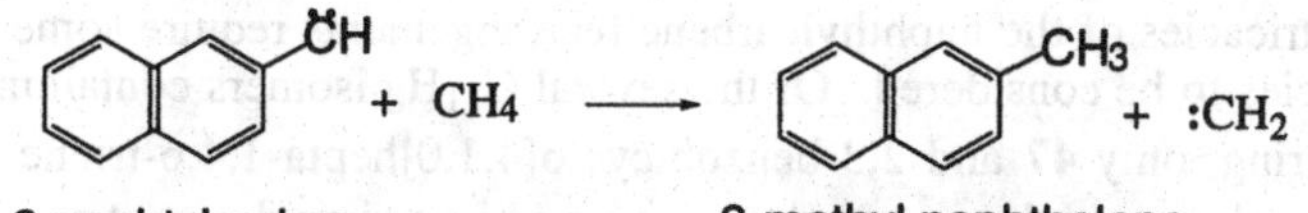

2-naphtylcarbene   2-methyl naphthalene

ΔH_R (singlet *anti*-2-**46**) = +26.2 kcal mol⁻¹;

ΔH_R (triplet *anti*-2-**46**) = +20.1 kcal mol⁻¹

Scheme 4.46 Stabilization energies of anti-2-**46** vs. methylene.

The small computed triplet rotational barrier (3.5 kcal mol⁻¹) between triplet 2-*syn*- and 2-*anti*-**46** agrees with experiment (4.4 kcal mol⁻¹).[333] Scheme 4.47 displays the most important orbital interactions of the naphthyl fragment and the *exo*-methylene.

π   p   π   p   π   p

sp²   sp²   sp²

triplet   singlet   triplet TS

Scheme 4.47 The interactions of the naphthylcarbene π-system with the carbene carbon in the singlet as well as the triplet ground states and in the triplet *exo*-methylene rotation transition structure.

The singlet and triplet states as well as the triplet transition structure benefit from p-π interactions. The energy difference between the ground state triplet and the triplet transition structure thus arises from the less favorable overlap of an sp^2 in the rotation transition structure $vs.$ a p orbital in triplet 2-syn-**46**. Since the exo-methylene bond angle widens more in the transition structure (from 135° to 144°), the activation energy is relatively small. This interpretation is consistent with the earlier analysis of the rotational barrier of triplet phenylcarbene.[335] Similar arguments can be put forward to explain why the singlet barrier (19.9 kcal mol^{-1}; not known experimentally) is much higher than the triplet barrier. While the π-system interacts favorably with the empty p-orbital in the singlet ground state, the interaction of the naphthyl-π-HOMO with the doubly occupied sp^2 orbital is repulsive in the singlet rotation transition structure.[333]

An in-plane least-motion-pathway (linearization) triplet transition structure for syn-$anti$ isomerization of the carbene hydrogen also exists, but it has two imaginary frequencies. However, this second order saddle point is only 1.3 kcal mol^{-1} higher in energy than the triplet rotation transition structure. By analogy, linear triplet methylene (rel. energy = 5.9 kcal mol^{-1}, B3LYP/6-31G*) is the transition structure (with a doubly degenerate imaginary mode of 987 i cm^{-1}) for linearization of bent methylene.

The intricacies of the naphthylcarbene rearrangements require some important related species to be considered. Of the several $C_{11}H_8$ isomers containing a three-membered ring, only **47** and 2,3-benzobicyclo[4.1.0]hepta-1,4,6-triene (**49**) have been observed and characterized to some extent.[203,246,247,253,258,263,279,280,283,285,286] 2,7-Benzobicyclo[4.1.0]hepta-1,3,5-triene (**50**) and 2,4-benzobicyclo[4.1.0]hepta-2,5,7-triene (**51**) were suggested[242-244,248,261,336] to participate in the rearrangements of **46**, but they have remained experimentally elusive.

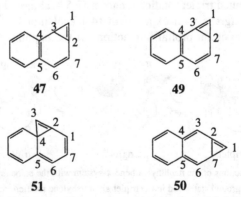

Scheme 4.48 Benzobicycloheptatrienes related to the naphthylcarbene PES.

Neither **50** nor **51** are minima at levels of theory which attempt to include electron correlation. These structures correspond to transition states for rearrangement of 1- and 2-**46** into benzocycloheptatetraenes. The instability of these molecules is due to loss of aromaticity in *both* rings. 2,3-Benzobicyclo[4.1.0]hepta-2,4-diene-6-ylidene (**52**), is a true minimum; the three-membered ring moiety is geometrically close to that of cyclopropylidene.

In line with the findings for **42**, benzannelated cycloheptatrienylidenes singlets do not exist at correlated levels of theory! All attempted optimizations of the carbenes led directly to their cycloallenes. Most strikingly, singlet benzobicyclohepta-2,4,6-triene-1-ylidene (**53**), suggested to be an intermediate in the carbene-carbene rearrangement of naphthylcarbene,[237] has two imaginary frequencies! The a_2 imaginary mode points towards the corresponding allene. Again, this is like the phenylcarbene PES, where **42** also is a transition structure for interconverting two enantiomeric **44** species (*vide supra*). Triplet benzobicyclohepta-2,4,6-triene-1-ylidene is a minimum with a ΔE_{ST} of 5.5 kcal mol^{-1}. This seems to agree well the experimental ΔE_{ST} estimate (only from the ESR characterization of the triplet, 3.5 - 4.0 kcal mol^{-1}),[14,305] but the allene may serves as the singlet species.

Benzobicyclohepta-1,4,6-triene-3-ylidene (**54**) has been proposed[282,331] as an intermediate resulting from rapid ring expansion of 1-**46**. However, the rearrangement to the more stable **48** predominates under the reaction conditions (600° C, 10^{-3} Torr), and **54** has not been observed experimentally. Not surprisingly, as found for the other cyclic conjugated carbenes, it is not a minimum (NIMAG=1; the imaginary mode leads to the corresponding cyclic allene) either! Thus, supposedly "aromatic" singlet seven-membered ring carbenes (benzannelated or not) do not exist, and the observed behavior derives from allenes. The corresponding triplet states, however, are stable; the ΔE_{ST}'s (with the cycloallenes as the singlets) are generally low (> 5 kcal mol^{-1}).

4,5-Benzocyclohepta-1,2,4,6-tetraene (**55**) has been prepared and characterized via various routes,[202,242-244,248,261,336,337] and it was recently identified as the second lowest minimum with a relative energy of 14.3 kcal mol^{-1} above the global $C_{11}H_8$ minimum **48**.[80] This finding is not surprising as **44** is the global minimum on the phenylcarbene PES, and the annelated benzene ring in **55** provides additional stabilization. It is quite conceivable that West *et al.* might have produced some **55** in the rearrangements of naphthylcarbenes.[280,338]

The global $C_{11}H_8$ minimum, **48**, is commonly found to be the final product in naphthylcarbene rearrangements at high temperatures, and it can be prepared on a preparative scale starting from naphthylcarbene precursors.[203] Despite the highly strained four membered ring bridging the *peri*-position of naphthalene, **48** is the only non-carbenoid structure which retains the aromaticity in *both* rings and has no additional unsaturation.

Both the similar relative energies of 2-**46**, **47** as well as 5,6-benzocyclohepta-1,2,4,6-tetraene (**56**) (usually depicted as **53**) and the ΔE_{ST} for 2-**46** (all energies within 5 kcal mol-1) are consistent with the experimentally observed facile interconversion of these structures at ambient temperatures. The barrier for ring-opening of **47** to **56** is also relatively small (9.4 kcal mol-1), allowing all species to equilibrate. Most notably, **53**, although only 9.5 kcal mol-1 above 2-**46**, is not a minimum (NIMAG=2) and can thus not participate. It should rather be ascribed to the allenic structure **56**, which is remarkably stable (2.7 kcal mol-1 lower in energy than 2-naphthylcarbene).

By analogy to the rearrangement of 2-**46**, 1-**46** also should rearrange with **49** and 3,4-benzobicyclohepta-1,2,4,6-tetraene (**57**). However, only 1-**46** and **49** have been observed experimentally. Since the barriers for rearrangement of 1-**46** to **55** and to **48** are higher than the barrier for rearrangement of 1-**46** to **57**, the latter is very likely to be an observable species at low temperatures (note, however, that the activation energy for the **57** —> **49** reaction is only 6.6 kcal mol[-1]).

Since the activation barriers for 2-**46** —> **55** (E_a = 21.4 kcal mol[-1]), 1-**46** —> **55** (E_a = 23.3 kcal mol[-1]), and 1-**46** —> **48** (E_a = 23.8 kcal mol[-1]) are very similar, the experimental difficulties in interpreting and identifying the interconversion pathways for 2-**46** into 1-**46** and further to **48** are easily understood. At low temperatures, 2-**46** only equilibrates with **47** and **56** (analogously, 1-**46** interconverts with **49**, possibly also with **57**), while much higher temperatures only give **48**. Although **55** is a minimum, its observation may require synthesis and characterization directly from precursors which do not require elevated temperatures or involve the intermediate formation of naphthylcarbenes.

Most of the qualitative features of the naphthylcarbene PES compare very well to phenylcarbene (*vide supra*). The rotational barriers of the *exo*-methylene group (triplet: around 4 kcal mol[-1], singlet: around 20 kcal mol[-1] for 2-naphthylcarbene and for phenylcarbene, respectively) and the ΔE_{ST}'s (both around 5 kcal mol[-1]) are quite comparable. More strikingly, the allenic species (**54** and **55** on the $C_{11}H_8$ PES) are generally low-lying minima (note that **44** is the cyclic C_7H_6 minimum).

5 Summary and Outlook

Carbenes continue to excite both experimental and computational chemists. This account attempts to summarize the *status quo* in theoretical studies on prototype carbenes and some of their typical reactions.

Carbene structures and properties can now be computed with "chemical accuracy," despite the difficulties associated with multi-reference species such as singlet carbenes. This is particularly encouraging since the determination of singlet-triplet energy separations and accurate structures of carbenes, which very often are, at best, fleetingly observable intermediates, is extremely difficult experimentally. Transition structures also can be computed apparently with good accuracy and complement experimental studies of the reactions as well as the complex rearrangements commonly found in carbene chemistry.

The advent of hybrid Hartree–Fock/density functional theory allows larger systems to be computed. While computational data at highly correlated levels of theory [CISD and CCSD(T)] only are available for carbenes with three carbon atoms, DFT studies on hundred atom systems are now possible and linear scaling will further increase the size of molecules which can be investigated. The accuracy of common hybrid HF/DFT methods (e.g., employing the B3LYP functional) comes reasonably close to that of multi-reference (e.g., CAS, MR-CI) and highly correlated methods [e.g., CCSD(T)] at much lower computational cost. Further improvements of the correlation functionals and computational algorithms will increase the applicability and the use of hybrid HF/DFT methods not only for characterizing carbenes (and related species like biradicals), but also for treating large chemical systems in general.

Acknowledgments

We acknowledge support by the U.S. National Science Foundation, Grant CHE-9527468, the Stiftung Volkswagenwerk, the Deutsche Forschungsgemeinschaft, the Fonds der Chemischen Industrie (Liebig-Fellowship for PRS), the DAAD and the Freistaat Bayern (Fellowships for HFB).

References

1. Schaefer, H. F. *Science* **1986**, *231*, 1100.
2. Schaefer, H. F. *Chimia* **1989**, *43*, 1.
3. Goddard, W. D. *Science* **1985**, *227*, 917.
4. Shavitt, I. *Tetrahedron* **1985**, *41*, 1531.
5. Foster, J. M.; Boys, S. F. *Rev. Mod. Phys.* **1960**, *32*, 305.
6. Herzberg, G.; Shoosmith, J. *Nature* **1959**, *183*, 1801.
7. Herzberg, G. *Proc. Roy. Soc. (London)* **1961**, *A262*, 291.
8. Jordan, P. C. H.; Longuet-Higgins, H. C. *Mol. Phys.* **1962**, 121.

158

9.   Padgett, A.; Krauss, M. *J. Chem. Phys.* **1960**, *32*, 189.
10. Pople, J. A.; Segal, G. A. *J. Chem. Phys.* **1966**, *44*, 3289.
11. Dixon, R. N. *Mol. Phys.* **1964**, *8*, 201.
12. Harrison, J. F.; Allen, L. C. *J. Am. Chem. Soc.* **1969**, *91*, 807.
13. Bender, C. F.; Schaefer, H. F. *J. Am. Chem. Soc.* **1970**, *92*, 4984.
14. Gaspar, P. S.; Hammond, G. S. In *Carbenes*; Moss, R. A. and Jones Jr., M., Ed.; Wiley Interscience: New York, 1975; Vol. 2, pp 207.
15. Bernheim, R. A.; Bernard, H. W.; Wang, P. S.; Wood, L. S.; Skell, P. S. *J. Chem. Phys.* **1970**, *53*, 1280.
16. Wassermann, E.; Yager, W. A.; Kuck, V. *J. Chem. Phys. Lett.* **1970**, *7*, 409.
17. Herzberg, G.; Johns, J. W. C. *J. Chem. Phys.* **1971**, *54*, 2276.
18. Harrison, J. F. *Acc. Chem. Res.* **1974**, *7*, 378.
19. McLaughlin, D. R.; Bender, C. F.; Schaefer, H. F. *Theor. Chim. Acta* **1972**, *25*, 352.
20. Bunker, P. R.; Jensen, P. *J. Chem. Phys.* **1983**, *79*, 1224.
21. Hay, P. J.; Hunt, W. J.; Goddard, W. A. *Chem. Phys. Lett.* **1972**, *13*, 30.
22. Bender, C. F.; Schaefer, H. F.; Franceschetti, D. R.; Allen, L. C. *J. Am. Chem. Soc.* **1972**, *94*, 6888.
23. Rowland, F. S.; McKnight, C.; Lee, E. K. C. *Ber. Bunsenges. Phys. Chem.* **1968**, *72*.
24. Hase, W. L.; Phillips, R. J.; Simons, J. W. *Chem. Phys. Lett.* **1971**, *12*, 161.
25. Frey, H. M. *J. Chem. Soc. Chem. Comm.* **1972**, 1024.
26. Zittel, P. F.; Ellison, G. B.; O'Neil, S. V.; Herbst, E.; Lineberger, W. C.; Reinhardt, W. P. *J. Am. Chem. Soc.* **1976**, *98*, 3731.
27. Engelking, P. C.; Corderman, R. R.; Wendolowski, J. J.; Ellison, G. B.; O'Neil, S. V.; Lineberger, W. C. *J. Chem. Phys.* **1981**, *74*, 5460.
28. Lucchese, R. R.; Schaefer, H. F. *J. Am. Chem. Soc.* **1977**, *99*, 6765.
29. Harding, L. B.; Goddard, W. A. *J. Chem. Phys.* **1977**, *67*, 1777.
30. Roos, B. O.; Siegbahn, P. E. M. *J. Am. Chem. Soc.* **1977**, *99*, 7716.
31. Shih, S.-K.; Peyerimhoff, S. D.; Buenker, R. J.; Peric, M. *Chem. Phys. Lett.* **1978**, *55*, 206.
32. Frey, H. M.; Kennedy, G. J. *J. Chem. Soc. Chem. Comm.* **1975**, 233.
33. Frey, H. M.; Kennedy, G. J. *J. Chem. Soc. Faraday Trans. I* **1977**, *73*, 164.
34. Lahmani, F. *J. Phys. Chem.* **1976**, *80*, 2623.
35. McCulloh, K. E.; Dibeler, V. H. *J. Chem. Phys.* **1976**, *64*, 4445.
36. Simons, J. W.; Curry, R. *Chem. Phys. Lett.* **1976**, *38*, 171.

37.   Hase, W. L.; Kelly, P. M. **1977**, *66*, 5093.
38.   Lengel, R. K.; Zare, R. N. *J. Am. Chem. Soc.* **1978**, *100*, 4795.
39.   Feldman, D.; Meier, K.; Zacharias, H.; Welge, K. H. *Chem. Phys. Lett.* **1978**, *59*, 171.
40.   Monts, D. L.; Dietz, T. G.; Duncan, M. A.; Smalley, R. E. *Chem. Phys.* **1980**, *45*, 133.
41.   Hayden, C. C.; Neumark, D. M.; Shobatake, K.; Sparks, R. K.; Lee, Y. T. *J. Chem. Phys.* **1982**, *76*, 3607.
42.   McKellar, A. R. W.; Bunker, P. R.; Sears, T. J.; Evenson, K. M.; Saykally, R. J.; Langhoff, S. R. *J. Chem. Phys.* **1983**, *79*, 5251.
43.   Sherrill, C. D.; Van Huis, T. J.; Yamaguchi, Y.; Schaefer, H. F. *submitted to Theochem* **1996**.
44.   Sears, T. J.; Bunker, P. R. *J. Chem. Phys.* **1983**, *79*, 5265.
45.   Leopold, D. G.; Murray, K. K.; Lineberger, W. C. *J Chem. Phys.* **1984**, *81*, 1048.
46.   Jensen, P.; Bunker, P. R. *J. Chem. Phys.* **1988**, *89*, 1327.
47.   Davidson, E. R.; Feller, D.; Phillips, P. *Chem. Phys. Lett.* **1980**, *76*, 416.
48.   Handy, N. C.; Yamaguchi, Y.; Schaefer, H. F. *J. Chem. Phys.* **1986**, *84*, 4481.
49.   Kirmse, W. *Carbene Chemistry*; 2. ed.; Academic Press: New York, 1971.
50.   *Carbenes*; Jones, M.; Moss, R. A., Ed.; Wiley: New York, 1973; Vol. 1.
51.   *Carbenes*; Moss, R. A.; Jones, M., Ed.; Wiley: New York, 1975; Vol. 2.
52.   Moss, R. A.; Jones, M. In *Reactive Intermediates*; Jones, M. and Moss, R. A., Ed.; Wiley: New York, 1978; Vol. 1, pp 69.
53.   Moss, R. A.; Jones, M. In *Reactive Intermediates*; Jones, M. and Moss, R. A., Ed.: New York, 1981; Vol. 2, pp 59.
54.   Moss, R. A.; Jones, M. In *Reactive Intermediates*; Jones, M. and Moss, R. A., Ed.; Wiley: New York, 1985; Vol. 3, pp 45.
55.   *Kinetics and Spectroscopy of Carbenes and Diradicals*; Platz, M. S., Ed.; Plenum: New York, 1990, pp Ch. 6.
56.   Moss, R. A. *Acc. Chem. Res.* **1989**, *22*, 15.
57.   Nickon, A. *Acc. Chem. Res.* **1993**, *26*, 84.
58.   Sander, W.; Bucher, G.; Wierlacher, S. *Chem. Rev.* **1993**, *93*, 1583.
59.   Liu, M. T. H. *Acc. Chem. Res.* **1994**, *27*, 287.
60.   *Methoden der Organischen Chemie (Houben-Weyl)*; Regitz, M., Ed.; Thieme Verlag: Stuttgart, 1989; Vol. E19b.
61.   Schaefer, H. F. *Acc. Chem. Res.* **1979**, *12*, 288.
62.   Davidson, E. R. In *Diradicals*; Borden, W. T., Ed.; Wiley-Interscience: New York, 1982, pp 73.

63. Walsh, A. D. *J. Chem. Soc.* **1953**, 2260.

64. Cotton, F. A. *Chemical Applications of Group Theory*; 3 rd. ed.; Wiley: New York, 1990.

65. Szabo, A.; Ostlund, N. S. *Modern Quantum Chemistry*; Macmillan: New York, 1982.

66. Bauschlicher, C. W.; Taylor, P. R. *J. Chem. Phys.* **1986**, *85*, 6510.

67. Anglada, J. M.; Bofill, J. M. *Theor. Chim. Acta* **1995**, *92*, 369.

68. Werner, H.-J. *Adv. Chem. Phys.* **1987**, *69*, 1.

69. Shepard, R. *Adv. Chem. Phys.* **1987**, *69*, 63.

70. Roos, B. O. *Adv. Chem. Phys.* **1987**, *69*, 399.

71. Siegbahn, P. E. M. *Faraday Symp. Chem. Soc.* **1984**, *19*, 97.

72. Yamaguchi, Y.; Sherrill, C. D.; Schaefer, H. F. *J. Phys. Chem.* **1996**, *100*, 7911.

73. Schreiner, P. R.; Karney, W. L.; Schleyer, P. v. R.; Borden, W. T.; Hamilton, T. P.; Schaefer, H. F. *J. Org. Chem.* **1996**, *61*, 7030.

74. Parr, R. G.; Yang, W. *Density Functional Theory of Atoms and Molecules*; Oxford University Press: Oxford, U.K., 1989.

75. *Density Functional Methods in Chemistry*; Labanowski, J. K.; Andzelm, J. W., Ed.; Springer Verlag: Berlin, 1991.

76. *Modern Density Functional Theory: A Tool for Chemistry*; Seminario, J. M.; Politzer, P., Ed.; Elsevier: Amsterdam, 1995.

77. *Top. Curr. Chem.*; Dunitz, J. D.; Hafner, K.; Houk, K. N.; Ito, S.; Lehn, J.-M.; Raymond, K. N.; Rees, C. W.; Thiem, J.; Vögtle, F., Ed.; Springer: Berlin, 1996; Vol. 180-183.

78. Becke, A. D. *J. Chem. Phys.* **1993**, *98*, 5648.

79. Lee, C.; Yang, W.; Parr, R. G. *Phys. Rev. B* **1988**, *37*, 785.

80. Xie, Y.; Schreiner, P. R.; Schleyer, P. v. R.; Schaefer, H. F. *J. Am. Chem. Soc.* **1997**, *119*, 1370.

81. Bettinger, H. F.; Schreiner, P. R.; Schleyer, P. v. R.; Schaefer, H. F. *J. Phys. Chem.* **1996**, *100*, 16147.

82. Curtiss, L. A.; Raghavachari, K.; Trucks, G. W.; Pople, J. A. *J. Chem. Phys.* **1991**, *94*, 7221.

83. Curtiss, L. A.; Raghavachari, K.; Redfern, P. C.; Pople, J. A. *J. Chem. Phys.* **1997**, *106*, 1063.

84. Foresman, J. B.; Frisch, Æ. *Exploring Chemistry with Electronic Structure Methods*; 2nd. ed.; Gaussian, Inc.: Pittsburgh, 1996.

85. Bauschlicher, C. W.; Langhoff, S. R.; Taylor, P. R. *J. Chem. Phys.* **1987**, *87*, 387.

86. Cole, S. J.; Purvis, G. D.; Bartlett, R. J. *Chem. Phys. Lett.* **1985**, *113*, 271.

87. Balkova, A.; Bartlett, R. J. *J. Chem. Phys.* **1995**, *102*, 7116.

88. Woon, D. E.; Dunning, T. H. *J. Chem. Phys.* **1995**, *103*, 4572.

89. Lee, Y. S.; Bartlett, R. J. *J. Chem. Phys.* **1984**, *80*, 4371.

90. Lee, Y. S.; Kucharski, S. A.; Bartlett, R. J. *J. Chem. Phys.* **1984**, *81*, 5906.

91. Lee, Y. S.; Kucharski, S. A.; Bartlett, R. J. *J. Chem. Phys.* **1985**, *82*, 5761(E).

92. Herzberg, G.; Johns, J. W. C. *Proc. Roy. Soc. London* **1966**, *A295*, 107.

93. Petek, H.; Nesbitt, D. J.; Darwin, D. C.; Ogilby, P. R.; Moore, C. B.; Ramsay, D. A. *J. Chem. Phys.* **1989**, *96*, 2118.

94. Hoffman, R.; Zeiss, G. D.; Van Dine, G. W. *J. Am. Chem. Soc.* **1968**, *90*, 1485.

95. Luke, B. T.; Pople, J. A.; Krogh-Jespersen, M.-B.; Apeloig, Y.; Karni, M.; Chandrasekhar, J.; Schleyer, P. v. R. *J. Am. Chem. Soc.* **1986**, *108*, 270.

96. Hehre, W. J.; Radom, L.; Schleyer, P. v. R.; Pople, J. A. *Ab Initio Molecular Orbital Theory*; Wiley Interscience: New York, 1986.

97. Raghavachari, K.; Frisch, M. J.; Pople, J. A.; Schleyer, P. v. R. *Chem. Phys. Lett.* **1982**, *85*, 145.

98. Ma, B.; Schaefer, H. F. *J. Am. Chem. Soc.* **1994**, *116*, 3539.

99. Evanseck, J. D.; Houk, K. N. *J. Phys. Chem.* **1990**, *94*, 5518.

100. Gallo, M. M.; Schaefer, H. F. *J. Phys. Chem.* **1992**, *96*, 1515.

101. Köhler, H. J.; Lischka, H. *J. Am. Chem. Soc.* **1982**, *104*, 5884.

102. Pople, J. A.; Raghavachari, K.; Frisch, M. J.; Binkley, J. S.; Schleyer, P. v. R. *J. Am. Chem. Soc.* **1983**, *105*, 6389.

103. Khodabandeh, S.; Carter, E. A. *J. Phys. Chem.* **1993**, *97*, 4360.

104. Sulzbach, H. M.; Bolton, E.; Lenoir, D.; Schleyer, P. v. R.; Schaefer, H. F. *J. Am. Chem. Soc.* **1996**, *118*, 9908.

105. Schleyer, P. v. R.; Maerker, C.; Buzek, P.; Sieber, S. In *Stable Carbocation Chemistry*; Prakash, G. K. S. and Schleyer, P. v. R., Ed.; Wiley: New York, 1997, pp 19.

106. Lambert, C.; Schleyer, P. v. R. *Angew. Chem. Int. Ed. Engl.* **1994**, *33*, 1129.

107. Reed, A. E.; Weinstock, R. B.; Weinhold, F. *J. Chem. Phys.* **1985**, *83*, 735.

108. Reed, A. E.; Curtiss, L. A.; Weinhold, F. *Chem. Rev.* **1988**, *88*, 899.

109. Richards, C. A.; Kim, S.-J.; Yamaguchi, Y.; Schaefer, H. F. *J. Am. Chem. Soc.* **1995**, *117*, 10104.

110. Armstrong, B. M.; McKee, M. L.; Shevlin, P. B. *J. Am. Chem. Soc.* **1993**, *117*, 3685.

111. Gano, J. E.; Wettach, R. H.; Platz, M. S.; Senthilnathan, V. P. *J. Am. Chem. Soc.* **1982**, *104*, 2326.

112. Hartzler, H. D. In *Carbenes*; Moss, R. A. and Jones, M., Ed.; Wiley: New York, 1975; Vol. 2.

113. Vacek, G.; Thomas, J. R.; DeLeeuw, B. J.; Yamaguchi, Y.; Schaefer, H. F. *J. Chem. Phys.* **1993**, *98*, 4766.

114. Ervin, K. M.; Ho, J.; Lineberger, W. C. *J. Chem. Phys.* **1989**, *91*, 5974.

115. Stanton, J. F.; Gauss, J. *J. Chem. Phys.* **1994**, *101*, 3001.

116. Yoshimine, M.; Pacansky, J.; Honjou, N. *J. Am. Chem. Soc.* **1989**, *111*, 2785.

117. Hutton, R. S.; Manion, M. L.; Roth, H. D.; Wassermann, E. *J. Am. Chem. Soc.* **1974**, *96*, 4680.

118. Palmer, G. E.; Bolton, J. R.; Arnold, D. R. *J. Am. Chem. Soc.* **1974**, *96*, 3708.

119. Bofill, J. M.; Bru, N.; Farràs, J.; Olivella, S.; Solé, A.; Vilarrasa, J. *J. Am. Chem. Soc.* **1988**, *110*, 3740.

120. Collins, C. L.; Davy, R. D.; Schaefer, H. F. *Chem. Phys. Lett.* **1990**, *171*, 259.

121. Maier, G.; Reisenauer, H. P.; Schwab, W.; Cársky, P.; Hess, B. A.; Schaad, L. J. *J. Am. Chem. Soc.* **1987**, *109*, 5183.

122. Gottlieb, C. A.; Killian, T. C.; Thaddeus, P.; Botschwina, P.; Flügge, J.; Oswald, M. *J. Chem. Phys.* **1993**, *98*, 4478.

123. Hehre, W. J.; Pople, J. A.; Lathan, W. A.; Radom, L.; Wasserman, E.; Wasserman, Z. R. *J. Am. Chem. Soc.* **1976**, *98*, 4378.

124. Kenney, J. W.; Simons, J.; Purvis, G. D.; Bartlett, R. J. *J. Am. Chem. Soc.* **1978**, *100*, 6930.

125. DeFrees, D. J.; McLean, A. D. *Astrophys. J.* **1986**, *308*, L31.

126. Jonas, V.; Böhme, M.; Frenking, G. *J. Phys. Chem.* **1992**, *96*, 1640.

127. Stanton, J. F.; DePinto, J. T.; Seburg, R. A.; Hodges, J. A.; McMahon, R. J. *J. Am. Chem. Soc.* **1997**, *119*, 429.

128. Bernheim, R. A.; Kempf, R. J.; Gramas, J. V.; Skell, P. S. *J. Chem. Phys.* **1965**, *43*, 196.

129. Maier, G.; Reisenauer, H. P.; Schwab, W.; Cársky, P.; Spirko, V.; Hess, B. A.; Schaad, L. J. *J. Chem. Phys.* **1989**, *91*, 4763.

130. Skell, P. S.; Woodworth, R. C. *J. Am. Chem. Soc.* **1956**, *78*, 4496.

131. Skell, P. S.; Garner, A. Y. *J. Am. Chem. Soc.* **1956**, *78*, 5430.

132. Hoffmann, R. *J. Am. Chem. Soc.* **1968**, *90*, 1475.

133. Zurawski, B.; Kutzelnigg, W. *J. Am. Chem. Soc.* **1978**, *100*, 2654.

134. Rondan, N. G.; Houk, K. N.; Moss, R. A. *J. Am. Chem. Soc.* **1980**, *102*, 1770.

135. Crabtree, R. H. *Chem. Rev.* **1995**, *95*, 987.

136. Hill, C. L. *Activation and Functionalization of Alkanes.*; John Wiley & Sons Inc.: New York, 1989.

137. Doering, W. v. E.; Knox, L. H. *J. Am. Chem. Soc.* **1956**, *78*, 4947.

138. Doering, W. v. E.; Buttery, R. G.; Laughlin, R. G.; Chaudhuri, N. *J. Am. Chem. Soc.* , *78*, 3224.

139. Richardson, D. B.; Simmons, M. C.; Dvoretzky, I. *J. Am. Chem. Soc.* , *83*, 1934.

140. Halberstadt, M. L.; McNesby, J. R. *J. Am. Chem. Soc.* **1967?**, 3417.

141. Baskin, C. P.; Bender, C. F.; Bauschlicher, C. W.; Schaefer, H. F. *J. Am. Chem. Soc.* **1974**, *96*, 2709.

142. Bauschlicher, C. W.; Haber, K.; Schaefer, H. F.; Bender, C. F. *J. Am. Chem. Soc.* **1977**, *99*, 3610.

143. Woodward, R. B.; Hoffmann, R. *Angew. Chem. Int. Ed. Engl.* **1969**, *8*, 781.

144. Bauschlicher, C. W.; Schaefer, H. F.; Bender, C. F. *J. Am. Chem. Soc.* **1976**, *98*, 1653.

145. Kollmar *Tetrahedron* **1972**, *28*, 5893.

146. Kollmar, H.; Staemmler, V. *Theor. Chim. Acta* **1979**, *51*, 207.

147. Jeziorek, D.; Zurawski, B. *Int. J. Quantum Chem.* **1979**, *16*, 277.

148. Bach, R. D.; Su, M.-D.; Aldabbagh, E.; Andrés, J. L.; Schlegel, H. B. *J. Am. Chem. Soc.* **1993**, *115*, 10237.

149. Jorgensen, W. L.; Salem, L. *The Organic Chemist's Book of Orbitals*; Academic Press: New York, 1973.

150. Gordon, M. S.; Gano, D. R. *J. Am. Chem. Soc.* **1984**, *106*, 5421.

151. Gordon, M. S.; Boatz, J. A.; Gano, D. R.; Friederichs, M. G. *J. Am. Chem. Soc.* **1987**, *109*, 1323.

152. Gano, D. R.; Gordon, M. S.; Boatz, J. A. *J. Am. Chem. Soc.* **1991**, *113*, 6711.

153. Gordon, M. S.; Gano, D. R.; Binkley, J. S.; Frisch, M. J. *J. Am. Chem. Soc.* **1986**, *108*, 2191.

154. Seburg, R. A.; McMahon, R. J. *J. Am. Chem. Soc.* **1992**, *114*, 7183.

155. Modarelli, D. A.; Platz, M. S. *J. Am. Chem. Soc.* **1993**, *115*, 470.

156. Modarelli, D. A.; Platz, M. S.; Sheridan, R. S.; Ammann, J. R. *J. Am. Chem. Soc.* **1993**, *115*, 10440.

157. Morgan, S. C.; Jackson, J. E.; Platz, M. S. *J. Am. Chem. Soc.* **1991**, *113*, 2783.

158. Chen, N.; Jones, M.; White, W. R.; Platz, M. S. *J. Am. Chem. Soc.* **1991**, *113*, 4981.

164

159. Modarelli, D. A.; Morgan, S.; Platz, M. S. *J. Am. Chem. Soc.* **1992**, *114*, 7034.
160. Modarelli, D. A.; Platz, M. S. *J. Am. Chem. Soc.* **1991**, *113*, 8985.
161. Miller, D. M.; Schreiner, P. R.; Schaefer, H. F. *J. Am. Chem. Soc.* **1995**, *117*, 4137.
162. Nobes, R. H.; L., R.; Rodwell, W. R. *Chem. Phys. Lett.* **1980**, *74*, 269.
163. Nickon, A.; Huang, F.; Weglein, R.; Matsuo, K.; Yagi, H. *J. Am. Chem. Soc.* **1974**, *96*, 5264.
164. Kyba, E. P.; John, A. M. *J. Am. Chem. Soc.* **1977**, *99*, 8329.
165. Press, L. S.; Shechter, H. *J. Am. Chem. Soc.* **1979**, *101*, 509.
166. Seghers, L.; Shechter, H. *Tetrahedron Lett.* **1976**, *23*, 1943.
167. Evanseck, J. D.; Houk, K. N. *J. Am. Chem. Soc.* **1990**, *112*, 9148.
168. Schleyer, P. v. R. *J. Am. Chem. Soc.* **1967**, *89*, 701.
169. Becke, A. D. *J. Chem. Phys.* **1992**, *98*, 1372.
170. Sulzbach, H. M.; Platz, M. S.; Schaefer, H. F.; Hadad, C. M. *submitted* **1997**.
171. La Villa, J.; Goodman, J. L. *J. Am. Chem. Soc.* **1989**, *111*, 6877.
172. Liu; Lynch; Truong; Lu; Truhlar; Garrett *J. Am. Chem. Soc* **1993**, *115*, 2408.
173. Storer, J. W.; Houk, K. N. *J. Am. Chem. Soc.* **1993**, *115*, 10426.
174. Dix, E. J.; Herman, M. S.; Goodman, J. L. *J. Am. Chem. Soc.* **1993**, *115*, 10424.
175. Dix, E. J.; Goodman, J. L. *Res. Chem. Intermed.* **1994**, *20*, 149.
176. Burnett, S. M.; Stevens, A. E.; Feigerle, C. S.; Lineberger, W. C. *Chem. Phys. Lett.* **1983**, *100*, 124.
177. Ervin, K. M.; Gronert, S.; Barlow, S. E.; Gilles, M. K.; Harrison, A. G.; Bierbaum, V. M.; DePuy, C. H.; Lineberger, W. C.; Ellison, G. B. *J. Am. Chem. Soc.* **1990**, *112*, 5750.
178. Chen, Y.; Jonas, D. M.; Kinsey, J. L.; Field, R. W. *J. Chem. Phys.* **1989**, *91*, 3976.
179. Kiefer, J. H.; Sidhu, S. S.; Kumaran, S. S.; Irdam, E. A. *Chem. Phys. Lett.* **1989**, *159*, 32.
180. Dykstra, C. E.; Schaefer, H. F. *J. Am. Chem. Soc.* **1978**, *100*, 1378.
181. Pople, J. A.; Krishnan, R.; Schlegel, H. B.; Binkley, J. S. *Int. J. Quantum Chem.* **1978**, *14*, 545.
182. Harding, L. B. *J. Am. Chem. Soc.* **1981**, *103*, 7469.
183. Frenking, G. *Chem. Phys. Lett.* **1983**, *100*, 484.
184. Sakai, S.; Morokuma, K. *J. Phys. Chem.* **1987**, *91*, 3661.
185. Osamura, Y.; Schaefer, H. F.; Gray, S. K.; Miller, W. H. *J. Am. Chem. Soc.* **1981**, *103*, 1904.

186. Carrington, T.; Schaefer, H. F.; Gray, S. K.; Miller, W. H. *J. Chem. Phys.* **1984**, *80*, 4347.

187. Krishnan, R.; Frisch, M. J.; Pople, J. A.; Schleyer, P. v. R. *Chem. Phys. Lett.* **1981**, *79*, 408.

188. Gallo, M. M.; Hamilton, T. P.; Schaefer, H. F. *J. Am. Chem. Soc.* **1990**, *112*, 8714.

189. Davidson, E. R. In ; Daudel, R. and Pullman, B., Ed.; Reidel: Dodrecht, 1974, pp 17.

190. Langhoff, S. R.; Davidson, E. R. *Int. J. Quantum Chem.* **1974**, *8*, 61.

191. Petersson, G. A.; Tensfeldt, T. G.; Montgomery, J. A. *J. Am. Chem. Soc.* **1992**, *114*, 6133.

192. Cheng, N.; Shen, M.; Yu, C. *J. Phys. Chem.* **1997**, *106*, 3237.

193. Hammond, G. S. *J. Am. Chem. Soc.* **1955**, *77*, 334.

194. Chang, K.-T.; Shechter, H. *J. Am. Chem. Soc.* **1979**, *101*, 5082.

195. Olah, G. A.; Kelly, D. P.; Jeuell, C. L.; Porter, R. D. *J. Am. Chem. Soc.* **1970**, *92*, 2544.

196. Olah, G. A.; Kelly, D. P.; Jeuell, C. L.; Porter, R. D. *J. Am. Chem. Soc.* **1972**, *94*, 146.

197. Staral, J. S.; Roberts, J. D. *J. Am. Chem. Soc.* **1978**, *100*, 8018.

198. Dewar, M. J. S.; Reynolds, C. H. *J. Am. Chem. Soc.* **1984**, *106*, 6388.

199. Saunders, M.; Siehl, H.-U. *J. Am. Chem. Soc.* **1980**, *102*, 6868.

200. Brittain, W. J.; Squillacote, M. E.; Roberts, J. D. *J. Am. Chem. Soc.* **1984**, *106*, 7280.

201. Hehre, W. J.; Hiberty, P. C. *J. Am. Chem. Soc.* **1974**, *96*, 302.

202. Barcus, R. L.; Hadel, L. M.; Johnston, L. J.; Platz, M. S.; Savino, T. G.; Scaiano, J. C. *J. Am. Chem. Soc.* **1986**, *108*, 3928.

203. Engler, T. A.; Shechter, H. *Tetrahedron Lett.* **1982**, *23*, 2715.

204. Bettinger, H. F.; Schreiner, P. R.; Schleyer, P. v. R.; Schaefer, H. F. , manuscript in preparation.

205. Horn, K. A.; Chateauneuf, J. E. *Tetrahedron* **1985**, *41*, 1465.

206. Trozzolo, A. M.; Wasserman, E.; Yager, W. A. *J. Am. Chem. Soc.* **1965**, *87*, 129.

207. Doering, W. v. E.; LaFlamme, P. M. *Tetrahedron* **1958**, *2*, 75.

208. Moore, W. R.; Ward, H. R. *J. Org. Chem.* **1960**, *25*, 2073.

209. Moore, W. R.; Ward, H. R.; Merritt, R. F. *J. Org. Chem.* **1961**, *83*, 2019.

210. Skattebøl, L. *Acta Chem. Scand.* **1963**, *17*, 1683.

211. Schleyer, P. v. R.; Dine, G. W. V.; Schoellkopf, U.; Paust, J. *J. Am. Chem. Soc.* **1966**, *88*, 2868.

212. Schleyer, P. v. R.; Su, T. M.; Saunders, M.; Rosenfeld, J. C. *J. Am. Chem. Soc.* **1969**, *91*, 5174.

213. Schleyer, P. v. R.; Sliwinski, W. F.; Van Dine, G. W.; Schoellkopf, U.; Paust, J.; Fellenberger, K. *J. Am. Chem. Soc.* **1972**, *94*, 125.

214. Radom, L.; Hariharan, P. C.; Pople, J. A.; Schleyer, P. v. R. *J. Am. Chem. Soc.* **1973**, *95*, 6531.

215. Valtazanos, P.; Ruedenberg, K. *Theor. Chim. Acta* **1986**, *69*, 281.

216. Valtazanos, P.; Elbert, S. T.; Xantheas, S.; Ruedenberg, K. *Theor. Chim. Acta* **1991**, *78*, 287.

217. Xantheas, S.; Valtazanos, P.; Ruedenberg, K. *Theor. Chim. Acta* **1991**, *78*, 327.

218. Xantheas, S.; Elbert, P.; Ruedenberg, K. *Theor. Chim. Acta* **1991**, *78*, 365.

219. Valtazanos, P.; Ruedenberg, K. *Theor. Chim. Acta* **1991**, *78*, 397.

220. Yoshimine, M.; Pacansky, J.; Honjou, N. *J. Am. Chem. Soc.* **1989**, *111*, 4198.

221. Honjou, N.; Pacansky, J.; Yoshimine, M. *J. Am. Chem. Soc.* **1985**, *107*, 5332.

222. Chapman, O. L. *Pure Appl. Chem.* **1974**, *40*, 511.

223. Lee, T. J.; Rice, J. E.; Scuseria, G. E.; Schaefer, H. F. *Theoret. Chim. Acta* **1989**, *75*, 81.

224. Lee, T. J.; Taylor, P. R. *Int. J. Quantum Chem. Symp.* **1989**, *23*, 199.

225. Lee, T. J.; Rendell, A. P.; Taylor, P. R. *J. Phys. Chem.* **1990**, *94*, 5463.

226. Fukui, K. *J. Phys. Chem.* **1970**, 4161.

227. Schaefer, H. F. *Chem. in Brit.* **1975**, *11*, 227.

228. Fukui, K. *Acc. Chem. Res.* **1981**, *14*, 363.

229. Bettinger, H. F.; Schleyer, P. v. R.; Schreiner, P. R.; Schaefer, H. F. , submitted for publication.

230. Johnson, R. P.; Daoust, K. J. *J. Am. Chem. Soc.* **1995**, *117*, 362.

231. Baumgart, H.-D.; Szeimies, G. *Tetrahedron Lett.* **1984**, *25*, 737.

232. Fitzgerald, G.; Saxe, P.; Schaefer, H. F. *J. Am. Chem. Soc.* **1983**, *105*, 690.

233. Carlson, H. A.; Quelch, G. E.; Schaefer, H. F. *J. Am. Chem. Soc.* **1992**, *114*, 5344.

234. Fitzgerald, G.; Schaefer, H. F. *Isr. J. Chem.* **1983**, *23*, 93.

235. Krebs, A.; Wilke, J. *Top. Curr. Chem.* **1983**, *109*, 189.

236. Jones, W. M.; Brinker, U. H. In *Pericyclic Reactions*; Marchand, A. P. and Lehr, A. E., Ed.; Academic: New York, 1977; Vol. 1, pp Ch. 3.

237. Jones, W. M. In *Rearrangements in Ground and Excited States*; De Mayo, P., Ed.; Academic: New York, 1980; Vol. 1, pp Ch. 3.

167

238. Jones, W. M. *Acc. Chem. Res.* **1977**, *10*, 353.
239. Wentrup, C. In *Reactive Intermediates*; Ambramovich, R., Ed.; Plenum: New York, 1980; Vol. 1, pp Ch. 4.
240. Wentrup, C. *Reactive Molecules*; Wiley-Interscience: New York, 1984.
241. Wentrup, C. In *Methoden der Organischen Chemie (Houben-Weyl)*; Regitz, M., Ed.; Thieme: Stuttgart, 1989; Vol. E19b, pp 824.
242. Vander Stouw, G. G.; Shechter, H. *Diss. Abstr.* **1965**, *25*, 6974.
243. Vander Stouw, G. G.; Kraska, A. R.; Shechter, H. *J. Am. Chem. Soc.* **1972**, *94*, 1655.
244. Baron, W. J.; Jones, M., Jr.; Gaspar, P. P. *J. Am. Chem. Soc.* **1970**, *92*, 4739.
245. Hedaya, E.; Kent, M. E. *J. Am. Chem. Soc.* **1971**, *93*, 3283.
246. Gaspar, P. P.; Hsu, J.-P.; Chari, S.; Jones, M. *Tetrahedron* **1985**, *41*, 1479.
247. Chapman, O. L.; Johnson, J. W.; McMahon, R. J.; West, P. R. *J. Am. Chem. Soc.* **1988**, *110*, 501.
248. Mayor, C.; Wentrup, C. *J. Am. Chem. Soc.* **1975**, *97*, 7467.
249. Gleiter, R.; Rettig, W.; Wentrup, C. *Helv. chim. Acta* **1974**, *57*.
250. Patterson, E. V.; McMahon, R. J. *J. Org. Chem.* **1997**, submitted for publication.
251. Wong, M.-W.; Wentrup, C., 61, 7022. *J. Org. Chem.* **1996**, *61*, 7022.
252. McMahon, R. J.; Abelt, C. J.; Chapman, O. L.; Johnson, J. W.; Kreil, C. L.; LeRoux, J.-P.; Mooring, A. M.; West, P. R. *J. Am. Chem. Soc.* **1987**, *109*.
253. Platz, M. S. In *Kinetics and Spectroscopy of Carbenes and Diradicals*; Platz, M. S., Ed.; Plenum: New York, 1990, pp Ch. 6.
254. Platz, M. S. *Acc. Chem. Res.* **1995**, *28*.
255. Savino, T. G.; Kanakarajan, K.; Platz, M. S. *J. Org. Chem.* **1986**, *51*, 1305.
256. Schuster, G. B. *Adv. Phys. Org. Chem.* **1986**, *22*, 311.
257. Trozzolo, A. M.; Wasserman, E. In *Carbenes*; Moss, R. A. and Jones, M., Jr, Ed.; Wiley: New York, 1975; Vol. 2, pp Ch. 5.
258. West, P. R.; Chapman, O. L.; LeRoux, J.-P. *J. Am. Chem. Soc.* **1982**, *104*, 1779.
259. Haider, K. W.; Platz, M. S.; Despres, A.; Migirdicyan, E. *Chem. Phys. Lett.* **1989**, *164*.
260. Higuchi, J. *J. Chem. Phys.* **1963**, *39*, 1339.
261. Joines, R. C.; Turner, A. B.; Jones, W. M. *J. Am. Chem. Soc.* **1969**, *91*, 7754.
262. Jones, W. M.; Ennis, C. L. *J. Am. Chem. Soc.* **1967**, *89*, 3069.

263. Jones, W. M.; Ennis, C. L. *J. Am. Chem. Soc.* **1969**, *91*, 6391.

264. Schissel, P.; Kent, M. E.; McAdoo, D. J.; Hedaya, E. *J. Am. Chem. Soc.* **1970**, *92*, 2147.

265. Harris, J. W.; Jones, W. M. *J. Am. Chem. Soc.* **1982**, *104*, 7329.

266. Saito, K.; Omura, Y.; Mukai, T. *Bull. Chem. Soc. Jpn.* **1985**, *58*, 1663.

267. Saito, K.; Ishihara, H. *Bull. Chem. Soc. Jpn.* **1985**, *58*, 2664.

268. Saito, K.; Ishihara, H. *Bull. Chem. Soc. Jpn.* **1986**, *59*, 1095.

269. Saito, K. *Bull. Chem. Soc. Jpn.* **1987**, *60*, 2105.

270. Saito, K.; Ishihara, H. *Bull. Chem. Soc. Jpn.* **1987**, *60*, 4447.

271. Saito, K.; Watanabe, T.; Takahashi, K. *Chem. Lett.* **1989**, 2099.

272. Saito, K.; Suzuki, S.; Watanabe, T.; Takahashi, K. *Bull. Chem. Soc. Jpn.* **1993**, *66*, 2304.

273. Chapman, O. L.; McMahon, R. J.; West, P. R. *J. Am. Chem. Soc.* **1984**, *106*, 7973.

274. Coburn, T. T.; Jones, W. M. *J. Am. Chem. Soc.* **1974**, *96*, 5218.

275. Billups, W. E.; Lin, L. P.; Chow, W. Y. *J. Am. Chem. Soc.* **1974**, *96*, 4026.

276. Billups, W. E.; Reed, L. E. *Tetrahedron Lett.* **1977**, 2239.

277. Billups, W. E.; E., R. L.; Casserly, E. W.; Lin, L. P. *J. Org. Chem.* **1981**, *46*, 1326.

278. Mykytka, J. P.; Jones, W. M. *J. Am. Chem. Soc.* **1975**, *97*, 5933.

279. Albrecht, S. W.; McMahon, R. J. *J. Am. Chem. Soc.* **1993**, *115*, 855.

280. West, P. R.; Mooring, A. M.; McMahon, R. J.; Chapman, O. L. *J. Org. Chem.* **1986**, *51*, 1316.

281. Tyner, R. L.; Jones, W. M.; Öhrn, Y.; Sabin, J. R. *J. Am. Chem. Soc* **1974**, *96*, 3765.

282. Balci, M.; Winchester, W. R.; Jones, W. M. *J. Org. Chem.* **1982**, 47.

283. Janssen, C. L.; Schaefer, H. F. *J. Am. Chem. Soc.* **1987**, *109*, 5030.

284. Radom, L.; Schaefer, H. F.; Vincent, M. A. *Nouv. J. Chem.* **1980**, *4*, 411.

285. Kuzaj, M.; Lüerssen, H.; Wentrup, C. *Angew. Chem. Int. Ed. Engl.* **1986**, *25*, 480.

286. McMahon, R. J.; Chapman, O. L. *J. Am. Chem. Soc.* **1986**, *108*, 1713.

287. Wentrup, C.; Mayor, C.; Becker, J.; Lindner, H. J. *Tetrahedron* **1985**, *41*, 1601.

288. Waali, E. E. *J. Am. Chem. Soc.* **1981**, *103*, 3604.

289. Kassaee, M. Z.; Nimlos, M. R.; Downie, K. E.; Waali, E. E. *Tetrahedron* **1985**, *41*, 1579.

290. Dewar, M. J. S.; Landman, D. *J. Am. Chem. Soc.* **1977**, *99*, 6179.

291. Waali, E. E.; Jones, W. M. *J. Am. Chem. Soc.* **1973**, *95*, 8114.

292. Matzinger, S.; Bally, T.; Patterson, E.; McMahon, R. J. *J. Am. Chem. Soc.* **1996**, *118*, 1535.

293. Bally, T.; Matzinger, S.; Truttmann, L.; Platz, M. S.; Morgan, S. *Angew. Chem. Int. Ed. Engl.* **1994**, *33*, 1964.

294. Roos, B. O. *Int. J. Quantum Chem. Symp.* **1980**, *14*, 175.

295. Andersson, K.; Malmqvist, P.-Å.; Roos, B. O. *J. Phys. Chem.* **1992**, *96*, 1218.

296. Andersson, K.; Malmqvist, P.-Å.; Roos, B. O.; Sadlej, A. J.; Wolinski, K. *J. Phys. Chem.* **1990**, *94*, 5483.

297. Balková, A.; Bartlett, R. J. *J. Chem. Phys.* **1994**, *101*, 8972.

298. Balková, A.; Bartlett, R. J. *Chem. Phys. Lett.* **1992**, *193*, 364.

299. Cramer, C. J.; Dulles, F. J.; Falvey, D. E. *J. Am. Chem. Soc.* **1994**, *116*, 9787.

300. Curtiss, L. A.; Redfern, P.; Smith, B. J.; Radom, L. *J. Chem. Phys.* **1996**, *104*, 5148.

301. Curtiss, J. A.; Jones, C.; Trucks, G. W.; Raghavachari, K.; Pople, J. A. *J. Chem. Phys.* **1990**, *93*, 2537.

302. Pople, J. A.; Head-Gordon, M.; Fox, D. J.; Raghavachari, K.; Curtiss, L. A. *J. Chem. Phys.* **1989**, *90*, 5622.

303. Smith, B. J.; Radom, L. *J. Phys. Chem.* **1995**, *99*, 6468.

304. McMahon, R. J. , personal communication.

305. Chateauneuf, J. E.; Horn, K. A.; Savino, T. G. *J. Am. Chem. Soc.* **1988**, *110*, 539.

306. Wentrup, C.; Wilczek, K. *Helv. Chim. Acta* **1970**, *53*, 1459.

307. Waali, E. E.; Lewis, J. M.; Lee, D. E.; Allen, E. W.; Chappell, A. K. *J. Org. Chem.* **1977**, *42*, 3460.

308. Mayor, C.; Jones, W. M. *Tetrahedron Lett.* **1977**, 3855.

309. Hackenberger, A.; Dürr, H. *Tetrahedron Lett.* **1979**, 4541.

310. Kirmse, W.; Loosen, K.; Sluma, H.-D. *J. Am. Chem. Soc.* **1981**, *103*, 5935.

311. Liehr, A. *J. Phys. Chem.* **1963**, *67*, 389.

312. Borden, W. T.; Davidson, E. R. *Acc. Chem. Res.* **1996**, *29*, 67.

313. Bunker, P. R.; Jensen, P.; Kraemer, W. P.; Beardsworth, R. *J. Chem. Phys.* **1986**, *85*, 3724.

314. Bunker, P. R.; Sears, T. J. *J. Chem. Phys.* **1986**, *85*, 4866.

315. Leopold, D. G.; Murray, K. K.; Miller, A. E. S.; Lineberger, W. C. *J. Chem. Phys.* **1985**, *83*, 4849.

316. McMahon, R. J. *Personal Communication.* **1996**.

317. Brudzynski, R. J.; Hudson, B. S. *J. Am. Chem. Soc.* **1990**, *112*, 4963.

318. Roth, W. R.; Ruf, G.; Ford, P. W. *Chem. Ber.* **1974**, *107*, 38.

319.   Pedash, Y. F.; Ivanov, V. V.; Luzanov, A. V. *Theor. Exp. Chem.* **1992**, *28*, 114.

320.   Rauk, A.; Bouma, W. J.; Radom, L. *J. Am. Chem. Soc.* **1985**, *107*, 3780.

321.   Staemmler, V. *Theor. Chim. Acta* **1977**, *45*, 89.

322.   Staemmler, V.; Jaquet, R. In *Energy Storage and Redistribution in Molecules*; Hinze, J., Ed.; Plenum: New York, 1983, pp 261.

323.   Seeger, R.; Krishnan, R.; Pople, J. A.; Schleyer, P. v. R. *J. Am. Chem. Soc.* **1977**, *99*, 7103.

324.   Dykstra, C. E. *J. Am. Chem. Soc.* **1977**, *99*, 2060.

325.   *Diradicals*; Borden, W. T., Ed.; Wiley–Interscience: New York, 1982.

326.   Borden, W. T. *Mol. Cryst. Liq. Cryst.* **1993**, *232*, 195.

327.   Borden, W. T.; Davidson, E. R. *J. Am. Chem. Soc.* **1977**, *99*, 4587.

328.   Borden, W. T.; Iwamura, H.; Berson, J. A. *Acc. Chem. Res.* **1994**, *27*, 109.

329.   Hund, F. *Z. Phys.* **1925**, *33*, 345.

330.   Hund, F. In *Linienspektren und periodisches System der Elemente*; Springer-Verlag: Berlin, 1927, pp 124.

331.   Becker, J.; Wentrup, C. *J. Chem. Soc. Chem. Comm.* **1980**, 190.

332.   Billups, W. E.; Haley, M. M.; Lee, G. A. *Chem. Rev.* **1989**, *89*, 1147.

333.   Senthilnathan, V. P.; Platz, M. S. *J. Am. Chem. Soc.* **1981**, *103*, 5503.

334.   Wentrup, C. *Top. Curr. Chem.* **1976**, *62*, 173.

335.   Chapman, O. L.; Sheridan, R. S.; LeRoux, J.-P. *Recl. Trav. Chim. Pays-Bas* **1979**, *98*, 334.

336.   Dorigo, A. E.; Li, Y.; Houk, K. N. *J. Am. Chem. Soc.* **1989**, *111*, 6942.

337.   Kirmse, W.; Sluma, H.-D. *J. Org. Chem.* **1988**, *53*, 763.

338.   Largan, J. G.; Sitzmann, E. V.; Eisenthal, K. B. *Chem. Phys. Lett.* **1984**, *110*, 521.

VIOLATIONS OF HUND'S RULE IN ORGANIC DIRADICALS -- WHERE TO LOOK FOR VIOLATIONS AND HOW TO IDENTIFY THEM

DAVID A. HROVAT AND WESTON THATCHER BORDEN*

Department of Chemistry, Box 351700
University of Washington
Seattle, WA 98195-1700, USA

Violations of Hund's rule are predicted to be found at the geometries of highest symmetry of [4n]annulenes, such as D_{4h} cyclobutadiene (CBD) and D_{8h} cyclooctatetraene (COT). Violations are also predicted for twisted alkenes, for planar allenes, for σ,π–dehydroaromatics that are related to planar allenes, and for non-Kekulé hydrocarbon diradicals that have disjoint NBMOs. In [4n]annulenes, twisted alkenes, and planar allenes, the equilibrium geometry of the triplet is a transition state on the potential energy surface for the lowest singlet state. Consequently, experimental confirmation of the predicted violations is very difficult. The role of calculations in assigning the photoelectron spectrum of COT·⁻ and thereby confirming experimentally that D_{8h} COT really does violate Hund's rule is described. In addition, calculations of UV spectra have been found to be useful for deciding whether the singlet or triplet state is responsible for the UV-Vis spectrum of non-Kekulé hydrocarbon diradicals. For example, comparison of the spectra calculated for the singlet and the triplet states of 1,2,4,5-tetramethylenebenzene (TMB), 4,5-dimethylenecyclopentane-1,3-diyl (DMCPD) and 2,3-dimethylenecyclohexane-1,3-diyl (DMCHD) with those observed indicates that TMB has a singlet ground state but that DMCPD and DMCHD both have triplet ground states.

1 Introduction

February 4, 1996 marked the 100th birthday of Professor Fredrich Hund. In honor of this event, a review of Professor Hund's contributions to chemistry and physics has been published.[1] Of these contributions, the most famous are certainly the rules that bear Professor Hund's name.

1.1 Hund's Rules for Atoms

Hund's rules were developed for predicting the energy orderings of the different electronic states that arise from the occupancy of the same set of orbitals in atoms. The rules state: (1) The state with the highest spin, S, is lowest in energy, and (2) Of the states with the same spin, the state with the highest orbital angular momentum, L, is lowest in energy.[2,3] A third rule, less often associated with Hund's name, was subsequently given for predicting the value of the total angular momentum, J, that affords the lowest spin-orbit coupling energy.[3,4]

Exceptions to Hund's rules are well-known in atoms.[1] For example, of the electronic states that arise when one electron occupies a 2p orbital and a second electron occupies a 3p orbital, the lowest energy state is not ^3D but ^1P. This represents a simultaneous violation of both the first and second of Hund's rules. In general, for electronic states with values of L that give them "unnatural parity"[1,5] [e.g., the P ($L = 1$) states that arise from the configuration $2p^1 3p^1$], it can be shown that the singlet is lower than the triplet.[5] Unnatural-parity states represent cases where angular momentum causes Hund's first rule always to be violated in atoms.

1.2 Hund's Rule(s) for Molecules

In linear molecules, such as O_2, electronic states can also be classified by angular momentum, in this case about the molecular axis. However, the states that arise ($^3\Sigma^-$, $^1\Delta$, and $^1\Sigma^+$) from the occupancy of two π orbitals by two electrons all have "natural" parity ($\lambda = 0$ or 2, but not 1), and Hund's rules should thus hold. Indeed, Hund himself appears to have applied his first rule to molecules when he asserted that the lowest energy state of π^2 molecules, such as O_2 should be $^3\Sigma^-$.[6] Hund's second rule also holds in O_2, since $^1\Delta$ has lower energy than $^1\Sigma^+$.

The vast majority of molecules are not linear; so considerations of electron angular momentum do not apply to them. For non-linear molecules it is thus correct to speak about "Hund's rule" for predicting the relative energies of electronic states. Moreover, in molecules, electron angular momentum effects cannot lead to violations of Hund's first rule, which predicts that the electronic state with the highest spin is lowest in energy. Therefore, one might expect that this rule should work very well when applied to molecules.

In a molecule, when two electrons are placed, one each, in two different MOs, Hund's rule predicts that the triplet will lie below the singlet in energy. If the two MOs, ψ_i and ψ_j have exactly the same energy, the lowest singlet wavefunction can be written as having one electron in each of two

MOs, either ψ_i and ψ_j or a linear combination of them. Therefore, Hund's rule further predicts that in diradicals, molecules in which two electrons occupy two degenerate MO's,[7] the triplet will be the ground state.

2 Where to Look for Violations of Hund's Rule in Molecules

The physical basis for Hund's rule is that, when two electrons have the same spin, the Pauli exclusion principle, which is embodied in the antisymmetrization of the electronic wavefunction, does not allow the electrons simultaneously to occupy the same region of space.[1,5,7] However, electrons of opposite spin are not prevented by the Pauli principle from appearing in the same region of space. Thus, two electrons of the same spin usually have a much lower mutual Coulomb repulsion energy than two electrons of the opposite spin that occupy the same pair of MOs.

However, if two MOs have no atoms in common and each MO is occupied by one electron, no matter whether the electrons have parallel or anti-parallel spins, the electrons will not appear in the same region of space. Therefore, if two MOs are *disjoint*; i.e., have no atoms in common, the lowest singlet and triplet should have nearly the same energy. Thus, violations of Hund's rule[8] are most likely to be found in those diradicals, in which the degenerate MOs are disjoint.[7,9]

2.1 [4n]Annulenes

For example, as shown in Fig. 1, the degenerate Hückel non-bonding (NB)MOs for square cyclobutadiene (CBD) can be chosen so that they are disjoint. In both the lowest singlet and triplet state of D_{4h} CBD one electron occupies each of these NBMOs. Because these MOs are disjoint, regardless of whether the electrons in them have the same or opposite spin, there is no probability that both electrons will simultaneously appear in the same AO. Thus, to a first approximation, the lowest singlet and triplet state of square CBD have the same energy.[10] This is also the case in D_{8h} cyclooctatetraene (COT) and in other [4n]annulenes, all of which have a pair of NBMOs that are disjoint.[10]

In contrast, in [4n ± 1]- and [4n ± 2]annulene ions with 4n π electrons, the two degenerate MOs have atoms in common. Consequently, because these MOs are *non-disjoint*, at the geometry of highest symmetry the triplet is calculated to lie well below the singlet in energy.[10] Thus, as predicted by Hund's rule, many antiaromatic annulene ions (e.g. cyclopentadienyl cation[11] and hexachlorobenzene dication[12]) have been found experimentally to have triplet ground states.

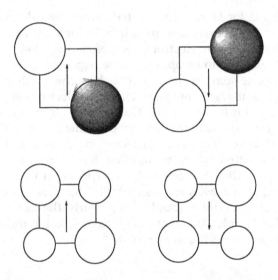

Figure 1: Bonding (bottom) and non-bonding (top) π MOs for the lowest singlet state of CBD. The bonding MOs have been drawn to show schematically the effect of dynamic spin polarization in the lowest singlet state. Only one lobe of each p-π AO is shown.

Ab initio calculations on [4n]annulenes, such as CBD[13,14b] and COT,[14] find that, when correlation is included between the electrons in the bonding and non-bonding π MOs, the singlet lies well below the triplet at the geometry of highest symmetry. Thus, D_{4h} CBD and D_{8h} COT are both predicted to violate Hund's rule.

For example, at the (4,4)CASSCF/6-31G* level of theory the singlet is calculated to be 10.6 kcal/mol lower in energy than the triplet in D_{4h} CBD.[14b] Correlation between the σ and π electrons selectively stabilizes the triplet, so that at the CASPT2N level of theory the singlet-triplet splitting is only $-\Delta E_{ST}$ = 4.1 kcal/mol. However, CASPT2N uses second-order perturbation theory to provide correlation between all but the four π electrons in the active space,[15] so it overestimates the selective stabilization of the triplet. SDCI calculations with a CAS reference space and the Davidson correction for the estimated effect of quadruple excitations[16] give $-\Delta E_{ST}$ = 6.3 kcal/mol.[14b]

Similarly, in D_{8h} COT the singlet is calculated to lie below the triplet by 15.8 kcal/mol at the CASSCF level,[14a] and this energy difference is reduced to 6.7 kcal/mol at the CASPT2N level.[14b] COT was too large to allow CAS-SDCI calculations to be performed, but it was possible to perform MR-(π-SD, σ-S)CI calculations. The reference space consisted of the three most important configurations in the CASSCF wave functions for the singlet and triplet -- the ROHF configuration and the two configurations in which one electron was excited from the degenerate pair of doubly occupied MOs into the degenerate pair of unoccupied MOs. Including the Davidson correction, these MR-(π-SD, σ-S)CI+(Q)/6-31G* calculations give $-\Delta E_{ST}$ = 8.5 kcal/mol in D_{8h} COT.[14b]

Zero-point energy (ZPE) corrections have a non-negligible effect on the value of $-\Delta E_{ST}$ in both D_{4h} CBD and D_{8h} COT. The "bond alternation" vibrational mode, which shortens half the C-C bonds and lengthens the other half, has a calculated frequency > 1600 cm^{-1} in both triplets but an imaginary frequency in the singlets, since the latter are transition states for bond shifting. Consequently, the triplet is calculated to have a higher zero-point energy than the singlet by 2.1 kcal/mol in D_{4h} CBD and by 2.1 kcal/mol in D_{8h} COT. Thus, after correcting for the difference in ZPEs, the singlet is predicted to lie below the triplet by 8.4 kcal/mol in D_{4h} CBD and by 10.6 kcal/mol in D_{8h} COT.

The physical reason why correlation between the bonding and non-bonding π electrons causes the singlet to fall well below the triplet in D_{4h} CBD and D_{8h} COT, is easy to understand.[17] Because in the lowest singlet state the two electrons in the disjoint NBMOs have opposite spin, it is possible for each of the electrons in the bonding π MO(s) to occupy preferentially the same set of AOs to which the non-bonding electron of the same spin is confined. This is shown schematically for CBD in Figure 1.

This type of correlation between the electrons in the bonding and non-bonding MOs in the singlet has been called "dynamic spin polarization".[18] It is energetically advantageous, because it confines bonding and non-bonding electrons of opposite spin to different regions of space, thus reducing their Coulombic repulsion energy. This type of correlation is unavailable to the triplet state, because both electrons in the NBMOs have the same spin. This is why including correlation between the bonding and nonbonding π electrons is calculated to drop the energy of the singlet well below that of the triplet in D_{4h} CBD and in D_{8h} COT.

Although violations of Hund's rule are predicted for D_{4h} CBD and for D_{8h} COT, these geometries are the transition states for bond shifting in the lowest singlet state of each of these [4n]annulenes.[13,14] Therefore, it is difficult to provide unequivocal experimental proof that at these geometries the singlet really does lie below the triplet in energy. However, as discussed later in this chapter, negative ion photoelectron spectroscopy has been used to show that, as predicted, D_{8h} COT really does have a singlet ground state and, therefore, does, indeed, violate Hund's rule.

2.2 Twisted Ethylene

In CBD the two NBMOs belong to the same π system. In twisted ethylene the two NBMOs lie in orthogonal planes. Nevertheless, calculations on twisted ethylene[18] give results that are qualitatively similar to the results for CBD.[7] The reason for this similarity has been pointed out by Mulder.[19]

If the AOs that form the σ and π C-C bonds in ethylene are replaced with the linear combinations of these AOs that provide a bent-bond representation of the molecule, the mathematical form of the MOs in twisted ethylene becomes equivalent to the form of the MOs in D_{4h} CBD. The C-C σ and σ^* MOs of twisted ethylene can thus be seen to correspond, respectively, to the bonding and antibonding π MOs of CBD; and the p-π AO at each carbon in twisted ethylene corresponds to one of the disjoint NBMOs in square CBD.

Like the disjoint NBMOs, in square CBD, the p-π AOs in twisted ethylene are localized in different regions of space. Therefore, to a first approximation, the lowest singlet and triplet states of twisted ethylene have the same energy. However, when correlation between the pair of electrons in the degenerate p-π AOs and the pair of electrons in the σ bond is included, the singlet falls below the triplet.

The singlet is stabilized by σ-π correlation, because dynamic spin polarization in the singlet allows the σ and π electrons of the same spin to become localized to the same carbon. In fact, at any geometry the singlet ground state of ethylene correlates with two triplet CH_2 groups. In contrast, because the spins of the electrons in the p-π AOs in the triplet state of ethylene are the same, one of the electrons in the σ MO cannot avoid being on the same carbon with an electron of opposite spin. In fact, triplet ethylene correlates with one CH_2 in the triplet ground state and one CH_2 in a singlet excited state.

2.3 Planar Allene

Calculations predict that planar allene (Figure 2) should also violate Hund's rule.[20] As in twisted ethylene, replacing the non-bonding σ AO and the p-π AO at the central carbon of planar allene with their sum and difference allows the MOs of planar allene to be written in the same form as the MOs of square CBD. The bonding and antibonding allyl π MOs of planar allene can thus be shown to correspond mathematically to the bonding and antibonding π MOs of CBD, and the allyl and the σ NBMOs of planar allene correspond to the disjoint NBMOs of square CBD.

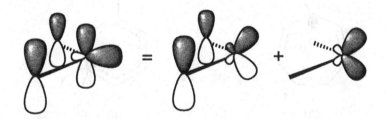

Figure 2: The π and σ AOs at the central carbon in planar allene, shown on the left, are mathematically equivalent, respectively, to the sum of the two out-of-phase hybrid AOs, shown in the center of the Figure, and their difference, which is shown at the right of the Figure.

Unlike the disjoint NBMOs of CBD, the allyl NBMO and the σ NBMO of planar allene are not degenerate; but they are disjoint. Since an electron in the allyl NBMO does not appear on the central carbon, to which an electron in the σ NBMO is largely confined, the lowest singlet and triplet states of planar allene have nearly the same energies. However, when correlation is included between the electrons in the bonding and nonbonding MOs, the singlet is calculated to be the ground state of planar allene. The reason is that the negative spin density which electron correlation causes to appear at the central carbon of the allyl π system is parallel to the spin of the electron in the σ NBMO in the singlet but antiparallel in the triplet.

2.4 2,4,6-Cyclohepatrienylidene (CHTD)

Depicted in Figure 3a is the $^1A_1$ state of 2,4,6-cyclohepatrienylidene, which is a possible transition state for racemization of the conjugated cyclic allene, 1,2,4,6-cycloheptatetraene. In the $^1A_1$ state of CHTD two electrons occupy the sigma NBMO, and there are two, low-lying, unfilled pi MOs, $2a_2$ and $3b_1$ that are nearly degenerate in energy. Excitation of one electron from the a_1 sigma NBMO can thus lead to two triplets, $^3A_2$ and $^3B_1$, and two open-shell singlets, $^1A_2$ and $^1B_1$.

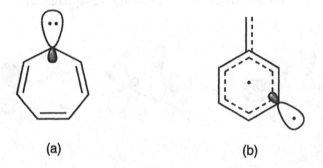

(a)   (b)

Figure 3: (a) $^1A_1$ cyclohepta-2,4,6-trienylidene (CHTD) and (b) *meta*-dehydrotoluene (MDHT).

As expected from the near degeneracy of $2a_2$ and $3b_1$ MOs, the two triplets are calculated to have nearly the same energies.[21] However, the two open-shell singlets have very different energies. $^1A_2$ is computed to be slightly lower in energy than both triplets and also lower in energy than the "aromatic" $^1A_1$ state of the carbene; whereas, $^1B_1$ is calculated to be much higher in energy than these other four states.

The reason for the low energy of $^1A_2$ is that $2a_2$, the π MO that is singly occupied in this state, has a node at the carbenic carbon, where the electron of opposite spin occupies the a_1 σ NBMO. The two singly occupied MOs in $^1A_2$ are thus disjoint, and the negative spin density in the pi AO at the carbenic carbon in this state causers it to lie lower than either triplet.

In contrast, $3b_1$, the π MO that is singly occupied in the $^1B_1$ state, has density at the carbenic carbon atom. Since in this singlet state the electron in $3b_1$ and the electron in the a_1 sigma NBMO are anti-correlated, their large mutual Coulomb replusion energy causes the $^1B_1$ state of CHTD to be very high in energy.

2.5 meta-Dehydrotoluene (MDHT)

Although violations of Hund's rule are predicted for twisted ethylene and planar allene, these geometries are transition states on the lowest singlet potential energy surface for these molecules. Therefore, as in the case of [4n]annulenes, such as square CBD, it is difficult to provide unequivocal experimental proof that at these geometries the singlet really does lie below the triplet in energy. However, experiments by Squires and coworkers have provided experimental evidence which supports their computational prediction that meta-dehydrotoluene (MDHT) diradical, shown in Figure 3b, has a singlet ground state and thus violates Hund's rule.[22]

MDHT can be regarded as a planar allene whose six-membered ring and highly delocalized π system make a planar geometry the equilibrium geometry of not only the triplet but also of the singlet. In the benzyl π system that MDHT contains, the unpaired p electron appears at the ortho and para carbons of the benzene ring; and, since the π NBMO has nodes at the meta carbons, negative spin density appears in the π AO at each of these atoms. Thus, it is easy to predict that when an ortho or para hydrogen is removed from benzyl radical, the resulting dehydrotoluene diradicals will have triplet ground states; but, when a meta hydrogen is removed, the MDHT diradical formed will have a singlet ground state.

The computational and experimental results of Squires and coworkers confirm these qualitative expectations.[22] Their CASSCF calculations predict that the $^1A''$ state of MDHT is lower in energy than the $^3A''$ state by 3.0 kcal/mol. In agreement with the prediction of a singlet ground state for MDHT, Squire's experiments find significant differences between formation of MHDT and its *ortho* and *para* isomers from singlet precursors. Both of the latter two dehydrotoluenes are qualitatively expected and quantitatively calculated to have triplet ground states; therefore the formation of each from a singlet precursor is a spin-forbidden reaction; whereas, formation of MDHT is not. Thus, MDHT represents an example of a diradical for which there exists experimental evidence for the violation of Hund's rule that is predicted by both qualitative theory and quantitative calculations.

2.6 Non-Kekulé Hydrocarbon Diradicals

As in twisted ethylene, planar allene, and MDHT, in non-Kekulé hydrocarbon diradicals there are two electrons that cannot be included in bonds in any resonance structure. However, in non-Kekulé hydrocarbon diradicals both electrons occupy 2p AOs that belong to the same π system. Violations of Hund's rule are predicted to occur in those non-Kekulé hydrocarbon diradicals in which, as is the case in all [4n]annulenes, the two NBMOs can be chosen to be disjoint.[7-9]

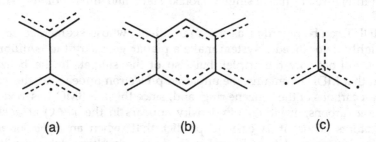

(a)   (b)   (c)

Figure 4: (a) tetramethyleneethane (TME), tetramethylenebenzene (TMB), and (c) trimethylenemethane (TMM).

Three non-Kekulé hydrocarbon diradicals are shown in Figure 4. The NBMOs in both tetramethyleneethane (TME) and tetramethylenebenzene (TMB) can be chosen to be disjoint; so both of these diradicals are candidates for having singlet ground states. In contrast, trimethylene-methane (TMM) has non-disjoint NBMOs. Hence, it is predicted to have a triplet ground state and a large singlet-triplet splitting, as has been found to be the case experimentally.[23]

The method of Ovchinnikov,[24] which is based on valence-bond, rather than on MO theory, makes the same predictions about the spin multiplicity of the ground states of diradicals such as CBD, COT, TME, TMB, and TMM. However, classifying the NBMOs of a diradical according to whether or not they are disjoint has several advantages, including the ability to predict that anti-aromatic annulene ions will generally be found to obey Hund's rule and to estimate whether the singlet-triplet splitting in a diradical is likely to be large or small.[25]

Ab initio calculations on TME at many different levels of theory all predict a singlet ground state for the planar diradical.[26] The predicted violation of Hund's rule at this geometry is a consequence of two effects. One is long-range bonding between the termini of the two allylic radicals in the singlet. Its magnitude can be assessed from calculations that correlate only the two non-bonding electrons, utilizing two configurations for the singlet (TCSCF) and one for the triplet (ROHF). At this level the singlet is calculated to lie below the triplet by about 1.8 kcal/mol with Dunning's SVP basis set.[26a,b]

The second effect that stabilizes the singlet, relative to the triplet, is interaction between the negative spin densities in the p-π AOs at the central carbons of the two allylic radicals in TME.[9] This interaction, like that between the electrons at the terminal carbons, is bonding in the singlet; but it only appears in calculations in which correlation is provided between all the π electrons. The size of this second effect can be assessed by comparing the results of CASSCF calculations, which correlate all six of the π electrons in TME, with the TCSCF/ROHF results. At the CASCSCF level the singlet is calculated to lie 3.8 kcal/mol below the triplet,[26b] an energy difference that is 2.0 kcal/mol larger that that computed by TCSCF/ROHF.

Confirming experimentally the predicted violation of Hund's rule in planar TME is very difficult, since the equilibrium geometry of neither the singlet nor the triplet is expected to be planar.[26,27] At the equilibrium geometry of the triplet long-range bonding interactions between the two allyl moieties, which are largest at dihedral angles of 0° and 90°, are minimal; and the overlap between the two p-π orbitals at the central carbons is also substantially reduced from that in planar TME. Consequently, calculations all agree that at the non-planar equilibrium geometry of the triplet state the magnitude of -ΔE_{ST} is very small.

The largest calculations to-date on TME are those of Nachtigall and Jordan.[26c] They correlated all the valence electrons at the SD-CI level; and, for the triplet, they included excited configurations that do not mix directly with the reference configuration. These CI calculations found that, after a correction for the effect of quadruple excitations, the triplet is calculated to lie slightly below the singlet in energy.

One might question whether these SD-CI calculations, based on just two reference configurations for the singlet and one for the triplet, are sufficiently accurate to predict correctly which state is lower in energy at the equilibrium geometry of the latter. However, Dowd's experimental findings, that TME has a triplet EPR spectrum[28a] and that a plot of the intensity of the spectrum versus 1/T is linear,[28b] can be taken as evidence that Nachigall and Jordan's SD-CI calculations are correct in finding that the triplet state is lower. Nevertheless, although the linear Curie plot obtained by Dowd could, indeed, indicate that the triplet state is lower in energy by several hundred cal/mol than any singlet state with which it is in equilibrium, the plot could also indicate that the two states have the same energy to within about 10 cal/mol.[29]

Planar TME should have a singlet ground state, and the problem of this not being the equilibrium geometry of either the singlet or triplet state is eliminated by the presence of the five-membered ring in 4,5-dimethylenecyclopentane-1,3-diyl (DMCPD), which is shown in Figure 5. Unfortunately, the CH$_2$ group that serves as a conformational lock in DMCPD also perturbs the electronic structure of the TME moiety via the hyperconjugative interaction between the C-H bonds of the methylene group and the adjacent radical centers. This interaction has been shown[30]

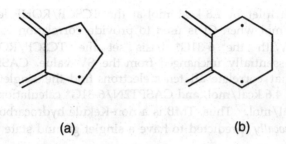

$$(a) \qquad\qquad (b)$$

Figure 5:   (a) 4,5-dimethylenecyclopentane-1,3-diyl (DMCPD) and   (b) 2,3-dimethylene-cyclohexane-1,4-diyl (DMCHD)

to be responsible for the triplet state being both calculated[31] and found[32] to be the ground state of cyclopentane-1,3-diyl, the diradical which lacks the pair of exocyclic methylene groups that are present in DMCPD.

Although calculations on DMCPD that correlated just the π electrons found the singlet to be slightly lower in energy than the triplet,[33] SD-CI calculations by Jordan and coworkers, which correlated all of the valence electrons, predicted the triplet to be the ground state.[34] Roth and coworkers have, in fact, found a triplet EPR spectrum and a linear Curie plot for the 2,2-dimethyl derivative of DMCPD.[35]

The ethano group in 2,3-dimethylenecyclohexane-1,4-diyl (DMCHD), which is also shown in Figure 5, should perturb the TME moiety less than the methylene group in DMCPD.[30] Were DMCHD planar, it would therefore be a better candidate than DMCPD for having a singlet ground state. However, the equilibrium geometry of DMCHD is computed to be non-planar, and at the SD-CI level this leads to a very small calculated preference for a triplet ground state.[34] This computational finding is consistent with the experimental results of Dowd, Roth, and their coworkers for DMCHD.[36]

Given the problems in devising an electronically unperturbed derivative of TME that is restricted to a planar geometry, TMB would appear to be a better candidate for a diradical in which a singlet ground state might be found experimentally. The two pentadienyl radical moieties in TMB are constrained to be coplanar by the two C-C bonds that join them. Therefore, unlike the case in TME, a planar geometry for TMB seems assured.

Both semiempirical and *ab initio* calculations predict a singlet ground state for TMB.[37] With Dunning's SV basis set the singlet is calculated to lie below the triplet by 1.8 kcal/mol at the TCSCF/ROHF level and by 5.0 - 6.6 kcal/mol when CI is used to provide correlation for all the π electrons.[37d] With the 6-31G* basis set the TCSCF/ROHF energy difference is essentially unchanged from the SV value. CASSCF/6-31G* calculations that correlate all ten π electrons find the singlet lies below the triplet by 4.6 kcal/mol, and CASPT2N/6-31G* calculations also give $-\Delta E_{ST} = 4.6$ kcal/mol.[37] Thus, TMB is a non-Kekulé hydrocarbon diradical that is *unequivocally* predicted to have a singlet ground state.

3   How to Identify Violations of Hund's Rule Experimentally

A diradical that obeys Hund's rule by having a triplet ground state is easy to identify experimentally, since it should have an EPR signal, and the temperature dependence of its intensity should obey the Curie law.[29] However, the failure to observe an EPR signal cannot be taken as proof that a diradical has a singlet ground state; since, in general, negative results provide only permissive, not conclusive evidence for a hypothesis.

3.1 The UV Spectrum of TMB -- Hoisted on Our Own Petard.

In order to facilitate the experimental verification of our unequivocal prediction that TMB has a singlet ground, we calculated the UV spectra of singlet and triplet TMB. Our π CI calculations on TMB led us to predict that, "Both states ... have an allowed absorption around or slightly below 3.0 eV, polarized along the long molecular axis. However, the triplet is expected also to have a weak absorption at lower energy, corresponding to dipole forbidden excitation [from $^3B_{1u}$] to $^3B_{3u}$."[37d] We fully expected that when TMB was prepared it would be found not to have an EPR spectrum and that the longest wavelength absorption would be intense, corresponding to an allowed excitation from the 1^1A_g ground state to the lowest $^1B_{3u}$ excited state.

Roth and coworkers were the first to prepare TMB.[38] They found that TMB has a triplet EPR signal whose intensity gives a linear Curie plot. Consequently, they assigned a triplet ground state to the diradical. Further, they noted, "Support for this assignment [of a triplet ground state for TMB] is provided by the UV/VIS spectrum ... with its [weak] long-wave[length] bands between 550 and 600 nm. The position of these bands is in good agreement with the long-wave[length], forbidden transition at 570 nm calculated for the triplet; these calculations do not lead one to expect any bands for the singlet in this region." Consequently, Roth and coworkers concluded, "Contrary to the theoretical expectations the singlet ... can lie 0.02 kcal/mol at most below the lowest triplet state, i.e. the states are [sic] degenerated, or -- what could be more likely -- the diradical has a triplet ground state."[38]

At the time that Roth *et al.* published their paper, Professor Jerome A. Berson and his coworkers at Yale had also generated TMB and were busy studying its spectroscopy and chemistry. These studies[39] led Berson and coworkers to the following conclusions: (a) The EPR spectrum observed when TMB is generated does not belong to TMB. (b) The UV-VIS spectrum does belong to TMB. (c) Since a sharp resonance is seen in the CPMAS ^{13}C NMR spectrum of TMB, enriched with ^{13}C, TMB has a singlet ground state. If the ground state were a triplet, dipolar broadening, due to the unpaired electrons, would have been observed.

Although these studies appeared to confirm the predictions of a singlet ground state for TMB,[37] Berson's findings also raised an important question. If the singlet is the ground state of TMB, why does the longest wavelength absorption in the UV-Vis spectrum of this state correspond to a forbidden transition, a spectral feature that our 1986 π CI calculations had predicted for the triplet but not the singlet.[37d] In order to understand why our attempt to predict the UV-Vis spectrum of TMB had given results that were apparently not even qualitatively correct, in 1994 we undertook additional calculatons of the UV-Vis spectra of the lowest singlet and triplet states of TMB.

3.2 CASPT2N Calculations of the UV-Vis Spectra of TMB

The size of TMB had made it impossible in 1986 to correlate more than just the π electrons. By 1994 not only had there been at least an order of magnitude improvement in the computer hardware available to us, but also Roos and coworkers had developed the CASPT2N method;[15] and they had shown that CASPT2N calculations give very good results for the UV spectra of conjugated molecules.[40] Therefore, we used the CASPT2N method to calculate the UV-Vis spectra of the lowest singlet and triplet states of TMB.[37e]

Our CASPT2N calculations found that inclusion of correlation between the σ and π electrons in TMB makes 2^1A_g, which is the third excited singlet state of TMB at both the π CI[37d] and π CASSCF levels,[37e] the lowest excited singlet state. The reason for the selective stabilization of 2^1A_g by inclusion of σ-π correlation is that, in the simplest MO description of this state, the wave function for 2^1A_g is largely comprised of the first two, zwitterionic, resonance structures in Figure 6.[37e] Dynamic correlation between the σ and π electrons in TMB and, more generally, between the inactive and active electrons in many molecules and transition states, stabilizes such zwitterionic structures.[41]

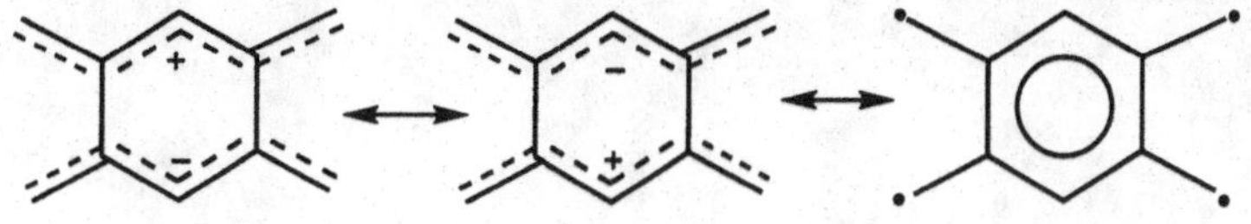

Figure 6: Resonance structures for the 2^1A_g state of TMB.

In valence-bond theory there is a purely covalent resonance structure for 2^1A_g, the third structure shown in Figure 6.[37e] The contribution of this structure to the 2^1A_g wave function suggested that 2^1A_g might have a very different geometry than the 1^1A_g ground state. Geometry optimization showed this to be the case. The large difference between the geometries of these two states means that the absorption corresponding to the dipole-forbidden excitation of TMB from 1^1A_g to 2^1A_g should not only be weak but should also show a large amount of vibrational structure. A great deal of vibrational structure is, in fact, present in the longest wavelength absorption in the UV-Vis spectrum published by Roth and coworkers.[38]

Not only the appearance of the long wavelength band in the spectrum of TMB but also its energy is in excellent agreement with that calculated at the CASPT2N level of theory for the $1^1A_g \longrightarrow 2^1A_g$ excitation. In fact, the first vibrational band in all four of the absorptions in the spectrum is within 0.1 - 0.2 eV of that calculated for an adiabatic excitation from the lowest singlet state of TMB; and, also as predicted by the CASPT2N calculations for the singlet, the first and third absorptions are weak, and the second and fourth are strong[37e] In contrast, the observed spectrum does not fit at all that computed at the CASPT2N level for excitations from the lowest triplet state. Therefore, the CASPT2N calculations of the UV spectrum provide strong support for assigning it to singlet TMB, thus confirming the computational prediction that TMB has a singlet ground state.

3.3 CASPT2N Calculations of the UV Spectra of DMCPD and DMCHD

Our success in using CASPT2N calculations to compute the UV spectra of singlet and triplet TMB and in thus assigning the observed spectrum to excitations from the 1^1A_g state, led us also to perform CASPT2N calculations of the UV spectra of two other non-Kekulé hydrocarbon diradicals, DMCPD and DMCHD (shown in Figure 5). Although the observation of triplet EPR signals that give linear Curie plots[35,36] is certainly consistent with a triplet ground state for each of these diradicals, it should be remembered that TMB also appeared to have a triplet EPR spectrum that gave a linear Curie plot.[38] Therefore, we compared the UV spectra, computed for the lowest singlet and triplet states of DMCPD and DMCHD, with the spectrum observed by Roth and coworkers for each of these diradicals,[35,36b] in order to confirm the assignment of a triplet ground state to both.

Comparison of the UV spectra computed for the lowest singlet and triplet states of DMCPD and DMCHD with that observed for each diradical by Roth and coworkers showed that in each case the observed UV spectrum fits much better that computed for the triplet than the singlet.[14b] Thus, our CASPT2N calculations provide very strong support for the assignment of a triplet ground state to both of these diradicals and further illustrate the utility of this type of calculation for identifying which spin state of a diradical gives rise to an observed UV spectrum.

4 Where to Look for Violations of the Strictest Version of Hund's Rule

It might seem that this question has already been answered; since, ^{13}C NMR and comparison of the UV-Vis spectrum of TMB with those calculated for the lowest singlet and triplet states show that, as predicted, this diradical has a singlet ground state.[37] However, although TMB certainly is a diradical with a singlet ground state, it does not really violate the strictest version of Hund's rule.

It should be recalled that Hund's rule can only be rigorously used to predict that the triplet will be the ground state of a diradical when the two orbitals that are occupied by a total of two electrons are degenerate.

Hund's rule cannot be said to be violated in the strictest sense when the orbitals have different energies, because, for a sufficiently large difference in orbital energies, having two electrons in the lower orbital will certainly be energetically more favorable than having one electron in each orbital with parallel spins.[7]

If the NBMOs of a diradical are non-degenerate, a violation of the strictest form of Hund's rule can still occur if the singlet state, in which one electron occupies each NBMO, is lower in energy than the corresponding triplet. This is the case in diradicals, such as planar allene, CHTD, and MDHT (Figures 2 and 3), which have one σ and one π NBMO.

However, in diradicals where both NBMOs are non-degenerate π orbitals, the lowest singlet state does not have one electron in each NBMO. Instead, the wave function for the lowest singlet state consists of two dominant configurations, each of which has one of the two NBMOs doubly occupied; and the configuration with the greatest weight is the one in which the lower energy of the two NBMOs contains two electrons.

In TMB the two NBMOs are not degenerate. Long-range interactions between the p-π AOs of nonbonded carbons make the in-phase combination of the two pentadienyl NBMOs lower in energy than the out-of-phase combination by about 12 kcal/mol. Consequently, in the TCSCF wave function for the lowest singlet state the electron occupation number of the in-phase combination (1.3) is almost twice as large as that (0.7) of the out-of-phase combination. Since the triplet must have exactly one electron in each MO, 1^1A_g is calculated to lie 1.8 kcal/mol lower than $^3B_{1u}$ at the TCSCF/ROHF level of theory.

If the strictest form of Hund's rule cannot be applied to TMB, where can a violation of the strictest form be found? In order for a molecule to have a pair of orbitals that are degnerate by symmetry, the molecule must have at least a three-fold axis of symmetry; and, in order for a degenerate pair of NBMOs to be disjoint, a four-fold symmetry axis is required.[10] Thus, experimental confirmation of a violation of the strictest form of Hund's rule in a diradical with two π NBMOs must be sought in a [4n]annulene at the geometry of highest symmetry. However, such an experiment must overcome the formidable difficulties associated with the fact that such a geometry is the transition state for bond shifting in the lowest singlet state.

5 How Can a Violation of Hund's Rule Be Identified in a [4n]Annulene?

In collaboration with Dr. Paul Wenthold and Professor Carl Lineberger, we succeeded in measuring the energy difference between the lowest singlet and triplet states of D_{8h} COT and thus confirmed experimentally the predicted violation of Hund's rule in this [4n]annulene.[42] The measurement of $-\Delta E_{ST}$ in D_{8h} COT was accomplished by obtaining the electron photodetachment spectrum of the radical anion of COT (COT$^{\cdot-}$). Theory not only motivated performing this experiment, but theory was also important in the design and interpretation of it.

Calculations have shown that COT$^{\cdot-}$ has a D_{4h} equilibrium geometry, with alternating long and short C-C bonds in the planar eight-membered ring.[44] The geometry calculated for the radical anion indicated that, upon electron photodetachment, (a) Franck-Condon factors would favor formation of the singlet and triplet at planar geometries; and (b) the differences in the degree of bond alternation between COT$^{\cdot-}$ and both the singlet and triplet states of planar neutral COT would lead to the observation of progressions in the bond-alternation vibrational mode in the band for each of the latter two states. Vibrational progressions were, in fact, seen in the bands for both states in the photodetachment spectrum of COT$^{\cdot-}$; and, with the help of calculations, this vibrational structure helped us to assign the lower energy band to the singlet and the higher energy band to the triplet.[42]

The higher energy band showed four major peaks with a spacing of 1635 ± 20 cm^{-1}. A second, weaker, vibrational progression was observed with a frequency of 735 ± 20 cm^{-1}. (8,8)CASSCF/6-31G* frequency calculations on the D_{8h} triplet, after scaling by 0.91 in order to account for anharmonicity and electron correlation effects, predict 1615 cm^{-1} for the bond-alternation vibrational mode and 735 cm^{-1} for the ring-breathing vibrational mode. The latter changes all the C-C bond lengths, while maintaining D_{8h} symmetry. Not only the frequencies but also the intensities of the vibrational peaks in the higher energy band were in excellent agreement with those calculated for the triplet. The intensities were obtained by calculating the Franck-Condon factors[44] from the

(8,8)CASSCF/6-31G* geometries and frequencies for COT$^{\cdot-}$ and for the $^3A_{2u}$ state of COT.

The first excited vibration at 740 cm^{-1} in the lower energy band of the electron photodetachment spectrum was assigned to the ring breathing mode in the D_{4h} singlet, which was calculated to have nearly the same frequency in $^1A_{1g}$ as in $^3A_{2u}$. However, the intense series of vibrational peaks, seen in the band for $^3A_{2u}$ and assigned to the bond-alternation mode, were missing in the band for $^1A_{1g}$. Instead, a pair of comparatively weak peaks at 1315 cm^{-1} and 1670 cm^{-1} was observed.

The scaled (8,8)CASSCF/6-31G* harmonic vibrational frequency, calculated for the bond alternation mode at the D_{4h} geometry of $^1A_{1g}$ COT, is 1535 cm^{-1}. However, at this energy (4.4 kcal/mol) above vibrationally unexcited D_{4h} COT, the first excited vibrational level of the bond-alternation mode is predicted to be very close to the energy of the D_{8h} transition state for bond shifting.[14] Therefore, the in-phase and out-of-phase combinations of the two D_{4h} wave functions for this vibrational level should be split by an observable amount. The appearance of two peaks, centered around 1500cm^{-1} above the first peak, is thus the expected signature of the band in the photoelectron spectrum that belongs to formation of planar singlet COT. Consequently, the peaks at 1315 cm^{-1} and 1670 cm^{-1} in the lower energy band of the spectrum of COT$^{\cdot-}$ provide strong evidence that this band should be assigned to the singlet.

The assignment of the lower energy band in the COT$^{\cdot-}$ photoelectron spectrum to the singlet and the higher energy band to the triplet leads to an experimental value of $-\Delta E_{ST} \approx 9$ kcal/mol in D_{8h} COT.[42] This experimental value is about 15% smaller than our MR-(π-SD,σ-S)CI+(Q)/6-31G* value of $-\Delta E_{ST} = 10.6$ kcal/mol but, probably fortuitously, the same as the value given by our CASPT2N calculations.[14b] Thus, we celebrated the 100th birthday of Professor Fredrich Hund by showing experimentally that theory correctly predicts not only the violation of the strictest form of Hund's rule in D_{8h} COT but also the amount by which the singlet lies below the triplet in energy.

Acknowledgements

This chapter is dedicated to the memory of Paul Dowd, a pioneer in the experimental study of non-Kekulé hydrocarbon diradicals, such as TMM[23] and TME.[28] As recounted elsewhere,[8] the interest of one of the authors of this chapter (W.T.B.) in diradicals that violate Hund's rule had its origin in an unsuccessful attempt, made in 1969 in collaboration with Professor Dowd, to observe triplet COT by EPR. We thank the National Science Foundation for support of the research performed in our group at the University of Washington.

References

(1)   Kutzelnigg, W. *Angew. Chem. Int. Ed. Engl.* **1996**, *35*, 4001.

(2)   Hund, F. *Z. Phys.* **1925**, *33*, 345.

(3)   Hund, F. *Linienspektren und periodisches System der Elemente*, Springer, Berlin, 1927

(4)   Hund, F. *Z. Phys.* **1925**, *33*, 855.

(5)   Kutzelnigg, W.; Morgan, J. D. III *Z. Phys.* D **1996**, *36*, 197.

(6)   Hund, F. *Z. Phys.* **1928**, *51*, 759.

(7)   See, for example, Borden, W. T. in *Diradicals*, Borden, W. T., Ed., Wiley-Interscience: New York, 1982, pp. 1-72.

(8)   Review: Borden, W. T.; Iwamura, H.; Berson, J. A. *Acc. Chem. Res.* **1994**, *27*, 109.

(9)   Borden, W. T.; Davidson, E. R. *J. Am. Chem. Soc.* **1977**, *99*, 4587.

(10) Borden, W. T. *J. Chem. Soc.,Chem. Comm.* **1969**, 1968.

(11) Saunders, M.; Berger, R.; Jaffe, A.; McBride, J. M.; O'Neill, J.; Breslow, R.; Hoffmann, J. M.; Perchonok, C.; Wasserman, E.; Hutton, R. S.; Kuck, V. J. *J. Am. Chem. Soc.* **1973**, *95*, 13017.

(12) Wasserman, E.; Hutton, R. S.; Kuck, V. J.; Chandross, E. A. *J. Am. Chem. Soc.* **1974**, *96*, 1965.

(13) (a) Buenker, R. J.; Peyerimhoff, S. D. *J. Chem. Phys.* **1968**, *48*, 354; (b) Kollmar, H.; Staemmler, V. *J. Am. Chem. Soc.* **1977**, *99*, 3583; (c) Borden, W. T.; Davidson, E. R.; Hart, P. *J. Am. Chem. Soc.* **1978**, *100*, 388; (d) Jafri, J. A.; Newton, M. *J. Am. Chem. Soc.* **1978**, *100*, 5012; (e) Agren, H.; Correia, N.; Flores-Riveros, A, Jensen, H. J. A. *Int. J. Quantum Chem.* **1986**, *19*, 237; (f) Nakamura, K.; Osamura, Y.; Iwata, S. *Chem. Phys.* **1989**, *135*, 67; (g) Balková, A.; Bartlett, R.J. *J. Chem. Phys.* **1994**, *101*, 8972.

194

(14) (a) Hrovat, D. A.; Borden, W. T. *J. Am. Chem. Soc.* **1992**, *114*, 5879; (b) Hrovat, D. A.; Borden, W. T. *Theochem*, in press.

(15) Andersson, K.; Malmqvist, P.-Å.; Roos, B. O. *J. Chem. Phys.* **1992**, *96*, 1218.

(16) Davidson, E. R. In *The World of Quantum Chemistry*; Daudel, R., Pullman, B., Eds.; Dordecht: Netherlands, 1974.

(17) Borden, W. T. *J. Am. Chem. Soc.* **1975**, *97*, 5968.

(18) Kollmar, H.; Staemmler, V. *Theor. Chim. Acta* **1978**, *48*, 223.

(19) Mulder, J. J. C. *Nouv. J. Chem.* **1980**, *4*, 283.

(20) (a) Borden, W. T. *J. Chem. Phys.* **1965**, *45*, 2512; (b) Staemmler, V. *Theor. Chim. Acta* **1977**, *45*, 89.

(21) (a) Matzinger, S.; Bally, T.; Patterson, E. V.; McMahon, R. J. *J. Am. Chem. Soc.* **1996**, *118*, 1535; (b) Schreiner, P. R.; Karney, W. L.; Schleyer, P. v. R.; Borden, W. T.; Hamilton, T. P.; Schaefer, H. F., III *J. Org. Chem.* **1996**, *61*, 7030

(22) Wenthold, P. G.; Wierschke, S. G.; Nash, J. J.; Squires, R. R. *J. Am. Chem. Soc.* **1994**, *116*, 7378.

(23) Dowd, P. *J. Am. Chem. Soc.* **1966**, *88*, 2587; Baseman, R. J.; Pratt, D. W.; Chow, M.; Dowd, P. *J. Am. Chem. Soc.* **1976**, *98*, 5726; Wenthold, P. G.; Hu, J.; Squires, R. R.; Lineberger, W. C. *J. Am. Chem. Soc.* **1996**, *118*, 475.

(24) Ovchinnikov, A. A. *Theor. Chim. Acta* **1978**, *47*, 297.

(25) Borden, W. T. *Mol. Cryst, Liq. Cryst.* **1993**, *232*, 195.

(26) (a) Du, P.; Borden, W. T. *J. Am. Chem. Soc.* **1987**, *109*, 930; (b) Nachtigall, P.; K. Jordan *J. Am. Chem. Soc.* **1992**, *114*, 4743; (c) Nachtigall, P.; K. Jordan *J. Am. Chem. Soc.* **1993**, *115*, 270.

(27) Dixon, D. A.; Foster, R.; Halgren, T. A.; Lipscomb, W. M. *J. Am. Chem. Soc.* **1978**, *110*, 1359.

(28) (a) Dowd, P. *J. Am. Chem. Soc.* **1970**, *92*, 1066; (b) Dowd, P. Chang, W.; Paik, Y. H. *J. Am. Chem. Soc.* **1986**, *108*, 7416.

(29) For a critical discussion of Curie plots see Berson, J. A. in *The Chemistry of Quinonoid Compunds*, Patai, S.; Rappoport Z., Eds.; Wiley: New York, 1988, Vol. 2, pp. 462-9.

(30) Goldberg, A. H.; Dougherty, D. A. *J. Am. Chem. Soc.* **1983**, *105*, 284.

(31) Conrad, M.; Pitzer, R.; Schaefer, H. F. III *J. Am. Chem. Soc.* **1979**, *101*, 2245; Sherrill, C. D.; Seidl, E. T.; Schaefer, H. F. III, *J. Phys. Chem.* **1992**, *96*, 3712; Xu, J. D.; Hrovat, D. A.; Borden, W. T. *J. Am. Chem. Soc.* **1994**, *116*, 5425.

(32) Buchwalter, S. L.; Closs, G. L. *J. Am. Chem. Soc.* **1975**, *97*, 3857; Buchwalter, S. L.; Closs, G. L. *J. Am. Chem. Soc.* **1979**, *101*, 4688.

(33) Du, P.; Hrovat, D. A.; Borden, W. T. *J. Am. Chem. Soc.* **1986**, *108*, 8086.

(34) Nash, J. J.; Dowd, P.; Jordan, K. D. *J. Am. Chem. Soc.* **1992**, *114*, 10071.

(35) Roth, W. R.; Kowalczik, U.; Maier, G.; Reisenauer, H. P.; Sustmann, R.; Muller, P. *Angew. Chem. Int. Ed. Engl.* **1987**, *26*, 1285.

(36) (a) Dowd, P.; Chang, W.; Paik, Y. H. *J. Am. Chem. Soc.* **1987**, *109*, 5284; (b) Roth, W. R.; Biermann, M.; Erker, G.; Jelich, K.; Gerhartz, W.; Görner, H. *Chem. Ber.* **1980**, *113*, 586.

(37) (a) Lahti, P. M.; Rossi, A. ; Berson, J. A. *J. Am. Chem. Soc.* **1991**, *113*, 2318; (b) Lahti, P. M.; Ichimura, A. S.; Berson, J. A. *J. Org. Chem.* **1989**, *54*, 958; (c) Lahti, P. M.; Rossi, A. ; Berson, J. A. *J. Am. Chem. Soc.* **1985**, *107*, 4362; (d) Du, P.; Hrovat, D. A.; Borden, W. T. Lahti, P. M.; Rossi, A. ; Berson, J. A. *J. Am. Chem. Soc.* **1986**, *108*, 5072; (e) Hrovat, D. A.; Borden, W. T. *J. Am. Chem. Soc.* **1994**, *116*, 6327.

(38) Roth, W. R.; Langer, R.; Bartmann, M.; Stevermann, B.; Maier, G.; Reisenauer, H. P.; Sustmann, R.; Müller, W. *Angew. Chem. Intl. Ed. Engl.* **1987**, *26*, 256; Roth, W. R.; Langer, R.; Ebbrecht, T.; Beitat, A.; Lennartz, H. -W. *Chem. Ber.* **1991**, *124*, 2751.

(39) (a) Reynolds, J. H.; Berson, J. A.; Kumashiro, K. K.; Duchamp, J. C.; Zilm, K. W.; Rubello, A.; Vogel, P. *J. Am. Chem. Soc.* **1992**, *114*, 763. (b) Reynolds, J. H.; Berson J. A.; Scaiano, J. C.; Berinstain, A. B. *J. Am. Chem. Soc.* **1992**, *114*, 5866. (c) Reynolds, J. H.; Berson, J. A.; Kumashiro, K. K.; Duchamp, J. C.; Zilm, K. W.; Scaiano, J. C.; Berinstain, A. B.; Rubello, A.; Vogel, P. *J. Am. Chem. Soc.* **1993**, *115*, 8073.

(40) Serrano-Andrés, L.; Merchán, M.; Nebot-Gil, I.; Roos, B. O.; Fülscher, M. *J. Am. Chem. Soc.* **1993**, *115*, 6184.

(41) Borden, W. T.; Davidson, E. R., *Acc. Chem. Res.* **1996**, *29*, 67.

(42) Wenthold, P. G.; Hrovat, D. A.; Borden, W. T.; Lineberger, W. C. *Science*, **1996**, *272*, 1456.

(43) Hammons, J. H.; Hrovat, D. A.; Borden, W. T. *J. Am. Chem. Soc.* **1991**, *113*, 4500.

(44) Chen, P. in *Unimolecular and Bimolecular Reaction Dynamics* Ng, C. Y.; Baer, T.; Powis, I., Eds., John Wiley & Sons, New York, 1994; Wenthold, P. G.; Polak, M. L.; Lineberger, W. C. *J. Phys. Chem.* **1996**, *100*, 6920.

AB INITIO METHODS FOR THE DESCRIPTION
OF ELECTRONICALLY EXCITED STATES:
SURVEY OF METHODS AND SELECTED RESULTS

ROBERT J. CAVE
Department of Chemistry, Harvey Mudd College,
Claremont, CA 91711, USA

An array of theoretical methods have been developed to describe the electronically excited states of atoms and molecules. This chapter reviews some of the general types of excited states of organic compounds and discusses several theoretical methods for their description. The strengths and weaknesses of the methods are discussed, and a comparison of the results of these methods is presented for ethylene and the three polyenes, butadiene, hexatriene, octatetraene.

1 Introduction

In many chemical reactions the reactants begin in their electronic ground states and, unless the reaction is highly exothermic, the products are produced in their electronic ground states. This is largely due to the fact that in most cases the lowest electronically excited states are at least 3 eV (about 70 kcal) above the ground state at its equilibrium geometry. However, this assumed separation between ground and excited states may break down at stages along a reaction path. Furthermore, it is not uncommon to photoinitiate reactions, in which case one makes direct use of an electronically excited state as a reactant, either for reasons of increased driving force or enhanced reactivity. Finally, in biological systems many important processes (e.g. in the visual pigments or the photosynthetic apparatus) are initiated by visual light and proceed at least for a portion of the time on an excited state surface. Thus, it is of primary importance to be able to obtain accurate theoretical descriptions of electronically excited states.

While this need has been acknowledged for as long as quantum mechanics has been applied to chemistry, it is fair to say that optimal methods for treating excited states are not nearly as refined, widely applicable, agreed upon, or accurate as those used for routine treatments of electronic ground states. For small to modest-sized problems (under 10 heavy atoms) treating electronic ground states, CCSD(T) (single and double excitation coupled cluster theory with noniterative inclusion of triple excitations[1]) is an attractive choice, combining accuracy with tractability.[2] For larger systems single reference perturbation theories (e.g.

second-order Møller Plessett (MP2) perturbation theory[3-5]), is the most common choice, allowing inclusion of some electron correlation. For excited states however, these methods are not directly applicable.

One reason for this greater difficulty in treating excited states is that both of the above methods are "single-reference based methods" - they include electron correlation relative to a single determinantal wavefunction (that is, an antisymmetric product of spin orbitals, either restricted Hartree-Fock (RHF) or unrestricted Hartree-Fock (UHF)) and many excited states are multi-configurational at zeroth-order (see below). In addition, even for excited states which are qualitatively well-described by a single configuration, it is not always possible to obtain accurate HF wavefunctions for these states (although the range of states for which accurate single configuration SCF descriptions can be obtained is significantly greater than is usually supposed[6,7]).

Thus, a wide range of tools has been developed over the past thirty years for the treatment of electronically excited states. Along with the concern to be able to treat multi-configurational zeroth-order states, a second overarching concern has been to develop methods which are "size-consistent."[8] That is, the energies should scale properly as system size is increased. While this may seem a technical point of modest interest, it turns out that "size-inconsistency" can lead to a significant decrease in accuracy for excitation energies and properties of excited states in some cases. It will come as little surprise that the most accurate methods tend to be the most costly (time consuming), but it also turns out that several methods developed in recent years combine quite reasonable accuracy with sufficient speed to allow one to treat systems containing up to, say, ten heavy atoms.

The present chapter will compare a number of these methods in a formal sense, and then examine the results of several of the methods when applied to the calculation of electronic excitation energies for the polyenes. In addition to a critical comparison of the methods, the discussion will focus on the question of "what has been learned *about the molecules*" from these time-consuming studies. These systems are frequently discussed in the context of simple Hückel theory and it is of interest to ask what new information is gleaned from application of more extensive computational approaches. Before foraying into the labyrinth of methods, a brief review is presented of some of the features of electronic spectroscopy and the qualitative description of electronically excited states.

2 Spectroscopy of Electronically Excited States

2.1 Qualitative Description of Excited States

The description of electronically excited states begins at essentially the same place as ground states. In principle, one would like to solve for the stationary states of the molecule of interest. The time-independent Schrödinger equation[9] (r and **R** indicating electronic and nuclear coordinates, respectively)

$$H\Psi(r,R)=E\Psi(r,R) \tag{1}$$

$$(-\sum_i^N \frac{h^2\nabla^2}{8\pi^2 M_i} - \sum_i^n \frac{h^2\nabla^2}{8\pi^2 m_e} + \sum_{i<j}^{N,N} \frac{Z_i Z_j e^2}{R_{ij}} + \sum_{i<j}^{n,n} \frac{e^2}{r_{ij}} - \sum_{ij}^{N,n} \frac{Z_i e^2}{r_{ij}^{en}})\Psi(r,R)=E\Psi(r,R) \tag{2}$$

contains the operators for nuclear and electronic kinetic energies (first and second terms), nuclear-nuclear repulsion (third term), electron-electron repulsion (fourth term) and electron nuclear attraction (fifth term). Born and Oppenheimer[10] realized that the great disparity between electronic and nuclear masses suggests an approximate solution to the above Schrödinger equation based on a product wavefunction of the form $\Psi(R,r) = \psi(r;R)\chi(R)$, where $\psi(r;R)$ depends explicitly on electronic coordinates and parametrically on the nuclear positions, while $\chi(R)$ depends only on nuclear coordinates. The approximation is valid when the electronic wavefunction varies little over the range of motion of the nuclei (thus the electrons see a nearly constant nuclear potential). (For cases where the approximation breaks down, see the accompanying chapter on Jahn-Teller interactions.) In order to obtain ψ and χ one solves a pair of coupled Schrödinger-like equations. The equation for ψ is

$$(-\sum_i^n \frac{h^2\nabla^2}{8\pi^2 M_i} + \sum_{i<j}^{N,N} \frac{Z_i Z_j e^2}{R_{ij}} + \sum_{i<j}^{n,n} \frac{e^2}{r_{ij}} - \sum_{ij}^{N,n} \frac{Z_i e^2}{r_{ij}^{en}})\psi(r;R)=E_e\psi(r;R) \tag{3}$$

which yields electronic wavefunctions and energies at fixed nuclear positions. The χ are then obtained by solving

$$(-\sum_i^N \frac{h^2 \nabla^2}{8\pi^2 M_i} + E_e(R))\chi(R) = E\chi(R) \tag{4}$$

yielding nuclear wavefunctions and total energies. Equation 4 demonstrates the nuclear wavefunction's dependence on the electronic energy. In many applications the electronic state of interest is the ground state, but it is clear from Eq. 3 that one could instead solve for excited electronic states at a given nuclear position. Variation of the nuclear positions than maps out not the ground but an excited state potential, and use of this excited state potential in Eq. 4 leads to nuclear wavefunctions (vibration, rotation) appropriate to the excited electronic state. The methods discussed below thus seek to obtain approximate solutions to Eq. 3 for states other than the electronic ground state of the system.

Experimental information about the excited states is most often obtained via spectroscopic investigations. A frequent starting point for discussions of the spectroscopy of these systems is Fermi's Golden Rule.[11] Derived using first-order time-dependent perturbation theory it yields an expression for the rate of transitions from a lower to an upper electronic state induced by incident light

$$k = \frac{4\pi^2}{h} \rho(E_{k0}) |<\Psi_k|H'|\Psi_0>|^2 \tag{5}$$

where $\rho(E_{k0})$ is the energy density of photons at the energy difference between states 0 and k, H' is the perturbation of the molecular energy due to the light (e.g the dipole moment operator ($\mu = \sum_i e r_i$) in the case of dipole allowed transitions) and Ψ_k and Ψ_0 are the excited state and ground state total wavefunctions, respectively. For the case of dipole-allowed transitions the "off-diagonal" dipole moment matrix element in Eq. 5 can be written as

$$<\Psi_k|\mu|\Psi_0> = <\chi_{kj}\psi_k|\mu|\psi_0\chi_{0l}> \tag{6}$$

for the j (electronic state k) and l (electronic state 0) nuclear wavefunctions. Franck and Condon suggested[12] that the inner integral over electronic coordinates (the electronic transition dipole moment) should be weakly dependent on nuclear coordinates (the electronic wavelength is much larger than the average nuclear displacement) leading to an approximate factorization of the above integral for electronic transitions:

$$<\Psi_k|\mu|\Psi_0> = <\chi_{kj}|<\psi_k|\mu|\psi_0>_e|\chi_{0l}>_n \approx <\psi_k|\mu(R_{eq})|\psi_0>_e<\chi_{kj}|\chi_{0l}>_n \qquad (7)$$

where the subscripts e and n indicate integration over electronic nuclear coordinates respectively. Note that the integral over nuclear coordinates (nuclear Franck-Condon factor) is not generally zero for $k{\neq}l$ because the excited electronic potential energy surfaces are usually somewhat displaced from the ground state surface. However, the size of the integral varies considerably with vibrational state and has a substantial effect on the intensity of the transition. Within the Franck-Condon approximation the greatest intensity will arise from the $0l{\rightarrow}kj$ transition having the largest overlap of vibrational wavefunctions. The practical import of the Franck-Condon approximation is that when the intensity maximum occurs at energies other than that of the 0-0 (or "adiabatic") transition (i.e. the transition from the 0 vibrational state of the ground electronic state to the 0 vibrational state of the excited electronic state, which is the lowest energy transition in absorption), one can infer that the geometries are different in the two electronic states. Analysis of the intensity distribution in the electronic spectrum of a molecule can therefore lead to a description of the excited state geometry.

Of course, one hopes to learn more from such calculations than the energy difference between a pair (or several) states at an arbitrary nuclear geometry. The electronic transition moment integral in part determines the intensity of the transition, and is chiefly responsible for the symmetry selection rules observed in the electronic spectra of symmetrical systems.[12] One normally calculates the so-called oscillator strength

$$f^{nm} = \frac{m_e hc^2 v_{nm}}{3he^2} |u_{nm}|^2 \qquad (8)$$

which is proportional to the absorption coefficient integrated over the entire band. It is clear from Eq. 8 that for transitions of similar energy the dominant factor affecting the oscillator strength (and therefore the intensity) is the transition dipole moment integral. Thus, it is common in many theoretical applications to merely quote μ_{nm}.

Other properties are also of interest. For example, significant charge rearrangement upon electronic excitation may lead to changes in dipole moment or polarizability (static and frequency-dependent). Expectation values of x^2, y^2, etc., while not readily measured, can yield useful information about the spatial extent of a state, which can be important in interpreting experimental spectra, and theoretical methods for their calculation have been pursued.

2.2 Qualitative Descriptions of Electronically Excited States

Molecular Orbital Theory. The most common starting point for discussing electronically excited states is qualitative Molecular Orbital (MO) theory. (By qualitative MO theory we mean here a simple "linear combination of atomic orbitals" approach.) In part MO theory's broad use is due to the average chemist's familiarity with it, as well as the success it has had in describing ground states. For diatomic molecules and a host of carbon containing compounds MO theory affords a simple means of rationalizing bond strength and bond length trends and one might assume similar success would be found for excited states. The results presented below show that this is a good assumption some of the time, but that the vast majority of electronically excited states do not fit into the neat category of "valence excited states." Having said which, when such valence excited states are low-lying they tend to be the most intense transitions, and at least some $\pi \rightarrow \pi^*$ transitions and $n \rightarrow \pi^*$ are well described in simple valence MO terms.[13-15] On the other hand, for the molecules considered here low-lying transitions into orbitals of σ^* symmetry are almost entirely non-valence-like. The origin and character of these states are considered in the following section.

In MO theory one combines atomic orbitals, localized on the various atomic centers in the molecule, into molecular orbitals of proper symmetry. The simplest such example occurs for ethylene. Mixing the π-like atomic orbitals on the left and right carbon atoms leads to bonding and anti-bonding orbitals

$$\pi = (p_l + p_r)/\sqrt{2}$$
$$\pi^* = (p_l - p_r)/\sqrt{2}$$

(9)

The ground state description of ethylene involves placing two electrons in the lowest (bonding) orbital. One predicts net overall bonding character and that the π molecular orbital will have similar spatial extent to the atomic orbitals which compose it. All of these predictions are borne out by more detailed calculations on the ground state of ethylene.

What about the excited states of ethylene? In MO theory, at least in the manifold of π excited states, it seems natural to assume that the lowest electronic excitation occurs from the π to the π^* orbital. In the simplest theories the splitting between the π MOs would be twice the "resonance" integral for the atomic p orbitals. When spin is taken into account two such states arise, one with overall triplet spin symmetry, the other a singlet. For planar ethylene, the triplet is thus lower in energy by twice the exchange interaction. However, for either the singlet or the triplet the spatial extent of the π^* orbital is expected to be similar to that of the π orbital, since it is formed from the same two "basis functions," p_l and p_r.

Within MO theory another class of ethylene excited states could be obtained from excitations of one or more of the π electrons into σ^* orbitals, and still other states could be obtained from excitations of the σ electrons into various anti-bonding orbitals. For excitations into orbitals formed from linear combinations of atomic valence AOs the spatial extents of the excited states are predicted to again be similar to that of the ground state.

One can easily extend this picture to other molecules in the series to be discussed below. We do so for here for butadiene to illustrate a second interesting feature of the polyenes, namely that of multi-configurational zeroth-order states.

The π orbitals of butadiene in the zero-overlap extended Hückel approximation are given by[16]

$$\psi_4 = 0.37\phi_1 - 0.60\phi_2 + 0.60\phi_3 - 0.37\phi_4$$
$$\psi_3 = 0.60\phi_1 - 0.37\phi_2 - 0.37\phi_3 + 0.60\phi_4$$
$$\psi_2 = 0.60\phi_1 + 0.37\phi_2 - 0.37\phi_3 - 0.60\phi_4$$
$$\psi_1 = 0.37\phi_1 + 0.60\phi_2 + 0.60\phi_3 + 0.37\phi_4$$

(10)

with energies

$$E_4 = \alpha - 1.62\beta$$
$$E_3 = \alpha - 0.62\beta$$
$$E_2 = \alpha + 0.62\beta$$
$$E_1 = \alpha + 1.62\beta$$

(11)

α and β (both less than zero) are the atomic orbital energies and resonance integrals between adjacent carbon atoms, respectively. The ground electronic state is obtained by placing the four π electrons in the two lowest energy orbitals (1 and 2) leading to net bonding for the π electrons. The lowest energy excited state is again expected to arise from excitation of an electron from the π HOMO to the π LUMO. As in ethylene, two such states are expected, due to triplet and singlet spin couplings, the triplet being lowest in energy. Either of these states is a "single excitation" both in electronic terms (one electron excited) and orbital terms (the electronic excitation is to the next highest energy orbital). In C_{2h} symmetry (the point group of *trans*-1,3-butadiene) the above orbitals are of a_u (1 and 3) and b_g (2 and 4) symmetry, making the ground state A_g symmetry and the two singly-excited states B_u (^3B_u and ^1B_u) symmetry.

A higher-lying $\pi \rightarrow \pi^*$ configuration could be obtained by exciting an electron from ψ_2 to ψ_4 (Φ_1). This configuration has the same overall symmetry as the ground state (A_g). Based on the simple orbital energies this is expected to be higher in energy than the single excitation discussed above. A distinct configuration of identical energy could be obtained by exciting from ψ_1 to ψ_3, again having A_g symmetry (Φ_2). In addition, for a state of singlet spin symmetry, one can generate another A_g configuration of similar energy by exciting both electrons in the ψ_2 orbital into the ψ_3 orbital (Φ_3). A more careful treatment than this one-electron Hückel model shows that these configurations actually have non-zero off-diagonal Hamiltonian matrix elements connecting them (analogous to the resonance integrals coupling orbitals in a one-electron model). Since they are of comparable energy one expects that they will interact and split, in complete analogy with the way atomic orbitals mix to yield molecular orbitals. The net result is that the lowest energy excited state of A_g symmetry is a mixture of all three excited A_g "configurations" of the form

$$\Psi(2^1A_g) = c_1\Phi_1 + c_2\Phi_2 + c_3\Phi_3 \tag{12}$$

with c_1, c_2 and c_3 of comparable size. For the ^3A_g state c_3 must be zero, but one again expects c_1 and c_2 to be of comparable size. The important point to make here is that in some cases one may find states that are *qualitatively* described as a sum of two or more electronic configurations, quite apart from any electron correlation effects ("dynamical correlation", see below). That is, at lowest order such states must be written as a sum of configurations. The reason one might not expect a similar "multi-configurational" description in the case of the B_u states is that other B_u symmetry excited configurations are much higher energy than the HOMO to LUMO excitation.

Of even greater interest than the fact that the A_g configurations interact and split is the question of the strength of their interaction. Resorting again to the analogy made with the bonding of atomic orbitals, when a pair of orbitals interacts, the new MOs turn out to be mixtures of the constituent atomic orbitals, at least one of which is lower in energy than the AOs. In like fashion, one also expects that the lowest energy ^1A_g excited state is multi-configurational, as in Eq. 12, but in addition is lower in energy than any of the constituent "zeroth-order" configurations out of which it is composed. A great deal of experimental and theoretical effort has been expended in attempting to answer "how much lower," especially in relation to the HOMO to LUMO (B_u) transition. While the answer is not known conclusively for butadiene and hexatriene, for octatetraene and higher polyenes this 2^1A_g state is actually lower than the ^1B_u state, at least in the 0-0 sense (see below).

Note that multi-configurational excited states occur quite naturally in many cyclic unsaturated systems (cyclobutadiene, benzene, etc.) due to symmetry. Two or more zeroth-order configurations are often exactly degenerate at points of high symmetry and mix in equal amounts. The case discussed above is different only in that the degeneracy of the configurations is not exact in all cases. The origin of the mixing is the same in either case.

Based on MO theory then, we might expect a variety of valence-like transitions of singlet and triplet character, and that some of the states will be multi-configurational even at the lowest possible level of description, due to near-degeneracies of low-lying configurations.

Valence Bond Description. A complimentary description of the low-lying states of simple systems is obtained using the valence-bond theory introduced by Pauling.[17] While somewhat less familiar than MO theory, it gives useful insights into the physical nature of some of the low-lying states of the polyenes and sheds light on the failures of many theoretical approaches used to describe them.

The valence bond treatment for the π electrons of ethylene forms a covalent bond from the left and right p_π orbitals, leading to (neglecting overlap of the atomic orbitals on different centers)

$$\Psi_{gs} = (p_l p_r + p_r p_l)(\alpha\beta - \beta\alpha)/2 \tag{13}$$

Electron number is implied by the order of the functions. Ψ_{gs} is, as it was in the MO description, of ^1A_g symmetry. This state is described as "covalent" since each atom in the bond retains one electron in the wavefunction for this state.

Excited states within the same set of atomic p orbitals are obtained via different "spin-couplings" and/or atomic orbital occupations. For example, an excited state of triplet spin symmetry is obtained from the wavefunctions

$$(p_l p_r - p_r p_l)\alpha\alpha/\sqrt{2}$$
$$(p_l p_r - p_r p_l)(\alpha\beta - \beta\alpha)/2 \tag{14}$$
$$(p_l p_r - p_r p_l)\beta\beta/\sqrt{2}$$

This excited state is of $^3B_{1u}$ symmetry and is covalent.

One can also form "ionic" states, by doubly occupying atomic orbitals. The lowest $^1B_{1u}$ state of ethylene is described in valence bond terms as

$$(p_lp_l - p_rp_r)(\alpha\beta - \beta\alpha)/2 \tag{15}$$

This state is qualitatively different from the previous two, in that the wavefunction involves a resonance between components that have both π electrons on one center. These ionic configurations would appear to be somewhat higher in energy than the covalent states; the ionic states can be thought of as crudely involving ionization of one carbon and electron attachment to the other. Of course, one could imagine lowering the energy of such configurations by shifting the σ electrons in response to the placement of π charge. That is, if the σ electrons shifted in the opposite direction to the π electrons, an overall energy lowering of the ionic configurations would be achieved. Note that the covalent states, at least to first order, require no such readjustment. In addition, the σ electrons cannot move in only a single direction (right or left) due to the symmetry of ethylene. In fact, their motion must be dynamic. A number of authors have discussed the need to "correlate" the σ and π electrons in such ionic states (see below), but a compact description has been recently presented by Borden and Davidson.[18] They show that to a first approximation the necessary $\sigma\pi$ correlation for the ethylene $^1B_{1u}$ state can be described by double excitations involving one σ electron and one π electron. While numerical calculations are not commonly performed using VB wavefunctions as starting points, it is nevertheless the VB representation that allows one to understand the qualitative difference between the ground state and such "ionic" states.

The polyenes can also be described in VB terms.[19] The most important points that arise are that in all cases the singlet $\pi \rightarrow \pi^*$ state is ionic in character, and that the 2^1A_g state can be described as a state with pairs of locally excited triplet ethylenes, spin-coupled to yield an overall singlet state. This is particularly useful in understanding the large geometrical rearrangement in the 2^1A_g state relative to the ground state.

Of course, as in the MO theory case, we have used a minimal atomic basis to describe the low-lying electronic states. One should ask whether this is such a severe approximation that one has forced an inaccurate picture on the electronically excited states. It is natural and most likely qualitatively accurate to assume that the bonding MOs in a molecule will be of similar character to those of their (ground state) atomic constituents. However, electronic excitations increase the energy in the molecule by from 3 to 10 eV. In atoms, excitation from the ground state to higher lying states lead to qualitatively different (larger spatial extent) electronic states. It should not be a surprise then that molecular excited states are often not well-described by a simple valence orbital picture. In fact, many molecular states are not at all valence-like, and this class of states is discussed in the following section.

Rydberg Excited States. The well known non-relativistic energy expression for the energy levels of the hydrogen atom is of the form[11]

$$E_n = -\frac{Ry}{n^2} = -\frac{13.6eV}{n^2} \quad n=1,2,3... \tag{16}$$

with Ry the "Rydberg constant." One notes that as n increases the energy levels quickly approach the ionization limit, and that the energies are independent of angular momentum (e.g. for $n=3$ the s, p, and d levels are degenerate in the non-relativistic limit). The average radial distance from the nucleus has also been worked out for hydrogen atom and is given by[20]

$$<r>_{nl} = \frac{1}{2}a_o(3n^2 - l(l+1)) \tag{17}$$

where a_o is the Bohr radius (=0.529Å). For the $n=3$ level for example, one finds that the energy is only 1.5 eV below the ionization limit, and the average radial distance from the nucleus is 7.1Å, 6.6Å, and 5.6Å for the 3s, 3p, and 3d levels, respectively. One can imagine that, were the nucleus distorted somewhat the effect might be to split the $n=3$ levels to some extent, but that given the average distance of the electron from the nucleus, elongation of even an Angstrom or two might not have a strong effect on the overall spectrum of the atom. That is, there would still be a series of ns levels, np levels, etc. following essentially the energy expression given in Eq. 16.

It turns out the an entire class of *molecular* excited states can be thought of in analogy to these hydrogenic states. Imagine the creation of a high-lying excited state by first ionizing the molecule and then restoring the electron from a large distance to a higher-lying orbital. At large distances the ionized molecule resembles a distorted positive point charge. (Even butadiene is only about 3.5Å from end carbon to end carbon, still small relative to the radius of the 3s orbital, and certainly modest compared to the radii of higher n hydrogenic-like orbitals.) Thus, at energies not far from the ionization energy of the molecule one might expect electronic orbitals of similar spatial extent and shape to hydrogenic orbitals. Molecular excited states could be formed by excitation of an electron from a valence orbital into these "diffuse" orbitals. These diffuse orbitals are commonly called "Rydberg orbitals," and the molecular excited states so formed are referred to as "Rydberg states."

Of course, the molecular core in such Rydberg states is not a point charge, and is certainly not spherically symmetric. Thus, deviations from the exact form of Eq. 16 are expected for their energies. From a spectroscopic point of view, it is natural to think about the excitation energy of these states from the molecular ground state, but for purposes of reinforcing the analogy with the hydrogenic states, it is common to quote the "term value" (*TV*) of such states defined as

$$TV_{nlm} = IP_j - E_{o,nlm} = \frac{Ry}{(n - \delta_{lm})^2} \tag{18}$$

where IP_j is the ionization energy of the jth MO and $E_{o,nlm}$ is the excitation energy from MO j to the Rydberg orbital with "quantum numbers" n, l, and m (to be taken loosely since the potential is not spherical). TV_{nlm} is thus the ionization energy of the Rydberg orbital to which the excitation has taken place. The "quantum defect," δ_{lm}, takes account of a) the effect of the non-spherical molecular potential on the energies of the Rydberg orbitals, and b) the differences in overlap of the given Rydberg orbital with core orbitals.[13-15] In case b), one expects that for a given l, if the molecule has orbitals of similar angular characteristics occupied in the core, the Rydberg orbital will overlap these (to a small extent, considering its diffuseness) and upon becoming orthogonal to the core, will pick up some small amount of core-like character.[13-15] This character leads to somewhat greater nuclear interaction, and lowered energy, relative to orbitals that do not have "precursors" in the core. Thus, it is generally expected that δ_{lm} decreases as l increases, and qualitative rules suggest that δ will be approximately 1.0, 0.6, and 0.1 for s, p, and d-like Rydberg orbitals respectively.[13-15] (However, these values can vary widely depending on how much valence-Rydberg mixing occurs in a given state.)

Given that the s and p quantum defects are appreciable, it seen from Eq. 18 that the $n=3$ Rydberg orbitals lie from 2 to 3 eV below the IP of the molecule. For molecules with low IPs this will mean that the lowest Rydberg states lie in the same region (or perhaps below) valence excitations. It should be noted that one could, of course, apply Eq. 18 to a valence-like transition, defining TV for it as well. One point that would distinguish the valence state from Rydberg states is that one would not expect to find another valence state at the energy obtained by increasing n by one in Eq. 18 however.

A second distinguishing feature of Rydberg states is their low intensity. While valence transitions can be quite intense (f values up to 1), Rydberg transitions are much less intense, and Robin has suggested an upper limit of 0.08 per degree of spatial degeneracy.[13-15] Purely Rydberg transitions also tend to have a sharpness similar to that seen for atomic transitions, whereas valence transitions are frequently broader. Finally, because of the large spatial extent, Rydberg transitions are easily perturbed, so much so that in most cases they disappear in condensed phases. Thus, a variety of diagnostics exist for distinguishing valence and Rydberg states, in addition to the theoretical diagnostic of the calculation of the average spatial extent of the state in question.

Of course, what has been described up to this point are the two limits of orbital type: valence- and Rydberg-like. It will come as no surprise that one can find intermediate states, where the excited state is of mixed character as well. Robin[13-15] discusses these cases and experimental evidence for such mixing. The polyenes turn out to have states that are excellent examples of valence-Rydberg mixing. In general one expects mixing when the two types of states occur near one another in energy (at "zeroth-order"), and this will more likely occur for molecules with relatively low ionization potentials. The question of the extent to which one can find vestiges of a valence-like transition, mixed in the dense manifold of Rydberg states has also been addressed.[13-15] In the case of transitions to valence-like σ^* states, it is more generally the case that these are either completely lost in the manifold of Rydberg states or observed only as autoionizing states (above the first ionization potential). The common exceptions to this rule are transitions to σ^* orbitals in bonds containing halides.[13-15] In these cases the σ and σ^* orbitals are lowered sufficiently to obtain valence-like excited states with σ^* character.

3 Theoretical Methods for the Treatment of Excited States

Single-Reference-Based Methods

3.1 Hartree-Fock methods

The starting point for almost all *ab initio* methods is some form of "self-consistent field" (SCF) wavefunction. When the SCF wavefunction is a single determinantal function one is using a Hartree-Fock description. In all such calculations one needs to choose a "one-electron basis" set in which to work. Conventional basis sets such as those from the Pople group[21](6-311G** etc.) or Dunning group[22] have typically been optimized to describe ground state systems. Rydberg states will require augmentation of these basis sets with quite diffuse basis functions, and even "valence" excited states may be somewhat more diffuse than those appropriate for valence ground states. In addition, although the energy of

Rydberg states may converge reasonably rapidly, the spatial extent can require extensive basis set augmentation. Polarization functions are required to treat energy differences accurately, as they are in calculations on ground states.

Once basis set limitations have been addressed, the question of which states can be explicitly treated at the SCF level arises. The HF wavefunction is a non-linear variational function,[23] and it is generally assumed that the variational principle only guarantees that the SCF energy is an upper bound to the lowest state of a given spatial and spin symmetry combination. Most electronic structure packages do not provide the option of solving for single-configurational SCF descriptions of excited states within a given symmetry, but in many cases such solutions can be obtained.[6,7] Given the ubiquity of multi-configurational SCF approaches and their ability to variationally treat excited states within a given symmetry this is not a great concern however. In fact, since at least some molecular electronically excited states are truly multi-configurational in character the restriction to a single-configurational description for excited states is generally too severe.

However, when the single-configurational SCF description is at least qualitatively appropriate, what are the advantages and drawbacks of this approach? The chief advantage is speed, since all of the methods to be discussed below are significantly slower than HF. Three real disadvantages exist however. First, since one would like to treat excited state potential surfaces, one needs to be able to treat regions of lowered symmetry. As mentioned above most HF implementations cannot treat excited states within a given symmetry and thus points of lowered symmetry can present problems. Second, the correlation energy (the difference between the HF or "mean-field" energy and the exact non-relativistic energy (full configuration interaction)) is generally larger for the ground state than for the excited state (since the excited electron is in a different region of space from the electron in the orbital out of which it originally came). Thus SCF excitation energies tend to be too low, often by on the order of 1 eV. Finally, since valence excited states will tend to have higher correlation energy than Rydberg states (the Rydberg electron is far removed from the molecule on average) the neglect of correlation in the HF calculation biases towards obtaining low-lying Rydberg states. While in many molecules Rydberg states may indeed be lowest-lying, it is a bias that one must be aware of in applications of HF wavefunctions to excited states. The "configuration interaction singles" approach[24] (see below) is a method not significantly more demanding than HF and, through a fortuitous cancellation of errors, manages to at times remedy some of these failures.

3.2 Full CI

The error in SCF approaches occurs because electron-electron repulsion is treated only in an average sense, rather than in a "correlated" or "dynamical" fashion. The error in the energy in the SCF description is called the "correlation error" ($\Delta E_{corr} = E_{exact} - E_{SCF}$). One type of wavefunction that includes dynamical correlation starts with the SCF solution for the given state, adding "excited determinants" (singly and higher excited configurations, taking one or more electrons from occupied orbitals, placing them in orbitals unoccupied in the SCF configuration (virtual orbitals)), each having a variable linear coefficient in the expansion of the total wavefunction. Application of the variational principle to the expectation value of the Hamiltonian obtained from this "configuration interaction" (CI) wavefunction leads to a matrix equation for the energies and the coefficients in the wavefunction. The Hylleraas-Undheim-MacDonald theorem[25-27] guarantees that not only is the ground state energy an upper bound to the true ground state energy, but that excited state energies are upper bounds to the successive excited state energies. When one constructs such a CI wavefunction out of all possible excitations (up to n-tuple for an n-electron system) into all possible virtual orbitals one has constructed a full-CI wavefunction. The wavefunction is then of the form

$$\Psi^{FullCI} = \psi_0 + \sum_i c_i^s \psi_i^s + \sum_j C_j^d \psi_j^d + \dots + \sum_m c_m^{n-tuple} \psi_m^{n-tuple} \qquad (19)$$

where ψ_0 is the zeroth-order configuration from which one generates excited determinants, and the ψ^s, ψ^d, on up to the $\psi^{n-tuple}$ represent the single, doubly, ... n-tupley excited determinants. Since all possible configurations (consistent with the given spin and spatial symmetry) are included in the full CI, the choice of ψ_0 is itself irrelevant (within the given space and spin symmetry). This is not true of any other method to be discussed here. Furthermore, since there is no greater variational flexibility in the given one-electron basis than the full CI wavefunction, one obtains "exact" (within the one-electron basis chosen) energies and energy differences based on the full CI.

Upon forming the expectation value of the Hamiltonian and applying the variational principle one obtains a set of coupled equations for the coefficients which also determine the energy eigenvalues. In matrix form, grouping all excitations of a given class into a single block one obtains

$$\begin{pmatrix} H_{00}-E & H_{0s} & H_{0d} & 0 & \cdots & 0 \\ H_{s0} & H_{ss}-E & H_{sd} & H_{st} & \cdots & 0 \\ H_{d0} & H_{ds} & H_{dd}-E & H_{dt} & \cdots & 0 \\ 0 & H_{ts} & H_{td} & H_{tt}-E & \cdots & 0 \\ \vdots & \vdots & \vdots & \vdots & \vdots & \vdots \\ 0 & 0 & 0 & 0 & \cdots & H_{nn}-E \end{pmatrix} \begin{pmatrix} c_0 \\ c^s \\ c^d \\ c^t \\ \vdots \\ c^n \end{pmatrix} = 0 \qquad (20)$$

(where s=single excitations relative to ψ_0, d=double excitations, etc.) Since the Hamiltonian contains at most two-electron operators, classes of configurations separated by more than double excitations from one another have zero Hamiltonian matrix elements between them. Furthermore, Brillouin's theorem[23] states that if ψ_0 is a Hartree-Fock wavefunction the block H_{0s} will be zero. This does not mean that c^s will be zero, but it is expected to be small when the HF description is qualitatively accurate.

Unfortunately, the full CI becomes computationally intractable (grows exponentially) as basis functions and electrons are added. Its growth is so rapid that for systems of even modest size, calculations in good quality basis sets are technically beyond reach. Thus, methods are required that yield accurate approximations to the full CI result (the "right" answer in the given basis) but which are computationally tractable.

3.2 Truncated CI

If one cannot perform the full CI because it grows too rapidly, perhaps reasonably accurate results can be obtained using only a subset of all excited configurations. A hierarchy of such methods suggests itself ordered by level of excitation, starting at CI singles (CIS), CI singles and doubles (CISD), CI singles, doubles and triples (CISDT), etc. One can perform calculations of this type using the ground state HF wavefunction (this is the common choice in the CIS approach) or using independently optimized HF wavefunctions for each state of interest when they can be obtained. Not surprisingly, the work rises rapidly as classes of configurations are included in the calculation, such that CISDTQ has been the largest CI-type wavefunction to see practical application,[28] with CISD being much more common. Considering CISD as an example, the CISD equations become

$$\begin{pmatrix} H_{00}-E & H_{0s} & H_{0d} \\ H_{s0} & H_{ss}-E & H_{sd} \\ H_{d0} & H_{ds} & H_{dd}-E \end{pmatrix} \begin{pmatrix} c_0 \\ c^s \\ c^d \end{pmatrix} = 0 \qquad (21)$$

Truncation of the full CI equation at a given order is akin to assuming that the excitations neglected have no effect on the correlation energy of the state $(E-H_{00})$ or the coefficients determining it (c^s and c^d). While the former is true in one sense (one only needs the *exact* c^s and c^d to determine E, even in a full CI calculation), it is nevertheless clear from Eq. 20 that c^s and c^d *are* determined, in part, by the interactions of the single and double excitations with higher-order excitations (*via* H_{st}, H_{dt}, and H_{dq}). It will not be a surprise then that truncated CI at any order is at best an approximation to the full CI result. While there is no rigorous means of defining why a given configuration is important or not important in the wavefunction, some general rules serve in guiding one's thinking concerning classes of configurations. For example, single excitations can be thought of as providing orbital shape changes, double excitations provide correlation for electron pairs, triple excitations serve to correlate singly excited configurations, and so on. Thus higher-than-double excitations are expected to have a non-negligible effect on the quantitative description of a state.

The quality of results obtained with the various truncated CI methods varies to a large extent with the types of states studied. Any single-reference CI other that full CI will do a poor job on states that are inherently multi-configurational at zeroth-order (since not all zeroth-order configurations are treated on equal footing; only ψ_0 is correlated). Confining consideration to excited states that are single-configurational at zeroth-order however, one can make some general comments. CI singles has been extensively used recently.[29-31] It allows excited state geometry optimizations,[24] is "size-consistent" (see below) and in many cases gives remarkably good excitation energies, even in at least one case where valence-Rydberg mixing is important.[31] It appears this accuracy arises in large part from a cancellation of errors; the ground state is biased against due to neglect of correlation, but the excited state is biased against by neglect of orbital relaxation (essentially the ground orbitals are used to describe the excited state). These two errors cancel to a reasonable degree in many cases, yielding better excitation energies than one would obtain at the HF level, (where only the error associated with neglect of correlation is made, leading to excitation energies that are too small).

On the other hand, the CISD approach, when used with the ground state HF wavefunction as ψ_0, is expected to do worse than CIS for excitation energies if all states are obtained as the eigenfunctions of the same CI matrix. In this case double excitations act to correlate ψ_0, but not the excited determinants (largely composed of the single and double excitations, which require triple and quadruple excitations relative to ψ_0 for correlation). The ground state is now artificially lower than the excited states leading to excitation energies that are too large. Better results are obtained with CISD when ψ_0 is chosen (when possible) as the HF wavefunction appropriate to the state of interest. Two separate calculations, based on two separate CI matrices are required, but given the balanced description of the two states one expects (and usually finds) better results for excitation energies.

For truncated CI beyond CIS, an additional problem known as "size-inconsistency" arises, due to the neglect of higher order excitations. In essence, CISD (and CISDT, and higher order non-full CI energies) do not scale linearly with system size. A simple system in which to investigate this behavior is a collection of non-interacting (infinitely separated) He atoms, as originally discussed by Davidson.[32] For simplicity we neglect single excitations, in a double zeta basis, such that the exact CID wavefunction for a single He atom is $c_0\Psi_0 + c_1\Psi_1$, where Ψ_1 is the doubly excited configuration relative to the HF wavefunction, Ψ_0. Since there are only two configurations in the wavefunction, the equation for c_1 becomes (assuming "intermediate normalization" i.e. $c_0=1$) (cf. Eq. 21)

$$c_1 = \frac{H_{01}}{E - H_{11}} = \frac{H_{01}}{E_{corr} - (H_{11} - H_{00})} \tag{22}$$

where $E_{corr} = E - H_{00}$. The total correlation energy is itself dependent on c_1, such that (again, cf. Eq. 21) $E_{corr} = c_1 H_{01}$ or

$$E_{corr} = \frac{|H_{01}|^2}{E_{corr} - (H_{11} - H_{00})} \tag{23}$$

If one now treats two He atoms at the CID level, two doubly excited configurations arise, one for each He. (The double excitations arising from a single excitation at each He can be neglected since the He are well-separated; the He matrix elements involving such cross-excitations are zero.) The expression for each c_1 is identical to that found in Eq. 22 (the He do not interact). The difference $H_{11} - H_{00}$ is the same as in the one He case. Thus, the only change in c_1 arises from the fact that E_{corr} is larger - the correlation error made in employing a RHF description for two He atoms should be twice as large as that for one He. As a result, the denominator in Eq. 22

is somewhat larger for two He than for one, c_1 is somewhat smaller for each double excitation than in the one-He case, and the total correlation energy, is somewhat smaller than that obtained from $2xE_{corr}(1-He)$. In short, the CID energy for two non-interacting He atoms is not twice the CID energy for a single He. As more He are added the problem will grow (as E_{corr} grows), and the overall energy will not scale linearly with system size; this is the origin of the "size-inconsistency" problem for truncated CI.

Of course, the proper analogue of the CID wavefunction for one He is the CIDQ (quadruple) wavefunction for two He, where the coefficient of the relevant quadruple excitation is the square of the one-He double excitation coefficient. This suggests that one can eliminate size-inconsistency effects by including higher excitations, and for small systems this is true. Inclusion of quadruple excitations in a CI description of small molecules does yield results in good agreement with full-CI. However, the work scales as N^8 rather than N^6 (as in CISD). Even if a CISDTQ calculation could be done for larger systems, one would merely be putting off the onset of size-inconsistency. Similar reasoning to that used above for the two-He case would show that the quadruple excitation coefficients (and all lower excitations) must deteriorate with system size, due to neglect of still higher excitations.

At first glance a discussion of size-inconsistency (improper scaling with system size) seems to have little to do with the accurate calculation of electronic excitation energies. Molecular ground and excited states have identical numbers of electrons and in no sense is the "number" of molecules altered upon electronic excitation. However, energy scaling with particle number is only a convenient means of observing a more general phenomenon, that is the correlation energy-based scaling of the CI coefficients. If one considers two states with differing correlation energies (for example, one is ionic, the other covalent) and examines the CISD coefficient for a double excitation not directly involved in the description of the transition of interest (e.g. correlation of core electrons), the coefficient will be different in the two states. Part of this arises because the electronic environment is different - however, a second difference arises due to the correlation-energy-dependent form of Eq. 22. Thus, when the correlation energy is bigger for a state, one expects a diminution of the contribution of all excitations when truncated CI is employed. Since this effect is really correlation-energy-dependent and not electron-number-dependent, states with larger correlation energy will be artificially raised in energy, and relative to the ground electronic state their associated excitation energies will be too high.

A perturbation theory analysis[33,34] suggests an approximate means of correcting the CISD energy for size-inconsistency effects. Several authors[32-36] have developed versions of such a size-inconsistency correction, but the first and most commonly used such correction is due to Davidson.[33,34] Analyzing the He case one can show that the principle contribution of quadruple excitations to the perturbation theory expression for the correlation energy is to cancel so-called "unlinked cluster" terms at fourth order, which are primarily responsible for the improper scaling discussed above. In the perturbation theory treatment the contribution from quadruple excitations in the non-interacting case is of the form

$$\Delta E_{SC} = \sum_i c_{i1}^2 E_{corr}(2) \qquad (24)$$

where ΔE_{SC} is the correction due to quadruple excitations which restores size-consistency and $E_{corr}(2)$ is the second order perturbation theory estimate of the correlation energy. Reasoning that a similar correction would arise in CI were quadruple excitations included, Davidson approximated $E_{corr}(2)$ with the CISD E_{corr} and the perturbation theory value of $\sum c_{i1}^2$ with the CISD value $(1-c_0^2)$, to obtain[33,34]

$$\Delta E_{SC} \approx (1-c_0^2) E_{corr}^{CISD} \qquad (25)$$

Later refinements of Eq. 25 treat the question of normalization more carefully and arrive at expressions better suited to use in systems with larger numbers of electrons, (e.g. N>10),[32,35,36] but the basis for all such treatments is the same - using a perturbation theory analysis of the effects of higher excitations to approximately cancel unlinked cluster effects in the CISD energy. In all cases the input to such corrections are the (size-inconsistent) CISD energies and coefficients, and as such are *a posteriori* attempts to correct the size-inconsistency problem in truncated CI. In following sections we discuss *a priori* attempts to correct truncated CI for size-inconsistency. The advantage of such methods is that they should not deteriorate significantly as system size increases, as Eq. 25 and its variants must (since their "input" is size-inconsistent). The "cost" of the *a. priori* approaches is that one sacrifices the variational upper bound property of CI, and can run the risk of collapse to lower states in the description of a particular excited state. While this is not a settled issue for the description of all excited states, sufficiently flexible approaches do exist that allow one to safeguard against variational collapse, and these will be discussed below.

It should also be noted that while the "Davidson Correction" (Eq. 25) approximately corrects CISD energies, it does not correct CI-based molecular properties for size-inconsistency. Thus, if size-inconsistency is an important limitation on the accuracy of a given property (e.g. dipole moment, transition moments), expectation values of the operator will still scale improperly with system size when a post-CI size-consistency correction is used for the energy.

3.3 Perturbation Theory Approaches.

Perturbation Theory (PT) offers another means to approximate the full CI results. While PT is not "variational," many perturbation theories are accurate enough so that the loss of the upper-bound property for a given state's energy is not a sacrifice. Most perturbation theories of general use for excited states are multi-reference in nature rather than single-reference. Here a pair of single reference perturbation theories are outlined for purposes of comparison with truncated CI, and different choices for zeroth-order Hamiltonians are discussed. In a following section they are generalized to the more important multi-reference approaches.

The full CI equations can be rewritten using "Löwdin Partitioning Theory"[37] as

$$H_{00} + H_{0e}(E - H_{ee})^{-1}H_{e0} = E$$
$$C_e = (E - H_{ee})^{-1}H_{e0}$$
(26)

Note that in Eq. 26 H_{0e} is a column vector, containing the coupling elements between the zeroth-order configuration and all other configurations, and $(E-H_{ee})^{-1}$ is the inverse of the matrix obtained by deleting the first row and column of the full CI matrix. Solving Eq. 26 is identical to solving Eq. 20. Any truncated CI approximation is obtained merely by truncating $E-H_{ee}$ and H_{0e} to span only the order of excitation desired.

Equation 26 seems to avoid the hard work of diagonalization involved in Eq. 20 or 21, but in fact entails the same work, recast in the need to obtain the inverse matrix $(E-H_{ee})^{-1}$. One can view use of perturbation theory as an attempt to avoid the inversion of $(E-H_{ee})$, approximating it and expanding the energy and wavefunction in an order-by-order fashion. A computationally simple approximation for $(E-H_{ee})$ is to take it as diagonal (although perturbation theories exist which employ non-diagonal forms), since diagonal matrices are particularly simple to invert. $(E-H_{ee})$ is relatively sparse, motivating such a choice. Thus, in perturbation theory one separates H into a zeroth-order part and a perturbation, i.e. $H = H^0 + V$ (V being the perturbation). In the order-by-order expansion $E = E^0 + E^1 + E^2 + ...$ Performing a double perturbation expansion of $(E-H_{ee})^{-1}$ leads to[37]

$$(E-H_{ee})^{-1}=(E^0-H^0_{ee})^{-1}$$
$$-(E^0-H^0_{ee})^{-1}(\Delta E-V)(E^0-H^0_{ee})^{-1} \tag{27}$$
$$+(E^0-H^0_{ee})^{-1}(\Delta E-V)(E^0-H^0_{ee})^{-1}(\Delta E-V)(E^0-H^0_{ee})^{-1}-...$$

Retaining terms order-by-order leads to the familiar Rayleigh-Schrödinger perturbation expressions:

$$E^0=H_{00}$$
$$E^1=V_{00}$$
$$E^2=H_{0e}(E^0-H^0_{ee})^{-1}H_{e0} \tag{28}$$
$$E^3=E^1H_{0e}(E^0-H^0_{ee})^{-2}H_{e0}-H_{0e}(E^0-H^0_{ee})^{-1}V_{ee}(E^0-H^0_{ee})^{-1}H_{e0}$$
$$E^4=....$$

Commonly used perturbation theories differ largely in the form chosen for H^0 and V.

Møller Plesset (MP) perturbation theory[3] chooses H^0 as the sum of the Fock-operators for the canonical Hartree-Fock orbitals. Any RHF or UHF single determinantal wavefunction is automatically an eigenfunction of H^0 with its associated eigenvalue being the sum of the one-electron orbital energies. $(E^0-H_{ee}{}^0)$ is a diagonal matrix and the difference $E^0 - (H_{ee}{}^0)_{ii}$ (for the ith configuration) equals the difference in orbital energies between orbitals that are vacated and occupied in forming configuration i from ψ_0. V is the difference between the exact Hamiltonian H and H^0, and for configurations that are double excitations relative to each other this difference reduces to a two-electron repulsion term. Møller-Plesset perturbation theories based on single-reference RHF or UHF determinants are size-consistent[8] order-by-order when all contributing higher-excitations are included at a given order.

Another choice for H^0 and V is found in Epstein-Nesbet (EN) perturbation theory.[38-40] Here H^0 is taken as the diagonal elements of the CI matrix, V comprises all off-diagonal terms. In this case $E^1=V_{00}=0$, and once again, for configurations differing by double excitations the elements of V are two-electron repulsion terms.

Other variants have been developed,[41] but the spirit is similar to that discussed above. The distinct advantage of PT is that it avoids diagonalization or inversion of large matrices, leading to theories that at second order are significantly faster than e.g CISD or CCSD (see below). Second order PT is generally not considered as accurate as CISD when size-inconsistency is not a problem, or CCSD (like CISD, another "infinite order" method), but where such infinite order approaches cannot be applied due to size limitations, second order PT usually yields the dominant correlation correction to the energy, especially when ψ_0 is a reasonably accurate zeroth-order description of the state.

In recent applications, a MP partitioning has been the more common one, in large part because at low-orders the energies tend to be closer to "exact" values than are EN results. They are also unaffected by orbital rotations in the virtual space, unlike EN results.[42]

3.4 Coupled-Cluster Theory

Based on ideas originally developed in the nuclear physics community and introduced to quantum chemistry by Cizek and Paldus,[43] coupled cluster theory applications were pioneered in the Bartlett[44] and Pople[45] groups, with extensive further application and development pursued in the Bartlett group.[8] In recent years a variety of coupled-cluster theories have been proposed for the treatment of electronically excited states. A few have been applied to the molecules of interest here, and the idea behind the excited state approaches will be discussed below. Here the single-reference theory is sketched, in order to differentiate it from the CI and perturbative approaches discussed above.

A motivation for the coupled cluster wavefunction can be found in an analysis of the separated He problem. The CISD wavefunction failed because it neglected quadruple and higher excitations, while the correlation energy increased with system size. Given the separated nature of the electron pairs, it is obvious that quadruple excitations should be formed as products of double excitation coefficients. The problem in CI is that the explicit inclusion of the quadruple (and higher) excitations is too costly; in CC an approximation is made for the higher excitations, essentially treating them as products of lower excitations. The approach achieves size-consistency,[46] in most cases yields quite high accuracy,[2,46] and at the CCSD level (coupled cluster singles and doubles) is about as computationally demanding as CISD. Augmented with perturbative and/or non-iterative corrections for triple excitations (CCSD(T), etc.[1,47] important in describing orbital shape changes under the influence of correlation) the CCSD(T) approache has become the current method of choice for the description of single configurational ground states.[3]

The CCSD wavefunction is written as[48]

$$\Psi_{CCSD} = e^{T_1 + T_2} \Psi_{RHF}$$

$$T_1 = \sum_{i,b} c_i^b a_b^\top a_i \qquad T_2 = \sum_{i>j,b>d} c_{ij}^{bd} a_b^\top a_i a_d^\top a_j \tag{29}$$

where b and d denote virtual orbitals, i and j denote orbitals occupied in Ψ_{RHF} and the a_j and $a_b{}^\tau$ respectively remove an electron from occupied orbital j and place an electron in unoccupied orbital b (relative to the reference wavefunction Ψ_{RHF}). One can expand the exponential operator yielding[48]

$$
\begin{aligned}
e^{T_1+T_2} &= 1+T_1+T_2+\frac{1}{2}(T_1+T_2)^2+\frac{1}{6}(T_1+T_2)^3+... \\
&= 1+T_1+T_2+\frac{1}{2}T_1^2+T_1T_2+\frac{1}{2}T_2^2+...
\end{aligned}
$$

(30)

Application of e^T to Ψ_{RHF} thus not only produces single and double excitations, but higher order excitations, expressed as products of the lower single and double excitations. The CCSD equations are given elsewhere[48] and will not be reproduced here but involve a set of nonlinear equations in the c_i^b and c_{ij}^{bd}, having the same dimension as CISD. Of interest to the present discussion are that 1) the equations are non-linear, containing up to quartic terms in the single excitation coefficients, 2) the method, while non-variational tends to be accurate enough so that the loss of the variational property of CI is more than compensated for by the gain in accuracy relative to truncated CI, 3) as is the case for truncated CI, full CCSDT or CCSDTQ prove to be so computationally demanding that they have only been applied to a few systems.[47] The quadratic CI (QCISD) approach of Pople and coworkers[49] is entirely analogous to CCSD, retaining only those terms in CCSD needed to restore size-consistency to CISD.

The success of these methods stems in large part from the fact that even in systems of interacting electron pairs, higher-order excitations can be expressed with reasonable accuracy as products of lower excitations. This is a good approximation for quadruple excitations (largely obtained as products of double excitations), it is a less accurate approximation for triple excitations (e.g. products of single and double excitations). This stems from the fact that to lowest order of perturbation theory single excitations make zero contribution to the correlation energy (relative to a HF wavefunction). Since the coefficients of the single excitations are expected to be small in general, the effects of triple excitations, where important, are poorly described by products of double and single excitations. For quantitative accuracy it has been important to include some independent (i.e. non-product) estimate of the effects of triple excitations, as has been noted above.[1,47]

Multi-Reference-Based Approaches

3.5 Multi-Configurational Self-Consistent Field Theory

Multi-Configurational Self-Consistent Field (MCSCF) wavefunctions are limited CI wavefunctions where the orbital and CI coefficients are optimized simultaneously. In most cases, for purposes of rapid convergence, some additional conditions are imposed on the classes of excitations included in the CI. For example, a common type of MCSCF wavefunction is the so-called "complete active space SCF"[50] where a full CI is performed for a subset of all electrons in a limited orbital space. Perfect pairing generalized valence bond (PP-GVB) wavefunctions are another type of MCSCF wavefunction, composed of products of locally correlated electron pairs. The orbitals and double excitation CI coefficients are obtained self-consistently.[51] Since for more than a single electron pair PP-GVB wavefunctions are not linear variational functions they have not been used as extensively as CASSCF for treatment of excited states.

In CASSCF one normally chooses a subset of the electrons in the system (the "active" electrons) and defines a set of orbitals in which these electrons can be distributed. All possible configurations of active electrons in active orbitals are obtained and the orbitals and CI coefficients optimized for the state of interest.[50] Since the CASSCF solution solves the full CI solution in the limited configuration space, the roots of the CASSCF CI matrix yield successive upper bounds to electronically excited states, and thus can be used to describe electronically excited states within a given symmetry, without "variational collapse." It is also common to perform "state-averaged" CASSCF calculations (SA-CASSCF) where the energy is optimized for an average of the n states of interest.

The CASSCF approach is a considerable step forward relative to RHF/UHF, in that it automatically includes a limited amount of correlation. It also allows one to design zeroth-order wavefunctions that explicitly treat multi-configurational states on a more-nearly equal footing with single configurational zeroth-order states. For example, in the case of the 2^1A_g state of butadiene the use of a CASSCF description leads to a dramatic change in the qualitative description of the state relative to the SCF description.[52] However, because of the need to include entire classes of configurations to improve convergence, CASSCF only allows limited correlation to be included (due to the rapid growth in size of full CI expansions). Difficulties in convergence can also occur when attempting to converge a given excited state having other close-lying states, where one encounters "root-flipping" by optimizing the orbitals for a single state. In these cases state-averaging often yields more rapid convergence. Because only limited correlation is included in CASSCF wavefunctions one does not expect quantitative accuracy for excitation energies. For example, only modest improvements relative to SCF values are obtained for Rydberg excitation energies.

3.6 Multi-Reference Truncated CI

One might reason that a CISD treatment includes the relevant excitations necessary for a multi-reference description of a state, should these configurations be important at zeroth-order (cf. the butadiene 2^1A_g example above). However, if these configurations *are* important at zeroth order, it seems natural to include double excitations relative to these configurations as well, in order to obtain a more balanced weighting of the configurations in the final CI. That is, since double excitations relative to a given configuration "correlate" it, lowering its effective energy, *neglect* of double excitations relative to one or more important zeroth-order configurations in the CI will lead to artificially small mixing with those zeroth-order configurations for which double excitations have been included. Thus, in Multi-Reference Singles and Doubles CI approaches (MRSDCI) single and double excitations are included relative to a set of configurations deemed important at zeroth-order. While the MRSDCI equations can be written exactly in the form given in Eq. 21, with the second through nth zeroth-order configurations and the single and double excitations relative to them included in the non-00 blocks, it turns out to be somewhat more transparent for comparison with other multi-reference approaches to rewrite the MRSDCI equations. Defining the "P" space as the set of zeroth-order configurations, the Q space as all single and double excitations relative to the P space, and the R space as all higher excitations relative to the P space, the full CI equations become

$$\begin{pmatrix} H_{PP}-E & H_{PQ} & 0 \\ H_{QP} & H_{QQ}-E & H_{QR} \\ 0 & H_{RQ} & H_{RR}-E \end{pmatrix} \begin{pmatrix} c^P \\ c^Q \\ c^R \end{pmatrix} = 0 \tag{31}$$

Clearly, since *all* excitations were included in the full CI to begin with, the definitions of P, Q and R can have no effect on the full CI energies. In the case of MRSDCI, however, there is a non-trivial effect due to expansion of the P space relative to CISD, leading to the MRSDCI equations

$$\begin{pmatrix} H_{PP}-E & H_{PQ} \\ H_{QP} & H_{QQ}-E \end{pmatrix} \begin{pmatrix} c^P \\ c^Q \end{pmatrix} = 0 \tag{32}$$

Note that H_{PP}, H_{PQ}, and H_{QQ} are now matrices, and augmentation of the P space leads to a concomitant increase in the size of the Q space. Using partitioning theory, one can rewrite Eq. 32 as a matrix equation of formal dimension spanning the P space yielding

$$(H_{PP} + H_{PQ}(E - H_{QQ})^{-1}H_{QP})c_P = Ec_P$$
$$c_Q = (E - H_{QQ})^{-1}H_{QP}$$

$$(33)$$

It should be noted that MRSDCI is still a size-inconsistent approach. One might have reasoned that the dominant triple and quadruple excitations are explicitly included in the MRSDCI by virtue of inclusion of single and double excitation relative to all zeroth-order configurations. Indeed this does often considerably lower the energy of even an essentially single-configurational state, leading to much more nearly size-consistent results. However, the roots of size-inconsistency are still present in the MRSDCI, since E continues to decrease with system size, while the configurations in the Q space remain largely uncorrelated, leading to a net increase in the effective denominators ($(E - H_{QQ})^{-1}$) in Eq. 33. Thus, there is a considerable gap between the MRSDCI energy and the estimated full CI result and it is common to employ a multi-reference version of the Davidson correction for MRSDCI as well.

One MRSDCI approach that has been shown to be quite accurate for the treatment of ground and excited states of small molecules is the "second-order CI," which includes all single and double excitations relative to a CASSCF wavefunction.[53,54] The difficulty with second-order CI is its rapid growth with number of active electrons/orbitals in the CASSCF, leading quite quickly to an intractable approach for the systems of interest here.

An alternative approach amenable to larger systems is to iteratively build up the zeroth-order space for the MRSDCI by performing successive CI calculations, augmenting the reference space (zeroth-order space) with any configurations that appear to be large contributors in the CI. Once a stable zeroth-order space is achieved (no new configurations with large coefficients in the final CI) one can use the CI energy for the calculation of MRSDCI excitation energies. Even with a limited reference space, for the systems of interest here MRSDCI tends to be too large to do, and some means of further selecting the CISD configuration space to be treated variationally is needed. Several groups have developed perturbation-theory-based selection and/or extrapolation schemes[55-57] that allow one to at least approach MRSDCI results for large systems, although the accuracy of the extrapolation schemes is an important question that has been addressed for some small systems.[58] It should be noted that once a perturbation theory based selection criterion has been applied to the singles and doubles CI space, the CI energy obtained for the given state is variational (i.e. is an upper bound to the state of interest) but the final extrapolated energy (accounting for the unselected single and double excitations) is not.

Rather than using perturbation theory selection, one can instead perform an *a priori* selected MRSDCI calculation, employing classes of excitations relative to a set of zeroth-order configurations. While less common, such *a priori* selected MRSDCIs have been used to describe the polyenes, as discussed below.

3.7 Size-Consistency Corrected MRSDCI Approximations

In order to describe potential energy surfaces in a global rather than local fashion using a CI-like approach, or to be able to treat excited states using a flexible multi-configurational reference space, several groups have attempted to approximately correct the MRSDCI equations for size-inconsistency effects.[59-70] An alternative means of deriving essentially identical methods is to view them as linearized multi-reference coupled cluster theories. The earliest approaches of this type did not allow reference-space readjustment upon correlation, however.[59-62] The two earliest methods that do, that also avoid "intruder state" effects, are the multi-reference averaged coupled-pair functional theory (MRACPF) of Gdanitz and Ahlrichs[67] and the Quasi-degenerate Variational Perturbation Theory (QDVPT) of Cave and Davidson.[66] All methods of this type can be viewed as approximations to the full CI (Eq. 20 or Eq. 31) leading to a set of equations of similar computational effort to the MRSDCI equations (Eq. 32). One can derive the QDVPT equations by assuming a) the main effect of the R space configurations is to correlate the Q space configurations, and b) that the average correlation energy for a given single or double excitation is similar to that for the zeroth-order wavefunction. This allows one to rewrite the Q space diagonal elements in Eq. 33 as

$$(H_{QQ}+H_{QR}(E-H_{RR})^{-1}H_{RQ})_{ii}-E \approx (H_{QQ})_{ii}-E_0 \qquad (34)$$

Using Eq. 34, Eq. 33 becomes

$$(H_{PP}+H_{PQ}(E_0-H_{QQ})^{-1}H_{QP})c_P=Ec_P$$
$$c_Q=(E_0-H_{QQ})^{-1}H_{QP} \qquad (35)$$

where E_0 is the zeroth-order energy of the state of interest in the P space. E_0 can be obtained as the appropriate eigenvalue of H_{PP}, or iteratively if there is significant readjustment in the reference space upon correlation. Note that $E_0-(H_{QQ})_{ii}$ is expected to be much less dependent on system size than is $E-(H_{QQ})_{ii}$. For example, if the zeroth-order state is a CASSCF wavefunction E_0 is size-consistent, and one expects significantly improved behavior as system size increases.[66] Even for a CASSCF reference function one can show, however, that the final equations are not exactly size-consistent.[66] The size-inconsistency appears to be far smaller than MRSDCI's size-inconsistency where they have been compared however.

The MRACPF equations of Gdanitz and Ahlrichs, can be derived similarly,[67] yielding equations of the form

$$\begin{pmatrix} H_{PP}-E & H_{PQ} \\ H_{QP} & H_{QQ}-E_0-2E_{corr}/n \end{pmatrix} \begin{pmatrix} c^P \\ c^Q \end{pmatrix} = 0 \qquad (36)$$

where n is the number of electrons correlated, and $E_{corr}=E-E_0$. Again, E_0 can be obtained as the appropriate eigenvalue of H_{PP} or iteratively. The additional term $2E_{corr}/n)$ in the lower right-hand block of Eq. 36 comes from an analysis of the separated Helium problem, and yields exact results in this case.

Examination of either Eq. 35 or 36 shows that one has assumed that the main role of higher excitations is to cancel the size-inconsistency effects in CISD that result from neglect of higher excitations. This is the same reasoning that leads to the post-CI Davidson size-consistency correction. Thus one could view QDVPT and MRACPF as methods in which one makes a Davidson-like correction to the MRSDCI equations, and *then* optimizes the wavefunction. In test cases on butadiene and butadiene radical cation, the total energies are somewhat different when comparing QDVPT/MRACPF and Davidson-corrected CI but the excitation energies obtained are in excellent agreement.[71,72]

Either of these approaches is in fact perturbative and not variational. Their accuracy stems from a) the same multi-configurational flexibility available in MRSDCI approaches, and b) the modifications that result in near size-consistency discussed above. A number of other more recent approaches have been developed that make a more careful size-consistency correction to the MRSDCI equations.[68-70] Many of these approaches have been reviewed by Szalay and Bartlett,[70] but have not been applied yet to the systems of interest here.

3.8 Multi-Reference Perturbation Theories

Work on multi-reference perturbation theories has a long history,[73-91] with much of the discussion aimed at the requirements of producing theories that preserve size-consistency while avoiding so-called "intruder" state problems. Crudely, intruder state problems arise when the reference space has configurations that are relatively high-lying in energy. Perturbation theory energy denominators involving these reference space configurations and low-lying single and double excitations can lead to small energy denominators, producing convergence problems or divergences. The natural solution to this problem would thus be to neglect high-energy P space configurations, but the route to rigorous size-consistency has generally involved reference spaces that are complete active spaces, virtually guaranteeing the presence of some high lying configurations in the reference space. Some work has attempted to preserve size-consistency while avoiding CAS P spaces.[86-88] For purposes of examining the polyenes and related molecules discussed below, one can largely avoid the details of these treatments and refer the interested reader to the literature.[73-91] Here the perturbation theories are classified according to whether they diagonalize "before" or "after" inclusion of correlation (see, e.g Ref. 89). (Most multi-reference perturbation theories have used a Moller-Plesset partitioning.[74-79, 84,85])

The differences between whether one diagonalizes before or after inclusion of correlation can be seen using the example below. A multi-reference perturbation theory approximation to the full CI equations can be developed in complete analogy with the single reference equations (Eqs. 27 and 28). Using the same partitioning as used for the analysis of the MRSDCI wavefunctions (P, Q, and R spaces), one obtains to second order

$$(H_{PP}+H_{PQ}(E^0-H_{QQ}^0)^{-1}H_{QP})c_P=Ec_P \tag{37}$$

If one further defines the P space zeroth-order states to be the eigenfunctions of H_{PP}, H_{PP} is diagonal. (One need not do so, but the differences between the two approaches are somewhat easier to see if one assumes this has been done.) The P space states thus represent multi-configurational zeroth-order states, each a linear combination of the individual configurations spanning the P space. In this case, the single and double excitations play two roles. Terms of the form $H_{iQ}(E^0-H_{QQ}^0)^{-1}H_{Qi}$ are added to each of the i diagonal elements of H_{PP}, accounting for the correlation of the ith state of the P space due to its interaction with the single and double excitations in Q. In addition, the zero off-diagonal elements in H_{PP} are augmented by terms of the form $H_{iQ}(E^0-H_{QQ}^0)^{-1}H_{Qj}$, arising from the interaction of P-space states i and j as mediated via the single and double excitations. In a perturbation theory that diagonalizes "before" correlation, the second order energies of the various states are just the diagonal elements of Eq. 37. This is equivalent to assuming that the additional coupling mediated via the Q space configurations is essentially zero. In a perturbation theory that diagonalizes "after correlation" one forms the *entire* matrix (diagonal and off-diagonal terms) in Eq. 37, and then diagonalizes the "dressed" P space H matrix (Eq. 37). In this case the weights of the various P space configurations are allowed to readjust in the final perturbation theory wavefunction in response to the effects of correlation.

The methods that diagonalize "before" correlation[74-79] are in essence single reference perturbation theories, where the single reference function happens to be multi-configurational in nature. If one defines excited configurations properly (using "excitation operators") it leads to work similar to that engendered in a single-configurational perturbation theory.[75] In addition, since one only considers specific (usually) low-lying combinations of the P-space configurations, intruder state effects are largely circumvented. Through use of CASSCF-type zeroth order states one can often obtain a qualitatively correct zeroth-order approximation leading to a general and flexible perturbation theory approach. All theories of this type have been applied at second-order.[74-79] While one can certainly develop formal expressions for higher order corrections, it appears that no implementations have been made to date.

The methods that diagonalize "after" correlation[80-91] have the advantage that they allow for P-space reorganization in response to correlation. This effect can play a role in getting an accurate qualitative description of a given state, since one expects that, for example, ionic-valence and Rydberg configurations should have very different correlation energies. The extent to which they mix will depend on their effective off-diagonal Hamiltonian matrix element and on their relative energies. Inclusion of correlation *before* diagonalizing the P-space Hamiltonian matrix will thus lead to very different weighting of the two-configurations in the final wavefunction, as compared with approaches that diagonalize and then correlate. This same effect can be responsible for large differences in theoretical predictions of the shapes of potential energy surfaces. The disadvantages of approaches that diagonalize "after" inclusion of correlation include the extra work (which grows considerably as the P space Hamiltonian increase in size) and the difficulties in obtaining an unambiguous choice for E^0.

Of the perturbation theories that have been used on the polyenes or similar systems, the CASPT2 method of Roos and coworkers,[73,74] the method due to Pulay, Wolinski, and Sellers,[78,79] the MRMP2 method of Murphy and Messmer,[77] and the method of Hirao[76] all diagonalize before correlating, and are all based on a MP partitioning of the Hamiltonian.

The perturbation theories that diagonalize after correlation include the B_k approach of Gershgoren and Shavitt,[91] a Rayleigh-Schrödinger B_k approach due to Nitzsche and Davidson,[90] the H_v approach of Freed and coworkers (Ref. 89, and references cited therein), the QD-PTs of Hoffman[82,83] and Kirtman,[81] and the CAS-MP2 method of Kowalski and Davidson.[84,85] Only the H_v method has been applied to the polyenes.

3.9 Multi-Reference Coupled-Cluster Theories.

Given the accuracy of the single reference coupled-cluster theories for ground states, it is no surprise that a number of workers have pursued multi-reference coupled cluster theories for the treatment of excited states.[92-106] One promising coupled-cluster-based approach for excited states is the "Equations of Motion" approaches of Stanton and Bartlett[104] and Jørgenson and coworkers.[105] These methods use a transformed Hamiltonian, based on the coupled cluster wavefunctions (up to CCSD with approximate inclusion of triple excitations) for the ground state of the system, and expand excited states in what amount to dressed single and double excitations. The accuracy of the approaches depends on the transferability of cluster amplitudes between states (an assumption that would on first pass seem dubious when the excited states are of different character than the ground state of the system). Nevertheless, the results on butadiene presented below are quite impressive. The method as currently implemented (dressed singles and doubles level) is best suited for excited states that are largely single excitations, and seems to do somewhat poorer for states possessing considerable doubly excited character.[104] However, the results for mixed-valence Rydberg states on ethylene and butadiene demonstrate the power of the method, even with the limitation to largely singly excited states. Work on using the approach to examine excited state potential energy surfaces is underway in Stanton's group,[107] again with quite encouraging results.

3.10 Summary

In this section a variety of *ab initio* approaches to the description of molecular ground and excited states were discussed. The chief factors considered in examining various techniques were whether the approach a) allows inclusion of electron correlation, b) is size-consistent, c) allows for a multi-configurational description of the state of interest, and d) allows for readjustment of the reference space configurations in response to inclusion of correlation. The table below summarizes the classes of methods discussed and the "performance" of a given approach in each category, Y indicating yes, N, indicating no, S indicating that the method does so in some cases. (In Table I, MR-PT indicates multi-reference PTs that diagonalize before correlation inclusion, QD-PT indicates PTs that diagonalize after.)

Table I. Comparison of the Characteristics of Several Theoretical Methods for the Treatment of Electronically Excited States.

Method	Correlation	Size-Cons.	Multi-Conf.	Ref. Readj.
HF	N	Y	N	-
CIS	N	Y	S	-
CISD	Y	N	N	-
MCSCF	Y	Y	Y	Y
MRSDCI	Y	N	Y	Y
QDVPT/ACPF	Y	Y	Y	Y
MR-PT	Y	Y	Y	N
QD-PT	Y	Y	Y	Y
EOM-CC	Y	Y	S	Y
MR-CC	Y	Y	Y	Y

4 Results

In this section several aspects of the theoretical results for the series ethylene through octatetraene will be surveyed. First, the extent to which the theoretical methods are able to reproduce and/or interpret the experimental results is considered. Given the importance of the polyenes as conceptual building blocks for conducting polymers, much experimental work has attempted to understand the spectroscopy of these systems, and it is important for theory to provide help in interpreting the spectra. Second, the "classes" of excited states that appear (valence, Rydberg, mixed valence-Rydberg) are discussed. Third, critical comparisons are made of the various theoretical methods that have been applied to these systems. In this regard, one is really asking the question "What does it take to describe state X"? It will turn out that in a number of cases most available methods do surprisingly well in reproducing the experimental excitation energies. Those for which discrepancies or difficulties persist point out the deficiencies in currently available methods for the treatment of excited states, and suggest what new features are needed to yield a truly general excited state method.

While space limitations do not allow detailed consideration of benzene, it too is an excellent case study in the application of many of the methods discussed above. The interested reader will find examples of these applications in Refs. 108-112 and references cited therein.

While obviously not a "poly-"ene, ethylene exhibits most of the interesting spectroscopic features of the longer members of the series, and almost all of the methods discussed above have been applied to it and butadiene, giving several points of comparison for the methods. For the longer polyenes fewer methods have been applied, but a sufficient number have been applied to hexatriene to still make useful comparisons. For octatetraene only two *ab initio* studies have been reported, but they both explore vertical and non-vertical excitation energies and the differences between the two suggest important features required to describe motion along the potential energy surfaces involved.

The first ionization potentials of the members of this series are: ethylene (10.51 eV),[113] butadiene (9.07 eV),[114] hexatriene (8.29 eV),[115] and octatetraene (7.8 eV).[116] As a crude estimate of the placement of the first Rydberg orbitals one can use Eq. 18, assuming average quantum defects for s, p, and d Rydberg states of $\delta_s = 0.8$, $\delta_p = 0.3$, and $\delta_d = 0.1$ and obtain approximate ranges for low-lying Rydberg transitions. In ethylene the first members of the n=3 Rydberg series should appear near 7.7 eV (3s), 8.6 eV (3p), and 8.9 eV (3d). Similar calculations place the n=3 Rydberg series for butadiene between 6.3 eV and 7.5 eV, for hexatriene between 5.5 eV and 6.7 eV, and for octatetraene between 5.0 eV and 6.2 eV. Thus the Rydberg states are expected to appear at lower energies as the molecule lengthens, since the IPs decrease with chain length. However, the valence transitions also shift to lower energy with increasing chain length, leading to a variation in valence-Rydberg mixing. The effects on the spectroscopy will be discussed below.

Ethylene Ethylene is one of the most well-studied polyatomic molecules, and an excellent summary of experimental and theoretical work on ethylene is contained in Robin's books.[13-15] As discussed above, two "valence"-like transitions are expected, one of triplet $\pi \rightarrow \pi^*$ character, the other of singlet $\pi \rightarrow \pi^*$ character. The triplet is expected to be below the singlet. The π^* orbital is of the same symmetry as a d-type Rydberg orbital, with which it may mix depending on the relative energy of the two zeroth-order states. The positions and assignments of several low-lying states are given in Table II. (Assignments from Ref. 15 and references cited therein.)

Table II. Experimental Positions and Assignments for Several Low-Lying States of Ethylene.[15]

State	Character	ΔE(eV)	
1^3B_{1u}	$\pi \to \pi^*$	4.36	vertical
1^1B_{1u}	$\pi \to \pi^*$	7.60 8.0 6.0	maximum vertical[118] 0-0
1^1B_{3u}	$\pi \to 3s$	7.11	adiab./vert.
1^1B_{1g}	$\pi \to 3py$	7.80	adiab./vert.
1^1A_g	$\pi \to 3px$	8.29	adiab./vert.
2^1B_{1u}	$\pi \to 3dxz$	9.33 9.13[117]	adiab./vert. adiab./vert.
2^1B_{3u}	$\pi \to 3d\sigma$	8.62	adiab./vert.

Triplet Rydberg states have also been observed, usually close in energy to the singlet Rydberg state reported above.[15] The Rydberg transitions tend to be quite sharp, while the valence transitions are significantly broader. The transition to the 1^1B_{1u} state is quite broad, since there is believed to be a large geometry change (twisted) in the excited state.[118] Note that the 1^3B_{1u} state is well removed from the region where one might expect to find Rydberg states, but the that 1^1B_{1u} is seen within 1 eV of the expected position of the 3d-like Rydberg state with which it could interact. Ethylene in solution still exhibits essentially the same structure and width for the transition to the 1^1B_{1u} state,[119] indicating it is largely valence-like. However, electron scattering experiments do indicate a variation in the character of the 1^1B_{1u} state with energy, appearing to have a greater Rydberg component at higher energies.[120]

In what follows, experiment and theory are compared, as well as various theoretical methods, for three classes of states: a) triplet valence states, b) Rydberg states, and c) singlet mixed valence-Rydberg states. A large number of publications have presented theoretical treatments of ethylene excited states, of which only a representative subset are considered here.[30,71,117,118,121-129] The basis sets used in the various studies are not identical, and tend to increase in size and flexibility with time. However, basis set questions do not appear to be the limiting factor on accuracy for the states examined here. Location of the lowest member of a Rydberg series implies the existence of higher-lying states, and many of these higher members have been identified for the Rydberg states listed above. For our purposes we will focus on the lowest members in comparisons with theory.

Consider first the triplet valence state. In Table III results are summarized from a variety of methods examining the 1^3B_{1u} state. For purposes of comparison one should note that the ground state is valence-like, and has a value of $<x^2>$ of from 11.5 to 12.0 a_0^2.[2,71,123,126]

Table III. 1^3B_{1u} Vertical Excitation Energies from Various Theoretical Approaches.

Method	$\Delta E(eV)$	$<x^2>$ (a_0^2)
CIS[30]	3.57	
$\sigma\pi$ CI[123]	4.49	12.0
CISD[122]	4.31	
$\sigma\pi$ MCSCF[126]	4.46	10.5
π CASSCF[128]	4.65	11.9
CASPT2[128]	4.39	
MRSDCI[118]	4.35	

The method designated $\sigma\pi$ CI is an *a priori* selected CI involving double excitations of the π electrons, along with all $\sigma\pi$ double excitations (no $\sigma\sigma'$ double excitations however). The $\sigma\pi$ MCSCF is similar to the $\sigma\pi$ CI, but in this case the orbitals were optimized self-consistently. The CASPT2 calculation uses the π CASSCF as its zeroth-order wavefunction. Aside from CIS and perhaps the π CASSCF, all methods do quite well in reproducing the vertical excitation energy for this state and the state is clearly valence-like. The single reference (CISD) or multi-reference (CASPT2, MRSDCI) approaches do essentially equally well here - this is a single configurational state at zeroth-order.

The failure of CIS is of interest, in that it exemplifies the necessity for error cancellation on which the method's accuracy is based. For a state of largely singly-excited character relative to the ground state, CIS optimizes an excited state orbital in the presence of essentially frozen ground state core orbitals. This can lead to an artificially high energy for the excited state if the core orbitals relax upon excitation. If the core orbitals do not relax significantly however, little error arises due to the (essentially) frozen core approximation. A second error arises from the neglect of electron correlation - in this instance biasing towards a too-low excited state. Since the triplet state is valence like, little core relaxation occurs upon excitation, so the first source of error is significantly smaller than the second, leading to a low estimate (by 0.8 eV) for the excitation energy.

Although the π CASSCF result is in reasonable agreement with experiment, it is seen that the CASPT2 calculation based upon it, which also includes $\sigma\sigma'$ and $\sigma\pi$ excitations in a multi-configurational single-reference second-order MP2-like perturbation theory, leads to a significant improvement in the excitation energy. This suggests that while $\sigma\sigma'$ and $\sigma\pi$ are not qualitatively important for the description of this state, they are nevertheless of quantitative importance to the excitation energy.

Consider next the low-lying Rydberg states of ethylene. Fewer calculations have focused on the Rydberg states, but enough exist so that one can make a critical comparison. In this case three Rydberg states are considered, those that involve excitations to the 3s orbital, the 3py orbital (in plane, perpendicular to the C=C bond), and the 3dσ orbital. Results from several theoretical treatments are presented in Table IV. The states are all of true Rydberg character, having much lower oscillator strengths than allowed valence transitions, and much larger values of $<x^2>$ than, for example, the ground state.

Table IV. Theoretical Vertical Excitation Energies (eV) to Several Low-Lying Singlet Rydberg States.

Method	$1^1B_{3u}(\pi\rightarrow3s)$	$1^1B_{1g}(\pi\rightarrow3py)$	$1^1B_{3u}(\pi\rightarrow3d\sigma)$
$\sigma\pi$ CI[129]	7.26	7.93	8.80
CIS[30]	7.13	7.71	8.93
CASSCF[128]	6.82	7.43	8.24
CASPT2[128]	7.17	7.85	8.66
MRSDCI[118]	7.13	7.89	8.73
EOM-CC[129]	7.24	7.91	8.75

In Table IV, the CIS, CASSCF, CASPT2, and MRSDCI treatments are the same as those of Table III. The entry listed as $\sigma\pi$ CI is an *a priori* selected CI, employing all double excitations involving the Rydberg electron. To obtain more accurate excitation energies they are referenced to the energy of the ion, rather than the neutral ground state. The CASSCF results use a similar π space to that used for the 1^3B_{1u} state, with σ orbitals added to the active space for the description of $\pi\to\sigma$ Rydberg states. The EOM-CC values are the EOM-CCSD(T) results of Ref. 129.

Except for the CASSCF results, the agreement of the various methods with each other *and* with the experimental data is excellent (within 0.2 eV) Of course, for higher-lying Rydberg states this type of accuracy would be insufficient, as the spacing of the states becomes less than or equal to 0.2 eV. However, for the low-lying states these results are sufficient to help in assignment of the states. The error in the CASSCF result (consistently low) likely stems from the neglect of $\sigma\pi$ correlation, artificially raising the ground state relative to the Rydberg excited states, since the ground state has one more valence-like π electron. Note also the accuracy of the CIS approximation for the Rydberg states.

Finally, consider the singlet $\pi\to\pi^*$ transition. The 1^1B_{1u} state has received the greatest attention theoretically, as it turns out to be the most difficult state to describe. The experimental spectrum extends over nearly a 2 eV region, with the intensity maximum occurring at 7.6 eV.[15] As discussed above, this intensity maximum is often indicative of the vertical transition energy, by reason of the Franck-Condon principle. However, the ionic nature of the valence state, coupled with the proximity of the valence configuration to the expected position of the 3dxz Rydberg state with which it can mix, have lead to serious difficulty in obtaining a converged description of this state. At the HF level, it was realized quite early that the energy of the 1^1B_{1u} state could be lowered considerably if diffuse functions were added to the atomic orbital basis. Closer examination of the wavefunction revealed that the 1^1B_{1u} state obtained in such a treatment was essentially Rydberg-like, in disagreement with experiment. All of the results presented below utilize basis sets with at least one set of diffuse functions, allowing for the treatment of valence-Rydberg mixing.

The methods of Table IV have been applied to the 1^1B_{1u} state (see Table V) along with some additional approaches. The $\sigma\pi$ RASSCF[130a] wavefunction is a π CASSCF augmented with $\sigma\to\sigma'$ excitations, all orbitals optimized self-consistently.[130b] The INO CIs are obtained using "iterative natural orbitals"[131] for the 1^1B_{1u} state from previous CIs. This is a non-self-consistent method for obtaining improved MOs based on the CI treatment of a state. The CAS-MP2(OP2) result[125] is a second order perturbation theory that "diagonalizes after correlation." Once diffuse functions have been added to the atomic orbital basis, almost any method achieves energy accuracy of similar quality (disregarding the CASPT2 result the spread is approximately 0.4 eV). Although the CIS result is closest to the experimental intensity maximum, later theoretical work suggests[118] that the intensity maximum is not the vertical transition, and that 7.9-8.0 eV is closer to the correct value. Having said which, the CIS result is still surprisingly good for such a simple method. All the other methods achieve quite good agreement with the expected 8.0 eV vertical transition as well.

Table V. Vertical Excitation Energies to the 1^1B_{1u} state of Ethylene from Various Theoretical Approaches.

Method	$\Delta E(eV)$	$<x^2>$
$\sigma\pi$ CI[123]	7.96	17.4
CISD (ROHF)[122]	8.25	42.9
CISD (INO)[122]	8.12	24.2
MRSDCI[127]	8.04	21.8
CIS[30,125]	7.74	17.8
$\sigma\pi$ MCSCF[126]	8.09	25.1
$\sigma\pi$ RASSCF[130b]	8.15	23.1
π CASSCF[128]	8.20	44.1
CASPT2[128]	8.40	
MRSDCI (INO)[121]	7.94	16.8
QDVPT[71]		20.5
CAS-MP2(OP2)[125]		17.8
EOM-CCSD(T)[129]	7.89	17.9

As expected based on the qualitative analysis discussed above the proper description of the state requires explicit inclusion of $\sigma\pi$ excitations. Some confusion in this regard has existed in the literature, but recent work by Davidson[125] or comparison of the $\sigma\pi$ RASSCF result[130b] (π CASSCF along with restricted $\sigma\rightarrow\sigma^*$ excitations) with the π CASSCF wavefunction dispels doubts on this matter. Neglect of the $\sigma\pi$ excitations (π CASSCF) has little effect on the excitation energy, but produces a much more diffuse 1^1B_{1u} state, almost indistinguishable from the ROHF result. Augmentation of the π CASSCF with σ excitations ($\sigma\pi$ RASSCF) leads to a similar excitation energy but a much less diffuse wavefunction. Thus, where ionic states are concerned one expects qualitative errors if $\sigma\pi$ correlation is neglected.

Further examples below will show that significant errors in excitation energies can also occur. The results in Table V also illustrate that limited treatments of correlation are affected by the starting orbitals they use. Consider the two CISD results, one beginning with (diffuse) ROHF orbitals, the other using "Iterative Natural Orbitals" (INO). Even though both CISD treatments include excitations from (formally) the same set of orbitals, the character of the occupied orbitals plays a large role in determining the nature of the final state. The more compact INOs yield a much less diffuse final CI wavefunction. The MRSDCI (INO) results use a similar procedure to describe the 1^1B_{1u} state based on a multi-reference CISD treatment. Extensive INO iterations starting from different SCF orbitals (1^3B_{1u}, 1^1B_{1u}, results not shown) illustrate the difficulties faced in converging the spatial extent using CI approaches focusing on a single state.[121] The QDVPT results[71] are based on averaged natural orbitals for the two lowest $^1B_{1u}$ states, which turn out to be largely valence and Rydberg-like respectively. Since this method allows for reference space readjustment upon correlation the final state is able to achieve a compromise valence-Rydberg mixing. The EOM-CCSD(T) method[129] also yields a compact description of the 1^1B_{1u} state, with an excitation energy in good agreement with experiment.

The CASPT2 method produces the highest excitation energy of all the approaches. The $<x^2>$ value is not reported but the value for the CASSCF wavefunction on which it is based is high, and there is no reference space readjustment in the CASPT2 method. It is likely that the CASSCF starting point is too Rydberg-like, leading to a excitation energy increase upon inclusion of correlation in analogy with true Rydberg states (see Table IV).

The variation in the character of the 1^1B_{1u} state arises from the combination of factors alluded to above: a) near-degeneracy of the "properly described" zeroth-order configurations $\pi \rightarrow \pi^*$ and $\pi \rightarrow 3dxz$, and b) the difficulty in achieving a balanced inclusion of $\sigma\pi$ correlation in order to describe the ionic $\pi \rightarrow \pi^*$ state. The variation in spatial extent is really an indication of the variable mixing between these two zeroth-order states obtained in the various calculations. The ideal approach would be to optimize the orbitals for a given state under the influence of *all* relevant types of correlation (e.g. $\sigma\pi$ RASSCF), but such methods are not generally applicable for molecules much beyond ethylene in size. Multi-configurational single-reference methods that diagonalize "before correlation" (e.g. CASPT2) do not allow for such optimal mixing to take place. Where such difficulties do arise, the methods of choice will allow reoptimization/readjustment of the zeroth-order wavefunction under the influence of correlation. It has also been shown that size-inconsistent approaches tend to bias towards a Rydberg-like description of the 1^1B_{1u} state.[71]

Although the results are not presented in tabular form, most of the methods applied in Table IV were also applied to the description of the 2^1B_{1u} state. This state is largely Rydberg-like but is found to occur at somewhat higher energy than the other 3d Rydberg states, in large part due to its interaction with the valence-like state.

Butadiene The spectroscopy of butadiene[132-142] is in many respects similar to ethylene. A strong broad $\pi \rightarrow \pi^*$ transition dominates the UV/VIS spectrum, with a number of Rydberg series also having been identified. At lower energies two valence-like triplet states have been identified, associated with a largely single configurational ^3B_u state, followed by a multi-configurational ^3A_g state. In analogy with what is observed for longer polyenes, a multi-configurational 2^1A_g state is expected. This is a "doubly" forbidden one-photon transition (symmetry and two-electron excitation) and has thus been difficult to observe. The ^3A_g and 2^1A_g states are the new features relative to ethylene that one encounters in considering the polyenes, and the position of the 2^1A_g state relative to the singlet $\pi \rightarrow \pi^*$ state (1^1B_u) has been the major focus of much of the theoretical work on this system. The $\pi \rightarrow \pi^*$ configuration has ionic valence bond character, and is of the same symmetry in C_{2h} as the 3p and 4f Rydberg states with which it is found to mix (to a greater or lesser degree depending on the correlation treatment). The experimental data and assignments are collected in Table VI. Where experimental evidence exists, the transitions are labelled (a) (adiabatic, 0-0) and (v) (vertical).

Table VI. Experimental Excitation Energies and Assignments for Low-Lying States of *trans* 1,3-Butadiene

State	Character	$\Delta E(eV)$
1^3B_u [132,135,140]	$\pi \rightarrow \pi^*$	2.59(a), 3.2(v)
1^3A_g [135]	$\pi \rightarrow \pi^*$	4.63(a), 4.95, 5.01(v)
1^1B_u [139]	$\pi \rightarrow \pi^*$	5.74(a), 5.92(v)
2^1A_g [137,138,141]	$\pi \rightarrow \pi^*$	5.4-5.8, 7.3
3^1A_g [133]	$\pi \rightarrow 3d\pi$	7.61
1^1B_g [114]	$\pi \rightarrow 3s$	6.20
1^1A_u [134,136]	$\pi \rightarrow 3p\sigma$	6.66
2^1B_u [134,136,137]	$\pi \rightarrow 3px,4pf$	7.07

Some information is also available on excited state geometries.[142]

Almost as many theoretical treatments have appeared for butadiene[19,31,52,57,71,89,143-149] as for ethylene, spanning the range from CIS and CASSCF to MRSDCI, EOM-CCSD(T), and two distinct multi-reference perturbation theories (CASPT2 and the H_v method). The theoretical results are again compared with the experimental results and each other. The results are compared not only for the three classes of states discussed for ethylene, but for the class of multi-configurational zeroth-order states (2^1A_g).

For the triplet states essentially all of the patterns observed for the methods when applied to ethylene carry over to butadiene.[19,31,57,71,89,143-146] Aside from CIS all of the CI, CASSCF, and perturbation theory approaches are within 0.2 eV of the triplet vertical excitation energies. CIS is too low for both triplets by about 0.6 eV, and π CASSCF tends to be slightly high. CASPT2 based upon the π CASSCF wavefunction is in excellent agreement with experiment. Essentially any method that includes electron correlation yields good agreement with experiment for the vertical excitation energies, as long as it is capable of treating the singly excited multi-reference 1^3A_g state.

Many of the Rydberg states are quite easily described as well. Considering the 3s, 3pσ, 3pσ', and the 3dxy states the results from CIS,[31] MRSDCI,[57,145] a priori selected GVB-CI,[19,146] CASPT2,[128] and EOM-CCSD(T)[129] are in excellent agreement with each other and experiment. No significant gain in accuracy is obtained by choosing one method over the other.

It is in the description of the 1^1B_u and 2^1A_g states that large differences between methods are found, significantly more pronounced in fact than those observed for the 1^1B_{1u} state of ethylene. Selected results are given in Table VII. All results are derived from AO basis sets with at least one set of diffuse functions, and the ground state value for $<x^2>$ is approximately 22 a_o^2. For comparison purposes, a π CI allowing only single excitations relative to the ground state in a basis *without* diffuse functions yields excitation energies of 9.03 eV and 10.99 eV for the 1^1B_u and 2^1A_g states, respectively.[143] In both cases these values are much higher than any experimental or later theoretical estimates, but for completely different reasons.

Table VII (see following page) is admittedly a dizzying array of data, but it is included here to emphasize that things are not even as simple as they seemed for ethylene. Certainly the 1^1B_{1u} state of ethylene presented some challenges in obtaining the correct spatial extent, but by and large the excitation energies varied little with method. In the cases of the two singlet "valence" states of butadiene, excitation energies and spatial extents show dramatic dependence on the method used to treat these states.

Many of the methods listed in Table VII were applied to ethylene (CIS,[31] $\sigma\pi$ CI(1^1B_u),[52] π CI(2^1A_g),[52] π CASSCF,[130b] $\sigma\pi$ RASSCF,[130b] MRSDCI,[52,144,145,149] QDVPT,[71] CASPT2,[128] and EOM-CCSD(T)[129]). The "new" methods applied to butadiene include GVB-CI[16,146] (*a priori* selected CI, involving limited $\pi\pi'$ and $\sigma\pi$ correlation), SAC-CI[148] (similar to EOM-CCSD), AV-CASSCF[128] (state averaged π CASSCF, the CASPT2 result is based on the third state) and H_v perturbation theory[89] (diagonalizes "after correlation"). Thus the range of methods considered spans essentially all approaches listed in Table I.

Table VII. Selected Theoretical Vertical Excitation Energies to the 1^1B_u and 2^1A_g states of *trans* 1,3-Butadiene.

Method	$1^1B_u \Delta E(eV)$	$1^1B_u <x^2>$	$2^1A_g \Delta E(eV)$	$2^1A_g <x^2>$
CIS[31]	6.21		7.19	"Ryd"
π CISD[143]	7.57		6.98	
π CASSCF[130b]	6.48	67.8	6.58	24.4
$\sigma\pi$ RASSCF[130b]	6.56	60.2	6.79	23.7
$\sigma\pi$ CI/π CI[52]	6.23	32.6	6.77	23.5
MRSDCI[57]	6.67	"Ryd"	7.02	
GVB-CI[146]	6.79	"Ryd"	7.0	
MRSDCI[147]	6.67		6.53	
QDVPT[71]	6.39	48.1		
MRSDCI[149]	6.70		6.78	
MRSDCI[52]	6.21	40.5	6.24	23.3
SAC-CI[148]	6.43		7.05	
AV CASSCF[128]	8.54 (3rd)	40.9		
CASPT2[128]	6.23		6.27	
H_v[89]	6.14		6.19	
EOM-CCSD(T)[129]	6.13	31.7	6.76	36.1

Consider first the 1^1B_u ($\pi \to \pi^*$) state. The excitation energies range over nearly 1.5eV, the highest value coming from a calculation that used ground state σ orbitals and included only π correlation (π CISD). Since this is an ionic state both of these factors bias towards a high excitation energy. The CIS result is in reasonably good agreement with experiment but the calculated oscillator strength is a factor of 2 too large, suggesting the state is too valence-like. Proceeding to methods that now allow optimization within the σ framework, one sees that a π CASSCF overestimates the experimental vertical excitation energy by 0.6 eV and produces a diffuse state. Similar effects were observed for ethylene, but in the case of ethylene the spatial extent of the $\pi \to \pi^*$ state was reduced when the MCSCF wavefunction also included single excitations of $\sigma \to \sigma'$ character ($\sigma\pi$ RASSCF). For butadiene however, there is little change in the diffuse nature of the 1^1B_u state when these excitations are included. This may be due to the fact that the $\sigma\pi$ RASSCF calculation includes only a limited subset of the possible $\sigma \to \sigma'$ excitations (9 occupied σ orbitals into 9 σ^* orbitals) due to computational limitations. A more tractable means of including $\sigma\pi$ correlation is to use an *a priori* selected CI ($\sigma\pi$ CI above) which allows all double excitations of π electrons with each other and with σ electrons and yields a state that is significantly more contracted. (However, in the $\sigma\pi$ CI calculation it is possible that too limited a set of diffuse functions was used, leading to an artificially contracted description of the state). One also notes significant variation of excitation energy considering only the MRSDCI results. All but the results of Ref. 149 are PT-selected MRSDCI results, the one from Ref. 52 including the largest set of configurations in the variational portion of the calculation for any PT-selected MRSDCI. Since a proper description of valence-Rydberg mixing requires a sensitive treatment of the correlation energy, part of the differences among the MRSDCI results may stem from this. Use of a size-consistency correction is also an important factor in lowering the 1^1B_u excitation energy, since the ionic state has significantly higher correlation energy than the ground state.[52] Overall the more recent results (CASPT2,[128] H_v[89] and EOM-CCSD(T)[129]) along with the MRSDCI result from Ref. 52 suggest excitation energies somewhat above the intensity maximum in the optical spectrum. The spatial extent of the 1^1B_u state certainly decreases upon inclusion of correlation, but there is no evidence to indicate that it is purely valence-like, on the basis of *ab initio* calculations.

Turning to the 2^1A_g state, it too shows great variation in character and excitation energy with method. Methods that are limited to descriptions of largely singly excited states (CIS,[31] EOM-CCSD(T)[129]) yield high energies and relatively diffuse characters for this state. The state at zeroth-order has a significant component from a true double excitation, and neglect of this configuration leads to state of significantly higher energy, which then mixes with Rydberg configurations of the proper symmetry, producing a diffuse state. On the other hand, any method capable of including π singly and doubly excited configurations relative to the ground state in a balanced fashion leads to a valence-like multi-configurational state. In the longer polyenes this state is unquestionably a valence state, and it is almost certain that this is the proper description for butadiene. Consider now its excitation energy. Early limited CI treatments centered on an excitation energy of approximately 7.0 eV.[57,146] CASSCF treatments correlating all π electrons drop the energy to approximately 6.6 eV, as do more extensive all π-electron CIs.[52] The two recent perturbation theory treatments, along with the largest PT-selected MRSDCI on the 2^1A_g state indicate the vertical transition is somewhat lower, around 6.2 eV.[52,89,128] On what basis might one choose amongst these results? Unfortunately the experimental data spans a 2 eV range, leading to the seductively comforting thought that almost any result for the 2^1A_g state could find a modicum of experimental support. Having said which, the high experimental estimate (7.3 eV) has less theoretical support than some of the more recent lower estimates. None of the theoretical estimates lend support to the experimental values in the range of 5.6 eV, but this may be nearer the 0-0 transition energy than the vertical transition energy. It should be noted that there is theoretical reason to believe that the π CASSCF value should overestimate the excitation energy for the 2^1A_g state.[150] It is well known that the SCF approximation tends to significantly overestimate vibrational force constants. This will play a differential role in the calculation of an excitation energy if the minima for a pair of states is well separated, and the geometry chosen corresponds to the minimum for one of the states. In the case of a vertical transition energy from the ground state, this leads to an artificially high excitation energy, since the excited state PES rises too steeply from its minimum. Both the π CASSCF and the π CIs include electron correlation in the π space, and it is known from semi-empirical calculations[151] that the 1^1A_g and 2^1A_g geometries are quite different, with essentially a switch in bond-length alternation. Thus, without significant correlation in the σ space as well, one would expect an overestimate of the excitation energy by these methods. The larger MRSDCI[52] and the two perturbation theory treatments[89,128] include $\sigma\sigma'$ correlation and yield reduced excitation energies. Whether this is the limit of the lowering for this excitation energy is not known at present, but results on octatetraene using the CASPT2 method (where the 0-0 and fluorescence maxima are known[152,153]) show it yields quite accurate results there.[154] Given the similar character of the two states it seems reasonable to expect similar accuracy here.

What about the experimental estimate of 5.6 eV for the 2^1A_g? Two sets of *ab initio* calculations have examined the nonvertical transition energies in butadiene,[149,155] using estimates of the excited state geometries from *ab initio*[149] or from semi-empirical calculations.[151] They predict a large difference between the vertical (6.78,[149] 6.77[155] eV) and 0-0 (5.57,[149] 5.66[155] eV) energy differences for the 2^1A_g state, using a limited MRSDCI[149] or π CI.[155] The 0-0 transition energy should be less affected by use of an SCF approximation for the σ electrons (see above) than the vertical excitation energy, and thus the agreement of the MRSDCI and π CI is reasonable in this case. Further calculations using methods that allow more extended reference spaces and correlate all electrons would be helpful in further probing the non-vertical excitation energy for the 2^1A_g state.

Recap Before addressing other examples, it is useful to summarize what can be concluded from the results on ethylene and butadiene. Low-lying $\pi \rightarrow \pi^*$ triplet states are unequivocally valence-like, but can be multi-configurational (beyond ethylene). These multi-configurational cases require some form of MRCI or MCSCF starting point to be accurately described. Many states are of Rydberg character and most methods do quite well in predicting the position of Rydberg states that are not mixed with low-lying valence states. For singlet valence states there is much greater variability in the results. For states where valence-Rydberg mixing is important the final outcome is sensitive to a) the "input orbitals" and b) the level of correlation. Since Rydberg states generally have lower correlation error, a method must recover a large fraction of the correlation energy in order to place the valence state properly relative to close-lying Rydberg states. This proves to be a difficult, and largely unsolved problem for many methods. For example, for perturbation theories that diagonalize before correlation inclusion (e.g. CASPT2) the wavefunction is unable to readjust its spatial extent in response to the more complete correlation treatment, retaining a too diffuse zeroth-order wavefunction. The EOM-CCSD(T) approach however, seems to do an excellent job of describing the spatial extent of the singlet $\pi \rightarrow \pi^*$ states, using what amount to "dressed" excited configurations. In regard to the multi-configurational singlet states, it is apparent that a true multi-configurational zeroth-order description is required to make progress (note the failure of EOM-CCSD(T) here), and that due to the large geometry changes relative to the ground state extensive correlation recovery is important in this case as well.

The conclusions drawn above apply equally well to hexatriene and octatetraene. In addition, hexatriene and octatetraene turn out to be somewhat easier to describe theoretically. Both the valence HOMO-LUMO gap and the IP decrease in the polyenes as the chain length decreases, but it appears that the HOMO-LUMO gap decreases somewhat faster, leading to a greater valence-Rydberg separation in the singlet manifold than is found in the smaller chain species. This leads to a much simpler description of the $\pi \rightarrow \pi^*$ state. Before examining the longer polyenes however, two related examples, *cis*-1,3 butadiene and the butadiene radical cation are discussed briefly.

cis-1,3-butadiene For *cis*-butadiene, only the intensity maximum for the intense $\pi \rightarrow \pi^*$ has been observed at 5.49 eV.[156] The second stable conformer of butadiene is most likely *gauche*,[157-159] but for computational simplicity the calculations have been performed at the low-energy C_{2v} *cis* saddle-point. In C_{2v} symmetry the $\pi \rightarrow \pi^*$ state is labeled 1^1B_2. All methods that have been applied (CIS,[31] MRSDCI,[52] *a priori* selected CI,[52] CASPT2,[160] H_v PT[161], SAC-CI,[148] and EOM-CCSD(T)[129]) find a much more nearly valence-like 1^1B_2 state, with vertical excitation energy within 0.15 eV of the experimental estimate, some low, some high. Note that the $\pi \rightarrow \pi^*$ transition is found to be approximately 0.6 eV lower than in the *trans* case. This is likely due[52] to a weak 1,4 interaction in the $\pi \rightarrow \pi^*$ state that stabilizes the valence component in *cis* isomer. By lowering the energy of the valence component the valence-Rydberg mixing is reduced, leading to a physically less diffuse state and a theoretically simpler description. This illustrates the subtle balance of interactions that lead to the complexity in the description of the *trans* 1^1B_u state. The 2^1A_1 state (corresponds to the 2^1A_g state in the *trans* isomer) is also lowered somewhat in energy, but its character remains that of a multi-configurational valence state.

butadiene radical cation The electronic spectroscopy of butadiene cation has been studied using a variety of experimental techniques.[162-164] The low-lying states are of interest due to the contribution of "non-Koopmans" configurations in the low-lying states, yielding essentially multi-configurational excited states similar to the 2^1A_g state of butadiene. A number of *ab initio* treatments have appeared[163,165,166] of the low-lying states, the most recent using MRSDCI and MRACPF, correlating all electrons.[165] Generally good agreement was found with the experimental excitation energies although one multi-configurational state tended to be somewhat high relative to experiment. Because the IP of the *cation* is considerable, no valence-Rydberg mixing is expected or observed. Thus, one essentially removes this difficulty in the description of the $\pi \rightarrow \pi^*$ excitations by moving the Rydberg states to much higher energy.

trans-1,3,5-hexatriene Aside from a shift of all transitions to lower energy, the spectroscopy of hexatriene is quite similar to that of butadiene. The lowest-lying states are a pair of triplets, the second being best described as multi-configurational. The dominant feature in the spectrum is a broad intense peak attributed to the 1^1B_u $\pi \rightarrow \pi^*$ transition, followed by a series of sharp Rydberg peaks. In the *cis* isomer unequivocal evidence has been found that the 2^1A_g-like state is the lowest excited state (at least in a 0-0 sense).[167] Assignments of some of the low-lying states are given in Table VIII.

The larger size of hexatriene has lead to fewer attempts to describe it theoretically, but a sufficient number of approaches (π-GVB-CI,[174] or $\sigma\pi$-GVB-CI[175] for σ Rydberg states, $\sigma\pi$ *a priori* selected CI,[176] CASSCF,[128] CASPT2,[128] and H_v PT[177]) have been applied to its spectroscopy to allow comparisons to be made. The two perturbation theory approaches (the only two which correlate all electrons) yield excellent results for the triplet and Rydberg states listed above.[128,177] The restricted CI approaches tend to be 0.2-0.3 eV higher in energy (nearly uniformly).[174-176] Data for the singlet valence states are presented in Table IX. For comparison purposes, the value of $<x^2>$ for the ground state is approximately 32 a_o^2.

Table VIII. Experimental Excitation Energies and Peak Assignments for *trans*-1,3,5-hexatriene.

State	Character	ΔE(eV)
1^3B_u[168]	$\pi\rightarrow\pi^*$	2.61
1^3A_g[168]	$\pi\rightarrow\pi^*$	4.11
1^1B_u[115,168,169]	$\pi\rightarrow\pi^*$	4.67(a),4.93(v),5.15(v)
2^1A_g[170,171]	$\pi\rightarrow\pi^*$	<4.22(a),5.21(v)
1^1A_u[168,172a]	$\pi\rightarrow3s$	5.67,6.0
1^1B_g[172b]	$\pi\rightarrow3p\sigma$	6.2
3^1A_g[173]	$\pi\rightarrow3p\pi$	6.2
2^1B_u[168,173]	$\pi\rightarrow3d\pi$	6.06,6.55

As has been observed before, neglecting $\sigma\pi$ correlation leads to significantly higher 1^1B_u excitation energies (π-GVB-CI, π-CASSCF). It should be noted that the π-CASSCF state used in Ref. 128 was not the lowest root of the CASSCF CI matrix, rather it was a higher root which was more nearly valence-like, and is the root that was use as the zeroth-order wavefunction for the CASPT2 calculation. The lowest ^1B_u CASSCF state was Rydberg-like and had a π-CASSCF excitation energy of 6.59 eV. The $\sigma\pi$-CI result nicely illustrates that the inclusion of $\sigma\pi$ correlation has a dramatic effect on the spatial extent of the 1^1B_u state (cf. the CASSCF result). For the 2^1A_g state, neglect of $\sigma\sigma'$ correlation (π GVB-CI, $\sigma\pi$ CI, π CASSCF) once again leads to a higher estimate of the excitation energy. The two methods that correlate all electrons are in good agreement with each other on the excitation energy to this state.

Table IX. Theoretical Excitation Energies and $<x^2>$ Values for the Singlet Excited Valence States of *trans*-1,3,5 Hexatriene.

Method	$\Delta E(eV)$ 1^1B_u	$<x^2>$ 1^1B_u	$\Delta E(eV)$ 2^1A_g	$<x^2>$ 2^1A_g
π-GVB-CI[174]	6.56	"valence"	5.87	
$\sigma\pi$-CI[176]	5.15	33.9	5.90	32.2
AV π-CASSCF[128]	7.36	40.4	5.65	31.8
CASPT2[128]	5.01	-	5.19	-
H_u[177]	5.21		5.20	

The "best" 1^1B_u excitation energies are still approximately 0.1 to 0.3 eV above the intensity maximum in the optical spectrum[115] but agree quite nicely with the electron energy loss values.[168] It is possible that the methods are consistently overestimating the vertical transition energy, but good agreement for the $\pi \rightarrow \pi^*$ state of *cis*-butadiene, coupled with the good agreement CASPT2 achieves for the same state in octatetraene (where essentially all of the Rydberg component is gone from the $\pi \rightarrow \pi^*$ state) indicate the methods are not fundamentally flawed in attempting to describe such valence states. It is possible that, as in ethylene, the intensity maximum in the absorption spectrum may not correspond to the vertical transition energy, perhaps due to increasing valence character in lower energy portions of the band. The $\sigma\pi$-CI results not only nicely reproduce the electron-energy loss vertical excitation energy, but also give excellent agreement with the optical 0-0 transition energy.[115] The perturbation theory results are in quite good agreement with the quoted vertical transition energy for the 2^1A_g state.[170] π-CI calculations using semi-empirical excited state geometries predict the vertical to adiabatic excitation energy shift is 1.2 eV.[155] However, the octatetraene results given below suggest the value might be somewhat large. The 0-0 transition energy is not known for *trans* hexatriene, but for *cis*-hexatriene it is 4.26 eV.[167] Further theoretical work is required using realistic excited state geometries and all-electron correlation methods to obtain an accurate theoretical estimate for the 0-0 transition energy for the 2^1A_g state.

all-trans octatetraene Octatetraene is the first of the polyenes for which fluorescence is observed.[152,178,179] Fluorescence is observed not from the 1^1B_u state but the 2^1A_g state, indicating that the 2^1A_g state is indeed the lower of the two valence states. Only two *ab initio* treatments have focused on all *trans* octatetraene, one using a $\sigma\pi$ *a priori* selected CI,[180] the other using CASPT2.[181] The former did not treat Rydberg states, and was somewhat high for the lowest triplet state (0.4 eV). The CASPT2[181] results are in excellent agreement with experiment for the low-lying triplet and Rydberg states, where comparison can be made. Results for the excited singlet valence states are presented in Table X. Both states are found to have spatial extents similar to the ground state.[180,181]

Table X. Theoretical and Experimental Excitation Energies to the 1^1B_u and 2^1A_g states of all *trans* Octatetraene.

Method	$\Delta E(eV)$ 1^1B_u	$\Delta E(eV)$ 2^1A_g
Expt.[152,178,179]	4.41(a), 4.41(v)	2.8-2.9(fm),3.59(a),4.33(v)*
$\sigma\pi$-CI/π-CI[180]	4.56(a), 4.79(v)	3.74(fm),4.15(a),5.21(v)
AV π-CASSCF[181]	6.67(v)	5.23(v)
CASPT2[181]	4.35(a), 4.42(v)	2.95(fm),3.61(a),4.38(v)

*using mirror rule based on fluorescence maximum and 0-0 transition.

These results are consistent with what was seen before for the shorter chain polyenes in regard to the performance of these two methods. Lack of $\sigma\pi$-correlation leads to an overestimate of the 1^1B_u excitation energy (π CASSCF), and lack of $\sigma\sigma'$ correlation leads to a too high estimate of the 2^1A_g state (π CI). However, here, for the first time, unequivocal experimental evidence for the 0-0 transition energy has been obtained, and the CASPT2 results are in quite good agreement with it. CASPT2 even does an excellent job predicting the positions of the other intensity maxima in the fluorescence and absorption spectra. (Note, the mirror rule asserts that for potential energy surfaces of similar curvature the fluorescence maximum and the absorption maximum should lie the same energy difference below and above the 0-0 transition, respectively.) The CASPT2 results also predict very little difference between the 0-0 and vertical excitation energies for the 1^1B_u state, in excellent agreement with experiment.

5 Conclusions and Opportunities

The above survey, admittedly over a limited set of molecules, nevertheless probes a wide variety of methods and essentially the entire array of bound molecular excited states. The polyenes are particularly demanding for theoretical methods, due to the presence of excited states that are multi-configurational at zeroth-order, as well as possessing another class of states that exhibit strong valence-Rydberg mixing. In the survey above, it was seen that most methods were able to do quite well in describing the lowest Rydberg states, and that in most cases similar accuracy could be expected for the valence triplet states. More demanding were the valence singlet states and here the results tended to be mixed. The method currently receiving the most use (and clearly demonstrating success) is the CASPT2 approach. Given the results described above, its general use is well deserved.

In terms of the qualitative description of the electronic states CASPT2 does not appear to be so successful. It is this fact that will require attention in the future, as methods such as CASPT2 are applied to explore excited state potential energy surfaces. Consider two excited states, A and B, near one another in energy but of different symmetry in the point group of the equilibrium geometry of the ground state of the molecule. If one now distorts the molecule, reducing the symmetry, A and B can end up having the same symmetry and will interact. Now, consider a case not unlike that discussed above for ionic states, where at zeroth order A is above B, but upon inclusion of correlation B is above A. As the molecule is distorted, one performs a CASSCF calculation, freezes the coefficients, and does a second order correlation treatment based on the CASSCF wavefunctions. However, these CASSCF zeroth-order wavefunctions behave incorrectly as the molecule is distorted. At the CASSCF level, B is below A, and as the molecule distorts and they interact, B is driven down in energy while A goes up in energy. The true ordering has A below B, and thus *A should go down in energy, while B goes up*. The correlation treatment will attempt to reverse this, but must do so on the basis of qualitatively incorrect zeroth-order wavefunctions (wrong sign for the mixing coefficients) since the zeroth-order wavefunctions mix on the basis of their CASSCF energies. Thus, if correlation reverses the order of the two states relative to their zeroth-order ordering serious problems can arise for multi-configurational single-reference perturbation theories. If on the other hand, one were to use a "diagonalize after correlation" perturbation theory, these effects would be largely avoided.

The H_v theory of Freed and coworkers[89] or the CAS-MP2 theory of Koslowski and Davidson[84,85] are theories of this type. The H_v theory has a potential drawback based on current prescriptions used to generate reference space orbitals which will make geometry optimizations difficult. In addition, while the third order results quoted above are in quite good agreement with experiment, the variation in quality from second order to third order results leads one to wonder whether fourth order results on these systems would be of similar quality. (Of, course the same could be said of any of the other PTs. Only for H_v PT are higher than second order results known.) However, results have been presented on a model system using H_v PT which help to understand the need for third order results, and suggest when the perturbation theory is expected to be rapidly convergent.[182]

Higher (infinite) order theories such as CC based approaches are appealing, but to date it has been a difficult task to produce a theory that is size-consistent, with a truly flexible zeroth-order wavefunction, that is tractable computationally. Recent attempts at multi-reference size-consistency corrected CI may provide a useful compromise along these lines.[69,70] However, for very large systems the speed of low-order perturbation theories suggests that they will continue to be very useful tools for theoretical treatments of excited states.

All of which only begins to scratch the surface of the challenges waiting for a truly general method for excited states. One would surely like to be able to characterize excited state surfaces, not merely near minima, but globally. Points and surfaces of intersection will certainly be sensitive to the zeroth-order description used as well as correlation effects. As energy increases the density of states increases as well, and one will need to be able to treat many states of a given symmetry simultaneously, not merely one or two. While such challenges might seem daunting, they also represent open doors for innovative research in the area of electronically excited states. Such advanced methods will certainly be brought to bare on the polyenes once again, where it will be of interest to accurately describe the interaction of the two low-lying singlet states as a function of single and/or double bond torsion. In visual chromophores exactly such a torsional motion stores light energy,[183] and for the shorter polyenes the interaction (or lack thereof) may be at the root of the lack of fluorescence.

Acknowledgements

The author wishes to acknowledge support from the Camille and Henry Dreyfus Foundation in the form of a Camille and Henry Dreyfus Teacher-Scholar Award (1993-98). Support from the NSF is also gratefully acknowledged, in the form of an NSF-ARI grant (CHE-9512467).

References

1. K. Raghavachari, G. W. Trucks, J. A. Pople, and M. Head-Gordon, *Chem. Phys. Lett.* **157**, 479 (1989).
2. M. Head-Gordon, *J. Phys. Chem.* **100**, 13213 (1996).
3. C. Møller and M. S. Plesset, *Phys. Rev.* **46**, 618 (1934).

4. R. J. Bartlett and D. M. Silver, *J. Chem. Phys.* **62**, 3258 (1975).

5. J. A. Pople, J. S. Binkley, and R. Seeger, *Int. J. Quantum Chem. Quantum Chem. Symp.* **10**, 1 (1976).

6. H. Hsu, E. R. Davidson, and R. M. Pitzer, *J. Chem. Phys.* **65**, 609 (1976).

7. E. R. Davidson and L. Z. Stenkamp, *Int. J. Quantum Chem. Quantum Chem. Symp.* **10**, 21 (1976).

8. R. J. Bartlett, *Ann. Rev. Phys. Chem.* **32**, 359 (1981).

9. J. C. Slater, *Quantum Theory of Molecules and Solids, Vol. 1, Electronic Structure of Molecules* (McGraw Hill, New York, 1963).

10. M. Born and J. R. Oppenheimer, *Ann. Physik*, **84**, 457 (1927).

11. L. I. Schiff, *Quantum Mechanics, Third Edition* (McGraw Hill, New York, 1968).

12. G. Herzberg, *Molecular Spectra and Molecular Structure, III. Electronic Spectra and Electronic Structure of Polyatomic Molecules* (Van Nostrand Reinhold, New York, 1966).

13. M. B. Robin, *Higher Excited States of Polyatomic Molecules, Vol. 1*, (Academic Press, New York, 1974).

14. M. B. Robin, *Higher Excited States of Polyatomic Molecules, Vol. 2*, (Academic Press, New York, 1975).

15. M. B. Robin, *Higher Excited States of Polyatomic Molecules, Vol. 3*, (Academic Press, New York, 1985).

16. W. T. Borden, *Modern Molecular Orbital Theory for Organic Chemists*, (Prentice Hall, Englewood Cliffs, New Jersey, 1975).

17. L. Pauling, *The Nature of the Chemical Bond*, (Cornell University Press, Ithaca, New York 1960).

18. W. T. Borden and E. R. Davidson, *Acc. Chem. Res.* **29**, 67 (1996).

19. M. A. C. Nascimento and W. A. Goddard III, *Chem. Phys.* **36**, 147 (1979).

20. A. Messiah, *Quantum Mechanics, Vol. I* (Wiley, New York, 1958).

21. R. Krishnan, J. S. Binkley, R. Seeger, and J. A. Pople, *J. Chem. Phys.* **72**, 650 (1980).

22. T. H. Dunning and P. J. Hay, in *Methods of Electronic Structure Theory*, Ed. H. F. Schaefer III (Plenum, New York, 1976).

23. A. Szabo and N. S. Ostlund, *Modern Quantum Chemistry* (McGraw Hill, New York, 1982).

24. J. B. Foresman, M. Head-Gordon, J. A. Pople, and M. J. Frisch, *J. Phys. Chem.* **96**, 135 (1992).

25. E. A. Hylleraas and B. Undheim, *Z. Phys.* **65**, 759 (1930).

26. J. K. L. MacDonald, *Phys. Rev.* **43**, 830 (1933).

27. I. Shavitt, in *Methods of Electronic Structure Theory*, Ed. H. F. Schaefer III

(Plenum, New York, 1976).

28. G. E. Scuseria and H. F. Schaefer III, *J. Chem. Phys.* **88**, 7024 (1988).

29. V. A. Walters, C. M. Hadad, Y. Thiel, S. D. Colson, K. B. Wiberg, P. M. Johnson, and J. B. Foresman, *J. Am. Chem. Soc.* **113**, 4782 (1991).

30. K. B. Wiberg, C. M. Hadad, J. B. Foresman, and W. A. Chupka, *J. Phys. Chem.* **96**, 10756 (1992).

31. K. B. Wiberg, C. M. Hadad, G. B. Ellison, and J. B. Foresman, *J. Phys. Chem.* **97**, 13586 (1993).

32. E. R. Davidson and D. W. Silver, *Chem. Phys. Lett.* **53**, 403 (1977).

33. S. R. Langhoff and E. R. Davidson, *Int. J. Quantum Chem.* **8**, 61 (1974).

34. E. R. Davidson, in *The World of Quantum Chemistry*, eds. R. Daudel and B. Pullman (Reidel, Dordrecht, 1974).

35. J. Simons, *J. Phys. Chem.* **93**, 626 (1989).

36. J. A. Pople, R. Seeger, and R. Krishnan, *Int. J. Quantum Chem. Quantum Chem. Symp.* **11**, 149 (1977).

37. P. O. Löwdin, *J. Mol. Spect.* **13**, 326 (1964).

38. R. K. Nesbet, *Proc. Roy. Soc.* **A230**, 312, 322 (1955).

39. P. S. Epstein, *Phys. Rev.* **28**, 695 (1926).

40. E. R. Davidson, L. E. Nitzsche, and L. E. McMurchie, *Chem. Phys. Lett.* **62**, 467 (1979).

41. Several examples are given in C. W. Murray and N. C. Handy, *J. Chem. Phys.* **97**, 6509 (1992).

42. R. B. Murphy and R. P. Messmer, *J. Chem. Phys.* **97**, 4170 (1992).

43. J. Cizek, *Adv. Chem. Phys.* **14**, 35 (1969), and references cited therein.

44. R. J. Bartlett and G. D. Purvis, *Int. J. Quantum Chem.* **14**, 561 (1978).

45. J. A. Pople. R. Krishnan, H. B. Schlegel, and J. S. Binkley, *Int. J. Quantum Chem.* **14**, 545 (1978).

46. R. J. Bartlett, *J. Phys. Chem.* **93**, 1697 (1989), and references cited therein.

47. G. E. Scuseria and T. J. Lee, *J. Chem. Phys.* **93**, 5851 (1990).

48. G. D. Purvis III and R. J. Bartlett, *J. Chem. Phys.* **76**, 1910 (1982).

49. J. A. Pople, M. Head-Gordon, and K. Raghavachari, *J. Chem. Phys.* **87**, 5968 (1987).

50. B. Roos, in *Ab Initio Methods in Quantum Chemistry II*, ed. K. Lawley (Wiley, New York, 1987).

51. F. W. Bobrowicz and W. A. Goddard III, in *Methods of Electronic Structure Theory*, Ed. H. F. Schaefer III (Plenum, New York, 1976).

52. R. J. Cave and E. R. Davidson, *J. Phys. Chem.* **91**, 4481 (1987).

53. C. W. Bauschlicher, Jr. and S. R. Langhoff, *Science*, **254**, 394 (1991).

54. C. W. Bauschlicher, Jr. and S. R. Langhoff, *Chem. Rev.*, **91**, 701 (1991).

55. D. C. Rawlings, E. R. Davidson, and M. Gouterman, *Int. J. Quantum Chem.* **26**, 251 (1984).

56. B. Huron, J. P. Malrieu, and P. Rancurel, *J. Chem. Phys.* **58**, 5745 (1973).

57. R. J. Buenker, S. Shih, and S. D. Peyerimhoff, *Chem. Phys. Lett.* **44**, 385 (1976).

58. R. J. Cave, S. S. Xantheas, and D. Feller, *Theor. Chim. Acta* **83**, 31 (1992).

59. A. Banerjee and J. Simons, *Int. J. Quantum Chem.* **19**, 207 (1981).

60. A. Banerjee and J. Simons, *J. Chem. Phys.* **76**, 4548 (1982).

61. H. Baker and M. A. Robb, *Mol. Phys.* **50**, 1077 (1983).

62. W. D. Laidig and R. J. Bartlett, *Chem. Phys. Lett.* **104**, 424 (1984).

63. K. Tanaka and H. Terashima, *Chem. Phys. Lett.* **106**, 558 (1984).

64. M. R. Hoffman and J. Simons, *J. Chem. Phys.* **88**, 993 (1988).

65. R. J. Cave and E. R. Davidson, *J. Chem. Phys.* **88**, 5770 (1988).

66. R. J. Cave and E. R. Davidson, *J. Chem. Phys.* **89**, 6798 (1988).

67. R. J. Gdanitz and R. Ahlrichs, *Chem. Phys. Lett.* **143**, 413 (1988).

68. J.-L. Heully and J. P. Malrieu, *Chem. Phys. Lett.* **199**, 545 (1992).

69. J. Meller, J. P. Malrieu, and R. Caballol, *J. Chem. Phys.* **104**, 4068 (1996).

70. P. G. Szalay and R. J. Bartlett, *J. Chem. Phys.* **103**, 3600 (1995), and references cited therein.

71. R. J. Cave, *J. Chem. Phys.* **92**, 2450 (1990).

72. R. J. Cave and M. J. Perrott, *J. Chem. Phys.* **96**, 3745 (1992).

73. B. H. Brandow, *Rev. Mod. Phys.* **39**, 771 (1967).

74. K. Andersson, P.-A. Malmqvist, B. Roos,A. Sadlej, and K. Wolinski, *J. Phys. Chem.* **94**, 5483 (1990).

75. K. Andersson, P.-A. Malmqvist, and B. O. Roos, *J. Chem. Phys.* **96**, 1218 (1992).

76. K. Hirao, *Chem. Phys. Lett.* **190**, 374 (1992).

77. R. B. Murphy and R. P. Messmer, *Chem. Phys. Lett.* **183**, 443 (1991).

78. K. Wolinski, H. L. Sellers, and P. Pulay, *Chem. Phys. Lett.* **140**, 225 (1987).

79. K. Wolinski and P. Pulay, *J. Chem. Phys.* **90**, 3647 (1989).

80. I. Shavitt and L. T. Redmon, *J. Chem. Phys.* **73**, 5711 (1980).

81. B. Kirtman, *J. Chem. Phys.* **75**, 798 (1981).

82. M. R. Hoffmann, *Chem. Phys. Lett.* **195**, 127 (1992).

83. M. R. Hoffmann, *J. Chem. Phys.* **100**, 6125 (1996).

84. P. M. Kozlowski and E. R. Davidson, *J. Chem. Phys.* **100**, 3672 (1994).

85. P. M. Kozlowski and E. R. Davidson, *Chem. Phys. Lett.* **222**, 615 (1994).

86. G. Hose and U. Kaldor, *J. Phys. B*, **12**, 3827 (1979).

87. S. Zarrabian and R. J. Bartlett, *Chem. Phys. Lett.* **153**, 133 (1988).

88. L. Meissner, S. A. Kucharski, and R. J. Bartlett, *J. Chem. Phys.* **93**, 1847 (1990).

89. R. L. Graham and K. F. Freed, *J. Chem. Phys.* **96**, 1304 (1992), and references

cited therein.

90. L. E. Nitzsche and E. R. Davidson, *J. Chem. Phys.* **68**, 3103 (1978).

91. Z. Gershgorn and I. Shavitt, *Int. J. Quantum Chem.* **2**, 751 (1968).

92. U. Kaldor, *J. Chem. Phys.* **87**, 467 (1987).

93. D. Mukherjee, *Chem. Phys. Lett.* **125**, 207 (1986).

94. S. Pal, M. Rittby, and R. J. Bartlett, *Chem. Phys. Lett.* **160**, 212 (1989).

95. B. Jeziorski and H. J. Monkhorst, *Phys. Rev. A*, **24**, 1668 (1988).

96. J. Geertsen, M. Rittby, and R. J. Bartlett, *Chem. Phys. Lett.* **164**, 57 (1989).

97. S. Pal, M. Rittby, R. J. Bartlett, D. Sinha, and D. Mukherjee, *J. Chem. Phys.* **88**, 4357 (1988).

98. A. Balkova and R. J. Bartlett, *Chem. Phys. Lett.* **193**, 364 (1992).

99. K. Jankowski, J. Paldus, I. Garowski, and K. Kowalski, *J. Chem. Phys.* **97**, 7600 (1992).

100. S. I. Kucharski and R. J. Bartlett, *Int. J. Quantum Chem. Quantum Chem. Symp.* **26**, 107 (1992).

101. A. Balkova, S. A. Kucharski, L. Meissner, and R. J. Bartlett, *J. Chem. Phys.* **95**, 4311 (1991).

102. S. A. Kucharski and R. J. Bartlett, *J. Chem. Phys.* **95**, 8227 (1991).

103. J. F. Stanton, R. J. Bartlett, and C. M. L. Rittby, *J. Chem. Phys.* **97**, 5560 (1992).

104. J. F. Stanton and R. J. Bartlett, *J. Chem. Phys.* **98**, 7029 (1993).

105. O. Christiansen, H. Koch, and P. Jørgensen, *J. Chem. Phys.* **105**, 1451 (1996).

106. H. Nakatsuji, *Chem. Phys. Lett.* **67**, 329 (1979).

107. J. F. Stanton and J. Gauss, *J. Chem. Phys.* **104**, 9859 (1996).

108. O. Kitao and H. Nakatsuji, *J. Chem. Phys.* **87**, 1169 (1987).

109. J. M. O. Matos, B. O. Roos, and P.-A. Malmqvist, *J. Chem. Phys.* **86**, 1458 (1987).

110. K. Hirao, H. Nakano, and T. Hashimoto, *Chem. Phys. Lett.* **235**, 430 (1995).

111. B. O. Roos, K. Andersson, and M. P. Fülscher, *Chem. Phys. Lett.* **192**, 5 (1992).

112. O. Christiansen, H. Koch, A. Halkier, P.Jørgensen, T. Helgaker, and A. Sanchez de Meras, *J. Chem. Phys.* **105**, 6921 (1996).

113. A. D. Baker, C. Baker, C. R. Brundle, and D. W. Turner, *Int. J. Mass Spect. Ion Phys.* **1**, 285 (1968).

114. T. Reddish, B. Wallbank, and J. Comer, *Chem. Phys.* **108**, 159 (1986).

115. R. M. Gavin, Jr., S. Risemberg, and S. A. Rice, *J. Chem. Phys.* **58**, 3160 (1973).

116. T. B. Jones and J. P. Maier, *Int. J. Mass Spect. Ion Phys.* **31**, 287 (1979).

117. M. H. Palmer, A. J. Beveridge, I. C. Walker, and T. Abuain, *Chem. Phys.* **102**,

63 (1986).

118. C. Petrongolo, R. Buenker, and S. D. Peyerimhoff, *J. Chem. Phys.* **76**, 3655 (1982).

119. E. Miron, B. Raz, and J. Jortner, *Chem. Phys. Lett.* **6**, 563 (1970).

120. M. Krauss and S. R. Mielczarek, *J. Chem. Phys.* **51**, 5241 (1969).

121. R. Lindh and B. O. Roos, *Int. J. Quantum Chem.* **35**, 813 (1989).

122. B. R. Brooks and H. F. Schaefer III, *J. Chem. Phys.* **68**, 4839, (1978).

123. L. E. McMurchie and E. R. Davidson, *J. Chem. Phys.* **66**, 2959 (1977).

124. L. E. McMurchie and E. R. Davidson, *J. Chem. Phys.* **67**, 5613 (1978).

125. E. R. Davidson, *J. Phys. Chem.* **100**, 6161 (1996).

126. K. K. Sunil, K. D. Jordan, and R. Shepard, *Chem. Phys.* **88**, 55 (1984).

127. R. J. Buenker, S.-K. Shih, and S. D. Peyerimhoff, *Chem. Phys.* **36**, 97 (1979).

128. L. Serrano-Andres, M. Merchan, I. Nebot-Gil, R. Lindh, and B. O. Roos, *J. Chem. Phys.* **98**, 3151 (1993).

129. J. D. Watts, S. R. Gwaltney, and R. J. Bartlett, *J. Chem. Phys.* **105**, 6979 (1996).

130. a) P. A. Malmqvist, A. Rendell, and B. O. Roos, *J. Phys. Chem.* **94**, 5477 (1990); b) R. J. Cave, unpublished results.

131. C. F. Bender and E. R. Davidson, *J. Phys. Chem.* **70**, 2675 (1966).

132. J. H. Moore, Y. Sato, and S. W. Staley, *J. Chem. Phys.* **69**, 1092 (1978).

133. P. H. Taylor, W. G. Mallard, and K. C. Smith, *J. Chem. Phys.* **84**, 1053 (1986).

134. K. B. Wiberg, K. S. Peters, G. B. Ellison, and J. L. Dehmer, *J. Chem. Phys.* **66**, 2224 (1977).

135. J. P. Doering, *J. Chem. Phys.* **70**, 3902 (1979).

136. W. M. Flicker, O. A. Mosher, and A. Kupperman, *Chem. Phys.* **30**, 307 (1978).

137. J. P. Doering and R. McDiarmid, *J. Chem. Phys.* **73**, 3617 (1980).

138. R. R. Chadwick, D. P. Gerrity, and B. S. Hudson, *Chem. Phys. Lett.* **115**, 24 (1985).

139. D. G. Leopold, R. D. Pendley, J. L. Roebber, R. J. Hemley, and V. Vaida, *J. Chem. Phys.* **81**, 4218 (1984).

140. P. Swiderek, M. Michaud, and L. Sanche, *J. Chem. Phys.* **98**, 8397 (1993).

141. R. R. Chadwick, M. Z. Zgierski, and B. S. Hudson, *J. Chem. Phys.* **95**, 7204 (1991).

142. R. J. Hemley, J. I. Dawson, and V. Vaida, *J. Chem. Phys.* **78**, 2915 (1983).

143. R. J. Buenker and J. L. Whitten, *J. Chem. Phys.* **49**, 5381 (1968).

144. R. P. Hosteny, T. H. Dunning, Jr., R. R. Gilman, A. Pipano, and I. Shavitt, *J. Chem. Phys.* **62**, 4764 (1975).

145. S. Shih, R. J. Buenker, and S. D. Peyerimhoff, *Chem. Phys. Lett.* **16**, 244 (1972).

146. M. A. C. Nascimento and W. A. Goddard III, *Chem. Phys.* **53**, 251 (1980).

147. L. Serrano-Andres, J. Sanchez-Mann, and I. Nebot-Gil, *J. Chem. Phys.* **97**, 7499 (1992).

148. O. Kitao and H. Nakatsuji, *Chem. Phys. Lett.* **143**, 528 (1988).

149. P. G. Szalay, A. Karpfen, and H. Lischka, *Chem. Phys.* **130**, 219 (1989).

150. M. Aoyagi, I. Ohmine, and B. E. Kohler, *J. Phys. Chem.* **94**, 3922 (1990).

151. A. Lasaga, R. J. Aerni, and M. Karplus, *J. Chem. Phys.* **73**, 5230 (1980).

152. R. M. Gavin, Jr., C. Weisman, J. K. McVey, and S. A. Rice, *J. Chem. Phys.* **68**, 522 (1978).

153. H. Petek, A. J. Bell, Y. S. Choi, K. Yoshihara, B. A. Tounge, and R. L. Christensen, *J. Chem. Phys.* **98**, 3777 (1993).

154. L. Serrano-Andres, R. Lindh, B. O. Roos, and M. Merchan, *J. Phys. Chem.*

155. R. J. Cave and E. R. Davidson, *Chem. Phys. Lett.* **148**, 190 (1988).

156. M. E. Squillacote, R. S. Sheridan, O. L. Chapman, F. A. L. Anet, *J. Am. Chem. Soc.* **101**, 3657 (1979).

157. I. L. Alberts and H. F. Schaefer, *Chem. Phys. Lett.* **161**, 375 (1989).

158. P. G. Szalay, H. Lischka, and A. Karpfen, *J. Phys. Chem.* **93**, 6629 (1989).

159. J. E. Rice, B. Liu, T. J. Leem C. M. Rohlfing, *Chem. Phys. Lett.* **161**, 277 (1989).

160. L. Serrano-Andres, B. O. Roos, and M. Merchan, *Theor. Chim. Acta*, **87**, 387 (1994).

161. S. Y. Lee and K. F. Freed, *J. Chem. Phys.* **104**, 3260 (1996).

162. J. H. Eland, *Int. J. Mass Spectrom. Ion Phys.* **2**, 471 (1969).

163. R. C. Dunbar, *Chem. Phys. Lett.* **32**, 508 (1975).

164. T. Bally, S. Nitsche, K. Roth, and E. Haselbach, *J. Am. Chem. Soc.* **106**, 3927 (1984).

165. R. J. Cave and M. G. Perrott, *J. Chem. Phys.* **96**, 3745 (1992).

166. T. Bally, W. Tang, and M. Jungen, *Chem. Phys. Lett.* **190**, 453 (1992).

167. W. J. Buma, B. E. Kohler, and K. Song, *J. Chem. Phys.* **94**, 6367 (1991).

168. W. M. Flicker, O. A. Mosher, and A. Kuppermann, *Chem. Phys. Lett.* **45**, 492 (1977).

169. M. F. Granville, B. E. Kohler, and J. B. Snow, *J. Chem. Phys.* **75**, 3765 (1981).

170. T. Fujii, A. Kamata, M. Shimizu, Y. Adachi, S. Maeda, *Chem. Phys. Lett.* **115**, 369 (1985).

171. R. M. Gavin and S. A. Rice, *J. Chem. Phys.* **60**, 3231 (1974).

172. a) A. Sabljic and R. McDiramid, *J. Chem. Phys.* **82**, 2559 (1985); b) J. P. Doering, A. Sabljic, and R. McDiarmid, *J. Phys. Chem.* **88**, 835 (1984).

173. D. H. Parker, J. O. Berg, and M. A. El-Sayed, *Chem. Phys. Lett.* **56**, 197 (1978).

174. M. A. C. Nascimento and W. A. Goddard III, *Chem. Phys. Lett.* **60**, 197 (1979).

175. M. A. C. Nascimento and W. A. Goddard III, *Chem. Phys.* **53**, 265 (1980).

176. R. J. Cave and E. R. Davidson, *J. Phys. Chem.* **92**, 614 (1988).

177. C. H. Martin and K. F. Freed, *J. Phys. Chem.* **99**, 2701 (1995).

178. M. F. Granville, G. R. Holtom, and B. E. Kohler, *J. Chem. Phys.* **72**, 4671 (1980).

179. W. G. Bouman, A. C. Jones, D. Phillps, P. Thibodeau, Ch. Friel, and R. L. Christensen, *J. Phys. Chem.* **94**, 7429 (1990).

180. R. J. Cave and E. R. Davidson, *J. Phys. Chem.* **92**, 2173 (1988).

181. L. Serrano-Andres, R. Lindh, B. O. Roos, and M. Merchan, *J. Phys. Chem.* **97**, 9360 (1993).

182. J. P. Finley, R. K. Chauduri, and K. F. Freed, *J. Chem. Phys.* **103**, 4990 (1995).

183. R. A. Mathies, C. H. B. Cruz, W. T. Pollard, and C. V. Shank, *Science*, **240**, 777 (1988).

LONG-RANGE INTRAMOLECULAR INTERACTIONS: IMPLICATIONS FOR ELECTRON TRANSFER

KENNETH D. JORDAN, DANA NACHTIGALLOVA[*]

Department of Chemistry, University of Pittsburgh, Pittsburgh, PA 15260 USA

MICHAEL N. PADDON-ROW

School of Chemistry, University of New South Wales, Sydney, 2052 Australia

In the weak coupling limit the rate of electron transfer in donor-bridge-acceptor molecules is proportional to the square of the electronic coupling between the donor and acceptor groups. The electronic coupling is usually dominated by through-bond interactions involving orbitals of the bridge. In this article model compounds and model Hamiltonians are used to illustrate key ideas about through-bond coupling through saturated hydrocarbon bridges.

1   Introduction

Chemists have long been fascinated by the problem of the extent to which one functional group influences the properties of other, remote functional groups. In recent years much of the interest in this problem has derived from the role that such interactions play in intramolecular electron, hole, and excitation transfer. These processes are of fundamental importance in photosynthesis and a host of other biological processes,[1a] electron transfer at semiconductor/electrolyte interfaces,[2] and in the design of molecular electronic devices.[3] Advances in our understanding of the factors that determine the strength of the coupling between remote functional groups, in particular, how it depends on the nature of the intervening media, have come largely as a result of: (1) the design, synthesis, and spectroscopic characterization of model chromophore-bridge-chromophore compounds[1,4] and (2) detailed theoretical studies tracing the "pathways" responsible for the coupling.[5-12]

In the nonadiabatic (weak-coupling) limit, the rate of electron transfer (ET) is given by

[*] J. Heyrovsky Institute of Physical Chemistry, Academy of Sciences of the Czech Republic, Dolejskova 3, 182 33 Prague 8

$$k = \left(\frac{2\pi}{\eta}\right) H_{el}^2 \{FCWD\}, \qquad (1)$$

where H_{el} is the relevant electronic coupling and $\{FCWD\}$ is the Franck-Condon weighted density of states.[13,14] Analogous expressions hold for the rates of hole transfer (HT) and electronic excitation transfer (EET). Along a series of closely related donor-bridge-acceptor (DBA) molecules with the same donor and acceptor groups and with similar bridges, differing primarily in length, e.g., the $I(n)$ series[4]

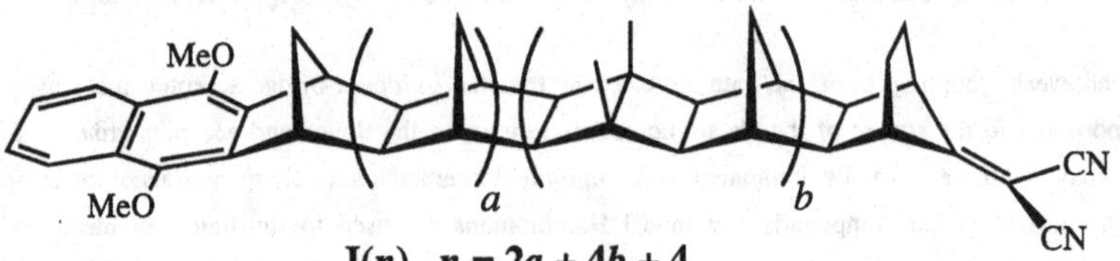

$$I(n) \quad n = 2a + 4b + 4$$

$$II(n) \quad n = 2m + 2 \qquad III(n) \quad n = 2m + 2 \qquad (a)$$

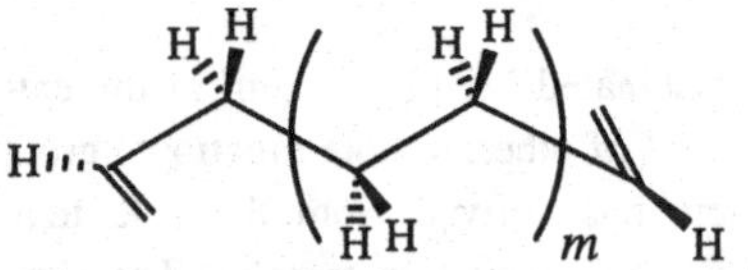

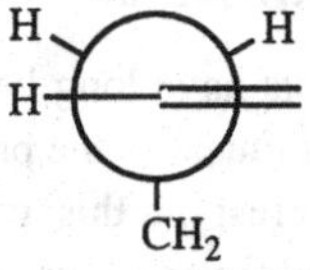

Figure 1. Bichromophoric species considered in this article. The length of the bridge in each species is given in terms of n, the number of C-C bonds that span the bridge termini. The orientations of the double bonds with respect to the alkane bridge are shown in diagram (a).

depicted in Fig. 1, the dependence of the rate of ET with bridge length is governed primarily by the variation of H_{el}. In elucidating the factors that control H_{el}, it has proven useful to focus on simpler model compounds, e.g., the $II(n)$ and $III(n)$ series.[4,6-12] In large measure this is because the model compounds are more accessible to theoretical study.

In the present article, the $II(n)$ and $III(n)$ series are used to illustrate the mechanisms for electronic coupling between remote chromophores in chromophore-bridge-chromophore species. The usefulness of orbital splittings and their decomposition into through-space and through-bond[15-18] contributions are discussed in Section 2. Approaches for decomposing the net through-bond coupling into contributions from individual pathways are presented in Section 3. Section 4 discusses the implications of our results for the ET problem, and Sections 5 and 6 will deal, respectively, with the construction of effective two-level model Hamiltonians and the role of solvent molecules in propagating electronic interactions. Section 7 addresses some of the issues connected with electronic excitation transfer.

2 π_+,π_- and π_+^*,π_-^* Splittings: Through-space and Through-bond Interactions

The molecules $\mathbf{II}(n)$, with two double bonds separated by "polynorbornyl" groups, have played a particularly important role in the development of our understanding of long-range intramolecular interactions. These molecules have two π cation states and two π^* anion states, the splittings between which, ΔIP and ΔEA, provide measures of the electronic couplings between the cation and anion states, respectively. In the Koopmans' theorem[19] (KT) picture, the ionization potentials (IP's) and electron affinities (EA's) are associated, respectively, with the negatives of the energies of the filled and unfilled Hartree-Fock orbitals of the neutral molecules. In this orbital picture ΔIP and ΔEA can be associated with the splittings between the filled π and unfilled π^* orbitals, respectively. As is well known, the KT approximation neglects relaxation and electron correlation contributions to the IP's and EA's. However, it has been shown for the $\mathbf{II}(n)$ series and for other bichromophoric systems that the KT approximation usually provides fairly reliable estimates of the ΔIP and ΔEA values. In the remainder of this article we will assume the validity of the KT approximation in discussing the splittings between cationic states and between anionic states of unsaturated bichromophoric systems.

In analyzing orbital interactions, it is convenient to adopt a localized orbital picture. Specifically, for the $\mathbf{II}(n)$ series we identify π_1 and π_2 as localized π orbitals on the two chromophores and π_1^* and π_2^* as the corresponding localized π^* orbitals. These localized orbitals are essentially unmixed with the σ orbitals of the bridge. It is also useful to form from these symmetry-adapted π_+,π_-,π_+^*, and π_-^* orbitals:

$$\pi_+ = \pi_1 + \pi_2 \qquad\qquad \pi_+^* = \pi_1^* + \pi_2^*$$
$$\pi_- = \pi_1 - \pi_2 \qquad\qquad \pi_-^* = \pi_1^* - \pi_2^* \tag{2}$$

The splittings between the π_+ and π_- orbitals and between the π_+^* and $\tilde{\pi}_-^*$ orbitals can be taken as measures of the direct or through-space (TS) coupling between the chromophores. These localized chromophore orbitals mix with the σ and σ^* orbitals of the bridge to give the canonical $\tilde{\pi}_+, \tilde{\pi}_-, \tilde{\pi}_+^*$, and $\tilde{\pi}_-^*$ orbitals. The splittings between the canonical orbitals, $\Delta E(\pi)$, and $\Delta E(\pi^*)$, contain contributions due to both TS and through-bond (TB) interactions.[15-18] By definition, TB interactions encompass all interactions between the chromophores that are "relayed" by bridge orbitals. An estimate of the TB contribution to the splitting can be obtained by subtracting the TS contribution, calculated as described above, from the net splitting.

In II(4), in which the two double bonds are separated by about 5 Å, photoelectron spectroscopy (PES) gives a splitting of 0.87 eV between the two π cation states,[20a,b,c] and electron transmission spectroscopy (ETS) gives a splitting of 0.90 eV between the two π^* anion states.[20d] These experimental state splittings are closely reproduced by the orbital splittings obtained from Hartree-Fock (HF) calculations using the split-valence 3-21G basis set[21] and invoking the KT approximation.[6,22,23] The $\Delta E(\pi)$, and $\Delta E(\pi^*)$ values for II(4) are about an order of magnitude greater than can be accounted for by TS interactions. This can be established by comparing the splittings for II(4) with the results of HF/3-21G calculations on a model ethylene dimer, with the two ethylene molecules the same distance apart and with the same relative orientation as the double bonds in II(4). For this model dimer, for which the interaction is necessarily TS, the π and π^* splittings are only about 0.1 eV. This leads us to conclude that TB interactions dominate the coupling between the two ethylenic groups in II(4). Moreover, the fraction of the coupling which is due to TB interactions grows with increasing length of the bridge.

3   Use of NBO's in analyzing TB interactions.

While the above approach has allowed us to separate the TB and TS contributions to the overall coupling, additional analysis is required in order to understand how the TB interactions depend on the length and other properties of the bridge. We,[6-10,24-26] Liang and Newton,[11] and Miller and coworkers,[12] have all adopted analysis schemes based on natural bond orbitals (NBO's[27]) for addressing this question. The NBO's are a set of orthogonal localized orbitals, obtained by a unitary transformation of the canonical Hartree-Fock MO's. They are close to the chemist's "intuitive" set of orbitals, and can be classified as two-center bonding, two-center antibonding, lone-pair, and "Rydberg" orbitals. (The "Rydberg" virtual orbitals are absent in a minimal basis set.)

The key to the NBO analysis procedure is the transformation of the Fock matrix to the basis of the NBO's (or symmetry-adapted NBO's). By zeroing out particular matrix elements in the NBO Fock matrix, followed by diagonalization, it is possible to establish the role of particular classes of interactions in the TB coupling. Alternatively, one can start with a blank Fock matrix and add in subsets of interactions. The strategy outlined above is close to that introduced by Heilbronner *et al.*[28a] and applied by Imamura and Ohsaku[28c] and by Gleiter and Schaefer.[28b]

3.1 Four-orbital four-site model

Consideration of a simple four-orbital, four-site model will suffice to illustrate the utility of the localized orbitals for analyzing TB interactions. The model consists of two localized π orbitals, π_1 and π_4, and two intervening localized σ orbitals, labeled, σ_2 and σ_3 (see Fig. 2). We further assume that the localized orbitals are NBO's, although the equations

derived, in fact, are valid for any set of orthogonal localized orbitals. If it is assumed that the TS coupling between the localized π orbitals is negligible and that only nearest-neighbor interactions are important, the NBO Fock matrix becomes

Figure 2. Four-orbital four-site model.

$$H = \begin{pmatrix} \varepsilon_\pi & T & 0 & 0 \\ T & \varepsilon_\pi & t & 0 \\ 0 & t & \varepsilon_\pi & T \\ 0 & 0 & T & \varepsilon_\pi \end{pmatrix} \tag{3}$$

where, t is the matrix element coupling σ_2 and σ_3, T is the matrix element describing the coupling of the π orbitals of the chromophores to the adjacent σ orbitals of the bridge, and ε_π and ε_σ are the self energies of the π and σ NBO's, respectively. By adoption of the symmetry-adapted orbitals π_+, π_-, σ_+ and σ_-, the Fock matrix can be block diagonalized to give:

$$H = \begin{pmatrix} \varepsilon_\pi & T & 0 & 0 \\ T & \varepsilon_\sigma + t & 0 & 0 \\ 0 & t & \varepsilon_\pi & T \\ 0 & 0 & T & \varepsilon_\sigma - t \end{pmatrix} \qquad (4)$$

The eigenvalues in the "+" and "-" symmetry blocks are:

$$E_+ = \left(\frac{\varepsilon_\pi + \varepsilon_\sigma + t}{2} \right) \pm \frac{1}{2} \sqrt{(\varepsilon_\pi - \varepsilon_\sigma - t)^2 + 4T^2}$$

$$(5)$$

$$E_- = \left(\frac{\varepsilon_\pi + \varepsilon_\sigma - t}{2} \right) \pm \frac{1}{2} \sqrt{(\varepsilon_\pi - \varepsilon_\sigma + t)^2 + 4T^2}$$

Taking the roots that correspond to the $\tilde{\pi}_+$ and $\tilde{\pi}_-$ eigenvalues, and using a Taylor series expansion of the square roots (with the assumption that $|t|$ and $|T| << \Delta \equiv \varepsilon_\pi - e_\sigma$), retaining terms through lowest order in the T and t interactions, one obtains:

$$\varepsilon_{\tilde{\pi}_+} = \varepsilon_\pi + \frac{T^2}{\Delta} + \frac{tT^2}{\Delta^2}$$

$$(6)$$

$$\varepsilon_{\tilde{\pi}_-} = \varepsilon_\pi + \frac{T^2}{\Delta} - \frac{tT^2}{\Delta^2}$$

Thus the interaction between σ_1 and σ_2 (which splits apart the localized σ orbitals to give σ_+ and σ_-.) causes both the π_+ and π_- orbitals to be destabilized, but to different extents. To lowest order in t the splitting between $\tilde{\pi}_+$ and $\tilde{\pi}_-$ is:

$$\Delta E(\pi) = -\left(\frac{2T^2}{\Delta} \right) \left(\frac{t}{2} \right). \qquad (7)$$

This expression for the splitting is just the result that is obtained from the McConnell model.[29,30] The generalization of this model to a bridge with n equivalent sites gives, again to lowest order,

$$\Delta E(\pi) = -\left(\frac{2T^2}{\Delta}\right)\left(\frac{t}{2}\right)^{n-1}. \qquad (8)$$

The most important predictions of the McConnell model are that the splitting due to TB coupling falls off exponentially with the number of bonds in the relay and that the net splitting is a product of two terms, one giving the coupling of the chromophore to the bridge (T^2/Δ) and the other $\{(t/\Delta)^{n-1}\}$ the propagation of the interaction along the bridge.

There is a number of assumptions and approximations in the derivation of the McConnell equation. These include:

1. restriction to nearest-neighbor interactions only
2. the restriction to a single relay between the chromophores
3. the restriction to only a single orbital on each site
4. the assumption that all bridge sites are identical
5. the assumption that the weak-coupling limit is valid (*i.e.*, $|t/\Delta| \ll 1$ and $|T/\Delta| \ll 1$).

In many chromophore-bridge-chromophore systems, *e.g.*, the $\mathbf{I}(n)$ and $\mathbf{II}(n)$ series, the first four of these criteria are not valid, and the validity of the last assumption is questionable. In light of this, it is amazing that both experiment and theory show for a wide range of chromophore-bridge-chromophore systems that the electronic coupling depends nearly exponentially on the length of the bridge. However, as will be discussed below, the McConnell model gives much too great an attenuation of the electronic coupling with increasing bridge length.

The assumption regarding the equivalence of the various bridge sites is readily relaxed, giving[30,31]

$$\Delta E(\pi) = \frac{-2T_{D,1}T_{n,A}}{\Delta_1}\prod_{i=1}^{n-1}\frac{t_{i,i+1}}{\Delta_{i+1}} \qquad (9)$$

where n is the number of bridge sites, $t_{i,i+1}$ is the coupling between the i and $i+1$ bridge sites, $T_{D,1}$ and $T_{n,A}$ are the matrix elements giving the coupling of the chromophores to the ends of the bridge, and $\Delta_i = \varepsilon - \varepsilon_i$ is the energy difference between the chromophore levels and the bridge level i. (The "D" and "A" denote "donor" and "acceptor" chromophores, respectively.)

Several extensions of the McConnell model have appeared.[32-36] But for the most part these have focused on removing the weak-coupling assumption and have retained the restriction to nearest-neighbor interactions only. An exception is Ref. 35, discussed below.

3.2 Interactions that skip over bonds

We now examine the role of next nearest-neighbor interactions. In doing so we consider a series of model compounds **III**(n) comprising two ethylenic groups connected by an *all-trans* alkane bridge as show in Fig 3. For each member of the series, HF/3-21G calculation[3,9] were carried out and used to generate the Fock matrix in the NBO representation. By deleting matrix elements from the full NBO Fock matrix, four different model systems A-D were generated:

A. Nearest-neighbor McConnell-type model (retaining only t interactions between adjacent C-C σ NBO's of the bridge and T interactions coupling the chromophores to the bridge).

B. Same as model *A*, except a longer-range coupling (T') between the chromophores and the bridge is included.

C. Same as *B*, except that next nearest-neighbor interactions (t') along the bridge are included.

D. Same as *C*, except that interactions (t'') that skip over two C-C bonds of the bridge are included.

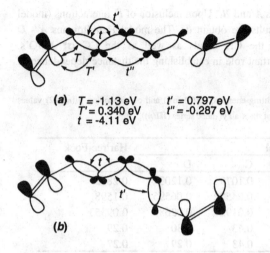

(a) $T = -1.13$ eV $t' = 0.797$ eV
$T' = 0.340$ eV $t'' = -0.287$ eV
$t = -4.11$ eV

(b)

Figure 3. (a) Definition of the matrix elements used in models A, B, C, and D of the through-bond coupling in **III**(n), illustrated for **III**(6). There are slight variations in the t, t', and t'' matrix elements depending on their location along the bridge. (b) Definition of the t and t' matrix elements at a *cis* linkage in an alkane bridge. From comparison of (a) and (b), it is seen that t' is of opposite sign at *cis* and *trans* linkages.

Table I reports the $\tilde{\pi}_+$, $\tilde{\pi}_-$ splittings calculated using each of these models as well as at the HF/3-21G level of theory. The various matrix elements are shown in Fig. 3 in the case of **III**(6).

Examination of the results of these four sets of calculations reveals that: (1) the splittings obtained from model A are much smaller than those from the full HF calculations, and (2) the inclusion of each of the T', t', and t'' interactions leads to sizable increases in the splittings.

Of more interest than the absolute splittings is how they depend on the separation between the chromophores. To examine the distance dependence, the splittings for each pair of consecutive members of the series were fit to an exponential,

$$\Delta E(\pi) = A \exp(-\beta \cdot n), \tag{10}$$

where, as before, n refers to the number of CC σ bonds along a side of the bridge. From examination of the resulting β values (also summarized in Table I), it is clear that the nearest-neighbor model, gives much too rapid a falloff of the π_+, π_- splitting with increasing bridge length. In fact, the β values obtained from this model are about a factor of three larger than those from the HF calculations. Moreover, adding a second coupling mechanism (T') between the chromophore and the bridge does not appreciably alter the distance dependence of the splittings which is to be expected on the basis of the McConnell model in Eq. 8. In contrast, inclusion of the next nearest-neighbor (t') interactions (model C), leads to a much weaker falloff of the coupling with bridge length, with the resulting β

values being about half those of models *A* and *B*. Upon inclusion of t'' interactions (model *D*), β values very close to the HF results are obtained. The model Hamiltonians *A - D* exclude the NBO's associated with the CH bonds as well as the CC σ* NBO's. Apparently, these do not play an important role in establishing the distance dependence of the π_+,π_- splittings.

Table 1: NBO/3-21G and HF/3-21G π_+,π_- splitting energies, $\Delta E(\pi)$ (eV) and corresponding $\beta_h(n,n+2)$ values (per bond) for the 6-, 8-, and 10-bond members of the *n*-alkyl diene series (**III**(*n*)).

n	NBO Model				Hartree-Fock
	A	B	C	D	
6	0.01860	0.05902	0.1079	0.1203	0.2882
8	0.003556	0.01113	0.04522	0.06595	0.1598
10	0.0006832	0.002108	0.01927	0.03689	0.09353
$\beta_h(6,8)^a$	0.83	0.83	0.43	0.30	0.29
$\beta_h(8,10)^a$	0.82	0.83	0.43	0.29	0.27

[a]The $\beta_h(6,8)$ values are calculated from the splittings for **III**(6) and **III**(8); the $\beta_h(8,10)$ values are calculated from the splittings for **III**(8) and **III**(10).

The importance of the t' and t'' interactions for determining the distance dependence of the π_+,π_- splitting can be traced to the rapidly growing number of ways that such interactions can appear with increasing bridge length.[7,9] For example, neglecting pathways that double back, a relay with *n* σ bonds has only one pathway involving *t* matrix elements, *n*-2 pathways with one t' interaction (and the remainder *t*-type interactions), and $(n-4)(n-3)/2$ pathways with two t' interactions.

Even with the large changes introduced by the inclusion of the t' and t'' interactions, the splittings, particularly for the longer bridges, still show a near exponential dependence on the bridge length. This implies that it should be possible to "revive" the McConnell model by using renormalized matrix elements and energy denominators. A renormalization incorporating next nearest-neighbor interactions has recently been described by Lopez-Castillo *et al.*[35]

Lopez-Castillo *et al.* also concluded that the inclusion of next nearest-neighbor t' interactions causes a weakening of the TB coupling (compared to that predicted in a nearest-neighbor McConnell model).[35] However, as shown above, for *trans*-alkane bridges the inclusion of next nearest t' neighbor interactions leads to an enhanced coupling. This apparent contradiction can be understood in terms of the signs of the matrix elements. For *trans*-alkane bridges, the t matrix elements giving the coupling between σ NBO's of adjacent C-C bonds are negative and the t' next nearest-neighbor matrix elements are positive. This can be clearly seen from Fig. 3A in which the nearest-neighbor or NBO adopt a trans conformation with respect to each other. In this case the next nearest-neighbor t' interaction involves an out-of-phase orbital relationship, whereas the nearest-neighbor t interaction involves on in-phase orbital relationship. (In a cis arrangement of the bonds, t and t' would be of the same sign as illustrated in Fig. 3B.) Because a t' interaction replaces two t interactions, the pathways involving next nearest-neighbor interactions have the same overall sign as that which includes only nearest-neighbor interactions. In contrast, in the model considered by Lopez-Castillo *et al.*, the nearest-neighbor and next nearest-neighbor matrix elements were both negative, thereby causing the destructive interference between the McConnell-type pathway and those that include a single t' interaction. The model considered by these authors would be appropriate for describing coupling through a series of bridge sites with s orbitals only (*e.g.*, a chain of H atoms), but is not appropriate for *trans*-alkane bridges.

The above extension of the McConnell model has retained a single relay between the chromophores. In bridges with multiple relays, *e.g.*, those comprising hydrocarbon rings as in $I(n)$ and $II(n)$, significant interference between different TB coupling pathways can result.[7,10,36] Indeed, the electronic coupling is less efficient in the $II(n)$ than in the $III(n)$ series due to the destructive interference between different coupling pathways in the former. The analysis of the contributions due to different pathways is somewhat more straightforward in the π than in the π^* manifold. The matrix element between a σ NBO on one relay and the opposite σ NBO on the parallel relay (t_1) is negative, as is the t-type

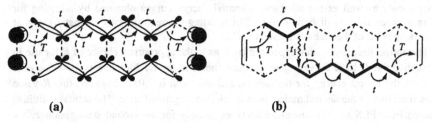

(a)   (b)

Figure 4. (a) A pathway proceeding along one side of the main relay in $II(8)$; (b) a pathway involving a single hop (t_1) between the two main relays. The expressions are those based on simple McConnell theory; the signs of the contributions to the electronic coupling are given in parentheses.

matrix element between adjacent σ NBO's on the same relay (see Fig. 4). Because a pathway that involves a single "jump" between relays involves one more interaction than one proceeding along a single relay, its contribution to the $\tilde{\pi}_+, \tilde{\pi}_-$ splitting is necessarily of opposite sign from a pathway proceeding along a single relay (see Fig. 4). Thus, the major source of the destructive interference in the $\mathbf{\Pi}(n)$ series stems from pathways that involve a single jump between the two main relays.

Although the above discussion has focused on TB coupling between occupied π orbitals, most of the conclusions also hold for TB coupling between π^* orbitals. However, while the TB coupling between π orbital is due primarily to interactions involving the σ orbitals of the bridge, interactions involving both σ and σ^* orbitals of the bridge are important for describing the TB coupling between π^* orbitals.[18,20c]

4   Connections with the Electron/hole Transfer Problem

As noted in the Introduction, in the weak-coupling limit, the rate constants for ET and HT depend on the square of H_{el}. It is customary to associate H_{el} with one-half of $\Delta E(\pi)$ or $\Delta E(\pi^*)$, in the case of hole or electron transfer, respectively. We now address the assumptions underlying this association.

Geometry changes can play an important role in ET and HT processes. In understanding the role of geometry changes, it is instructive to adopt a diabatic picture and to plot as a function of the "transfer" coordinate the potential energy curves for the electron (or hole) localized on one or the other chromophore. For the $\mathbf{\Pi}(n)$ dienes, the main geometrical change accompanying transfer of an electron from one ethylenic group to the other is the lengthening of one ethylenic CC bond and a shortening of the other. In this case, it is convenient to denote the two ethylenic bond lengths as $R + \delta$ and $R - \delta$, where δ is the reaction coordinate. Figure 5 sketches as a function of δ the relevant diabatic potential energy curves as well as the adiabatic potential energy curves obtained by allowing for mixing between the two diabatic states. The splitting between the two adiabatic states at the crossing point of the diabatic curves is $2H_{el}$.

The above discussion sidestepped the issue as to the appropriate choice of R for use in constructing the diabatic curves. The standard choice in the ET literature is the value of R that gives the lowest energy for the delocalized ion (with $\delta = 0$). However, this R value differs from that of the neutral molecule in its electronic ground state. The orbital splittings extracted from PES and ETS measurements are usually for the ground state geometry, as are most calculations of the $\tilde{\pi}_+, \tilde{\pi}_-$ and $\tilde{\pi}_+^*, \tilde{\pi}_-^*$ splittings. Fortunately, the splittings generally do not depend strongly on R, and the error incurred by associating $2H_{el}$ with the splittings determined at the geometry of the neutral molecule is small.

There is a second assumption in using orbital (or state) splittings as the measure of the electronic coupling relevant for ET or HT, namely, that the problem is well described as an effective 2-level Hamiltonian.[29,30] For our "ethylene-bridge-ethylene" model systems, this means that the mixing of the localized π (or π^*) orbitals with localized σ orbitals of the bridge is weak in the sense that most of the π (or π^*) character resides in two canonical MO's. In the limit of strong mixing

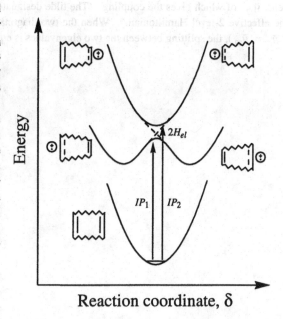

Reaction coordinate, δ

Figure 5. Diabatic and adiabatic potential energy curves for hole transfer in the radical cation of a hypothetical ethylene-bridge-ethylene system. The lowest curve corresponds to the ground state of the neutral molecule, assumed to possess C$_{2v}$ symmetry. The diabatic and adiabatic curves are represented by dashed and solid lines, respectively. Similar potential energy curves exist for electron transfer in the corresponding radical anion.

between the chromophore and bridge levels, this is not the case, and the association of the electronic coupling responsible for electron or hole transfer with an orbital (or state) splitting is no longer appropriate.[30,32-34,37] In such cases it is necessary to solve a time-dependent equation involving a multi-level Hamiltonian in order to deduce an effective coupling responsible for electron transfer. In the remainder of this article we will assume that we are dealing with systems for which a two-level model is appropriate. In such systems one can construct an effective 2x2 Hamiltonian matrix, the off-diagonal matrix element, $\tilde{H}_{12}$, of which gives the coupling. (The tilde designates that the matrix element is for the effective 2-level Hamiltonian.) When the two diagonal matrix elements are equal (*i.e.*, $\tilde{H}_{11} = \tilde{H}_{22}$), the splitting between the two eigenvalues is equal to twice $\tilde{H}_{12}$.

4.1 Asymmetric Bichromophoric Systems

For asymmetrical donor-bridge-acceptor (DBA) molecules, e.g., $I(n)$, the electronic coupling appropriate for describing the ET or HT processes should be determined under conditions that the two relevant diabatic states cross. The crossing of the diabatic states can be the result of an intramolecular distortion, the application of an external field, or a solvent fluctuation, for processes occurring in solution.[30,38] To illustrate procedures for calculating the relevant diabatic states and for estimating the couplings in asymmetrical molecules, we consider as an example the exo-5-chloro-2-norbornene molecule (IV) shown in Fig. 6. This species has been selected because in the gas-phase it has been found to produce Cl⁻ upon electron capture at the energy of the π^* anion state.[39]

ETS measurements locate the π^* and σ^* anion states of IV at 1.10 and 2.78 eV, respectively. (The energies of the anion states are reported relative to the ground state of the neutral molecule.) Rough estimates of the energies of the diabatic π^* and σ^* anion states

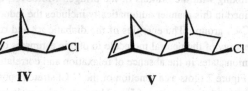

IV **V**

Figure 6. Two rigid chloroalkenes used as examples of asymmetric donor-bridge-acceptor systems.

of IV at the equilibrium geometry of the ground state of the neutral molecule can be obtained from the experimental electron attachment energies of norbornene (π^* anion at 1.70 eV) and chloronorbornane (σ^* anion at 2.3 eV). The splitting between the π^* and σ^* anion states of IV is 1.1 eV greater than that between the two diabatic states, as estimated above. Can half this value (0.55 eV) be taken as a measure of the electronic coupling governing electron transfer from the π^* to the σ^* anion state? The answer is no on two counts. First, part of the shift in the energy of the π^* orbital is due to the electric field imposed by the C-Cl group.[40] Because this is not caused by orbital overlap, it should not be counted as part of the electronic coupling. Second, the electron attachment energies determined from the ETS measurements are at the geometry of the neutral molecule, at which the diabatic π^* and σ^* anion states are not degenerate.

In order to determine the electronic couplings appropriate for describing ET in IV and other chloroalkenes, it is necessary to calculate diabatic π^* and σ^* anionic potential energy curves as a function of the reaction coordinate, which, to good approximation, can be taken to correspond to the C-Cl stretch degree of freedom. Here we describe an approach to this problem that we refer to as the "reduced Fock matrix" method.[39] In this approach, in keeping with the spirit of Koopmans' theorem, Hartree-Fock calculations are carried out on the ground state of the neutral molecule for a range of C-Cl bond lengths (R_{C-Cl}). At each R_{C-Cl} value, the Fock matrix is transformed to the NBO representation and two "reduced" Fock matrices are formed. In one all interactions with the π^* NBO are zeroed out, and in the other all interactions with the C-Cl σ^* NBO are zeroed out. This assures that σ^*/π^* mixing is "turned off". Diagonalization of these reduced Fock matrices give "diabatic" σ^* and π^* orbitals. These diabatic orbitals are "dressed" or "renormalized" in that they allow for mixing with the orbitals of the bridge. Moreover, the energy of the diabatic π^* orbital obtained in this manner automatically includes the inductive shift due to the electric field of the C-Cl group. The energies of the diabatic π^* and σ^* orbitals can be added to the total HF energy of the neutral molecule to obtain estimates of the energies of the diabatic σ^* and π^* anion states in the absence of relaxation and correlation effects.

Figure 7 plots as a function of the C-Cl stretch coordinate the diabatic potential energy curves for the two anion of **IV**. The figure also includes the corresponding adiabatic anion potentials generated by adding the energies of the canonical σ^* and π^* orbitals to the energy of the neutral molecule. All anion curves have been shifted downward by the amount needed to bring the calculated vertical attachment energy for formation of the π^* anion state into agreement with experiment. The two diabatic curves cross at $R_{C-Cl} = 1.9$ Å, which is about 0.1 Å greater than the equilibrium bond length in the neutral molecule. The splitting between the adiabatic curves at this crossing point is 1.4 eV, giving an electronic coupling of 0.7 eV. An alternative approach to obtaining diabatic states and to calculating the electronic coupling in DBA molecules involves use of the partitioning method to construct an effective two-level problem. Application of the partitioning method to **IV** gives nearly the same value for the π^*/σ^* coupling as obtained by the "reduced Fock matrix" approach.[39b] The partitioning method will be described in Section 5 below.

The yield of Cl⁻ upon electron attachment to **IV** is about 71 times greater than that upon electron attachment to exochloronorbornane. Moreover, the peak in the Cl⁻ current occurs at the energy of the π^* anion state as determined from ETS. This has led to the interpretation of the Cl⁻ production in **IV** and in other chloroalkenes in terms of a two-state diabatic model in which the electron is captured into the π^* orbital, and then jumps to the σ^* orbital, with the dissociation occurring from the resulting σ^* anion state. In the weak-coupling limit, the Cl⁻ yield depends quadratically on the coupling between the π^* and σ^* anion states, which implies that measurements of the Cl⁻ yield can be used as a probe of the distance dependence of the electronic coupling.

In concluding this section, a few comments about the applicability of the "weak-coupling limit" are in order. The π^*/σ^* coupling calculated for **IV** is actually too large for the weak-coupling limit to be valid. This means that for this system the dissociation most likely occurs directly on the lower adiabatic surface, rather than as a two-step process as described above. However, for **V** (see Fig. 6) and molecules with the ethylenic and C-Cl groups even further separated, it is probably valid to view the Cl⁻ production as physically occurring in a two step process.[39b] It should also be noted that in spite of the very large π^*/σ^* coupling, the chromophore levels of **IV** are relatively weakly coupled to the bridge. This is a consequence of the fact that roughly half of the coupling in **IV** is TS in origin.

274

4.2 Thermal and Photo-assisted Electron Transfer

Two types of ET can be distinguished: (1) thermal electron transfer in radical anions[4], and (2) photo-assisted ET in neutral molecules. In the former, an electron is added to the donor group (giving D⁻BA) and this is followed by electron transfer to the acceptor group (giving DBA⁻). In the latter, the DBA system is first electronically excited to give D*BA.[4,41,42] Electron transfer from D* to A then gives a charge-separated product D⁺BA⁻. In both photo-assisted ET and radical anion ET, an electron is transferred from a normally unfilled orbital of a donor group to a normally unfilled orbital of an acceptor group. Even though the energy gaps between the chromophore and bridge levels are different in these two processes, the electronic

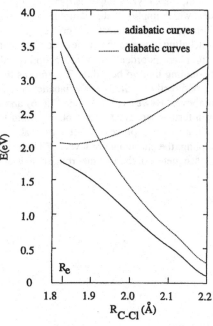

Figure 7. Potential energy curves for the π* and σ* anion states of **IV** in both the adiabatic (solid curves) and diabatic (dashed curves) representations as a function of the C-Cl bond length (reproduced with permission from ref 39a). The equilibrium bond length of the neutral molecule is denoted by ᴅ

coupling tends to have a similar distance dependence in the two cases. For example, the distance dependence of the electronic coupling in the **I**(n) series, as deduced from the experimental rate constants for photo-assisted ET, is very close to that of the calculated π_+^*, π_-^* splittings in the **II**(n) series.[4j]

Most studies of thermal and photo-assisted ET have been carried out in the condensed phase. The work on the chloroalkenes, described in the preceeding section, is unique in that it provides a probe of thermal ET in DBA systems in the gas phase.

5 Effective two-level Hamiltonians

Probably the most straightforward way of turning a multilevel problem into a two- level system is via Löwdin partitioning.[43] The application of this approach to electron transfer problems was pioneered by Larsson.[31] It is also closely related to the Greens function approaches.[37,44,45] In illustrating the partitioning approach, it is useful to rewrite the Fock matrix over NBO's in terms of four submatrices,

$$H = \begin{pmatrix} H_{cc} & H_{cb} \\ H_{bc} & H_{bb} \end{pmatrix} \tag{11}$$

where, H_{cc} is a 2x2 matrix involving the self energies of the two relevant orbitals of the two chromophores, H_{bb} describes the couplings among the orbitals localized on the bridge, and H_{cb} (and H_{bc}) describes the chromophore-bridge coupling. (If TS coupling between the relevant orbitals of the two chromophores is non-negligible, its effect can be incorporated through the off-diagonal matrix elements of H_{cc}.) For a system with n bridge orbitals, H_{bb} and H_{cb} are nxn and 2xn matrices, respectively. The matrix eigenvalue equation may be written as:

$$H_{cc}C_c + H_{cb}C_b = EIC_c \tag{12a}$$

$$H_{bc}C_c + H_{bb}C_b = EIC_b \tag{12b}$$

where I is the identity matrix, C_c is a two-component vector giving the coefficients of the chromophore orbitals, and C_b is an n-component vector giving the coefficients of the orbitals localized on the bridge. Solving Eq. 12b for C_b, and substituting this into Eq. 12a gives:

$$\{H_{cc} + H_{cb}(EI - H_{bb})^{-1}H_{bc}\}C_c = EIC_c, \tag{13}$$

and corresponding determinant to be solved for the eigenvalues is:

$$\left| H_{cc} - EI - H_{cb}(EI - H_{bb})^{-1}H_{bc} \right| = 0. \tag{14}$$

Thus far no approximations have been made, and the effective 2-level problem contains all the information present in the original $(n+2)$x$(n+2)$ problem. The price paid for turning the $n+2$ level problem into an effective two-level Hamiltonian is that the matrix elements have become energy dependent. In practice, an iterative approach is usually adopted for finding the eigenvalues of Eq. 14.

In order to illustrate the application of the partitioning method, we consider again the four-site four-orbital model. In this case, we adopt the numbering scheme so that "1" and "2" refer to the two chromophore orbitals, and "3" and "4" to the two bridge orbitals. We again assume that TS coupling between the two chromophore orbitals is negligible and that only nearest-neighbor interactions are important. The Hamiltonian matrix is then

$$
H = \begin{pmatrix} H_{11} & 0 & H_{13} & 0 \\ 0 & H_{22} & 0 & H_{24} \\ H_{31} & 0 & H_{33} & H_{34} \\ 0 & H_{42} & H_{43} & H_{44} \end{pmatrix}
\tag{15}
$$

The three submatrices needed for the partitioning are:

$$
H_{cc} = \begin{pmatrix} H_{11} & 0 \\ 0 & H_{22} \end{pmatrix} \qquad
H_{cb} = \begin{pmatrix} H_{13} & 0 \\ 0 & H_{24} \end{pmatrix} \qquad
H_{bb} = \begin{pmatrix} H_{33} & H_{34} \\ H_{43} & H_{44} \end{pmatrix}
\tag{16}
$$

The determinant of the 2-level system obtained by Löwdin partitioning is:

$$
\left| \begin{pmatrix} H_{11} - E & 0 \\ 0 & H_{22} - E \end{pmatrix} + \frac{\begin{pmatrix} H_{13} & 0 \\ 0 & H_{24} \end{pmatrix} \begin{pmatrix} H_{44} - E & -H_{34} \\ -H_{43} & H_{33} - E \end{pmatrix} \begin{pmatrix} H_{31} & 0 \\ 0 & H_{42} \end{pmatrix}}{D} \right| = 0,
\tag{17}
$$

where

$$
D = (H_{33} - E)(H_{44} - E) - H_{43}H_{34}
\tag{18}
$$

The determinant may be rewritten as:

$$\begin{vmatrix} H_{11}+\dfrac{H_{13}H_{31}}{H_{33}-E-H_{43}H_{34}\,/\,(H_{44}-E)}-E & \dfrac{-H_{13}H_{34}H_{42}}{(H_{33}-E)(H_{44}-E)-H_{43}H_{34}} \\[4mm] \dfrac{-H_{24}H_{43}H_{31}}{(H_{33}-E)(H_{44}-E)-H_{43}H_{34}} & H_{22}+\dfrac{H_{24}H_{42}}{(H_{44}-E)-H_{43}H_{34}\,/\,(H_{33}-E)}-E \end{vmatrix}=0 \quad (19)$$

The diagonal entries, $H_{11} + H_{13}H_{31}/\{(H_{33}\text{-}E) -H_{43}H_{34}/(H_{44}\text{-}E)\}$ and $H_{22} +H_{24}H_{42}/\{(H_{44}\text{-}E) - H_{43}H_{34}/(H_{33}\text{-}E)\}$ give the energies of the two diabatic states, and the off-diagonal term gives the coupling between the diabatic states. At the crossing point of the diabatic curves, the energy of the crossing point can be substituted for E in the off-diagonal matrix element in order to determine the coupling. Away from the crossing point, the appropriate choice of E has been the matter of some debate.[30,45,46] A common choice is to simply take the mean of H_{11} and H_{22}, or, better yet, the mean of $H_{11} + H_{13}H_{31}/(H_{33}\text{-}H_{11})$ and $H_{22} + H_{24}H_{42}/(H_{44}\text{-}H_{22})$.

With the definitions of the matrix elements introduced in Section 3.1, Eq. 19 becomes

$$\begin{vmatrix} \varepsilon_\pi +\dfrac{T^2}{\varepsilon_\sigma-E-t^2\,/\,(\varepsilon_\sigma-E)}-E & \dfrac{-T^2 t}{(\varepsilon_\sigma-E)(\varepsilon_\sigma-E)-t^2} \\[4mm] \dfrac{-T^2 t}{(\varepsilon_\sigma-E)(\varepsilon_\sigma-E)-t^2} & \varepsilon_\pi +\dfrac{T^2}{(\varepsilon_\sigma-E)-t^2\,/\,(\varepsilon_\sigma-E)}-E \end{vmatrix}=0 \quad (20)$$

If in the denominators the t^2 terms are neglected and E is replaced by ε_π, the off-diagonal matrix element reduces to $-T^2 t/\Delta^2$, precisely the coupling obtained from the McConnell model. (Recall that the coupling is one half the splitting.)

Solving the coupling through terms that are fifth order (in t and T) gives:

$$H_{el} = -\left(\frac{T^2 t}{\Delta^2}\right)\left(1+\frac{2T^2}{\Delta^2}+\frac{t^2}{\Delta^2}\right). \quad (21)$$

The three contributions are shown diagramatically in Fig. 8, from which it is seen that the two higher order terms are due to "retracing" pathways.

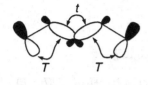

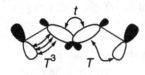

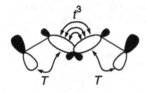

Figure 8. Three pathways for coupling the chromophore orbitals in a simple four-orbital four-site model. The second and third diagrams represent retracing pathways.

also in van der Waals contact with each other. For both model systems HF/3-21G calcu-lations give β values for hole transfer similar to those obtained for the covalently linked **IV**(*n*)

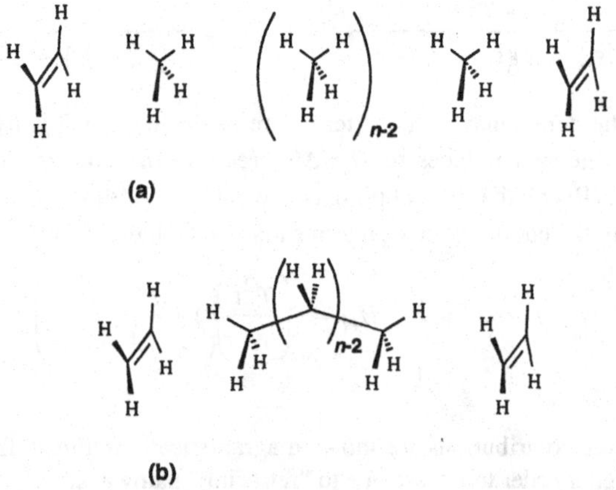

6 The Role of Solvent in Mediating Electronic Coupling.

The discussion thus far has focused on electronic coupling between covalently linked chromophores. Calculations on various model systems show that molecules not covalently linked to the chromophores can also be very effective at relaying the coupling.[9,18] The $C_2H_4 \cdot (C_nH_{2n+2}) \cdot C_2H_4$ and $C_2H_4 \cdot (CH_4)_n \cdot C_2H_4$ model systems, shown in Figure 9, illustrate this point. These model systems consist of two ethylene molecules, arranged in a face-to face manner, with either an all-*trans* alkane chain or a series of methane molecules placed between the two ethyelene molecules. The spacing between the ethylene molecules is chosen so that the ends of the alkane chain or the terminal methane molecules are in van der Waals contact with the ethylenes. In the case of $C_2H_4 \cdot (CH_4)_n \cdot C_2H_4$ com-plex, adjacent methane groups are

(a)

(b)

Figure 9. Model $C_2H_4 \cdot (CH_4)_n \cdot C_2H_4$ and $C_2H_4 \cdot C_nH_{2n+2} \cdot C_2H_4$ systems.

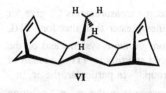

VI

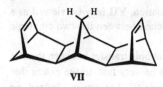

VII

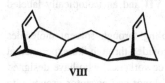

VIII

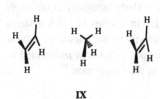

IX

Figure 10. Systems used to elucidate electronic coupling through a methane "solvent" molecule. **VIII**, an ethylene-bridge-ethylene system with relatively weak coupling between the ethylenic groups. **VI**, derived from **VIII**, by "insertion" of a methane molecule between the two double bonds. The closely related system **VII** has a bridging CH_2 group between the two double bonds. **IX** is a simplified model system with a methane molecule sandwiched between two ethylene molecules separated the same distance as in **VIII**.

species, discussed above. (For the geometries used for the model systems, coupling of the π^* orbitals via the σ and σ^* orbitals of the C_nH_{2n+2} and $(CH_4)_n$ chains is negligible. This is on account of the fact that the π^* orbitals are antisymmetric with respect to the plane containing the carbon atoms of the alkane spacers. However for other geometries, the HF/3-21G calculations give β values for coupling of the π^* orbitals similar to those obtained for the $IV(n)$ species.)

It follows from the results for the $C_2H_4 \cdot (C_nH_{2n+2}) \cdot C_2H_4$ and $C_2H_4 \cdot (CH_4)_n \cdot C_2H_4$ model systems that in certain situations solvent molecules can be very effective at relaying through-bond interactions in chromophore-bridge-chromophore systems. In general there will be a multitude of pathways, passing through different numbers of solvent molecules, that couple the two chromophores. As a result, some pathways will contribute to the coupling with a plus sign and others with a minus sign, resulting in considerable destructive interference. Consequently, through-solvent electronic coupling is likely to be most effective when (1) the coupling through the bridge covalently linking the chromophores is weak, and (2) the solvent molecules in the vicinity of the DBA are structured so as to minimize destructive interference.

An example of a system (**VI**) for which solvent-mediated electronic coupling should be important is shown in Figure 10. This is an ideal system in the sense that the covalently linked bridge has two cis linkages, making it a relatively poor relay, and the methane solvent molecule just "fits into" the region between the two chromophores. Although experiments have not been carried out for this system, results are available for the closely related species **VII**, with a methylene bridge between the two ethylenic groups. Photoelectron spectroscopic measurements[20b,c] give for **VII** a π_+,π_- splitting of 0.52 eV, approximately four times greater than that for **VIII**, which lacks the methylene bridge. Hartree-Fock calculations on these systems and on the ethylene-methane-ethylene model **IX** show that the enhanced splitting in **VII** is a result of the pathway proceeding through the bridging methylene group.[18] In particular, the π_+,π_- splitting calculated for **IX** is nearly the same as the enhancement in the π_+,π_- splitting found in going from **VIII** to **VII**. Thus to a good approximation, **VII** may be viewed as a model of **VI** which has a methane solvent molecule sandwiched between the two ethylenic groups.

The substantial π_+,π_- splitting found for **VII** suggests that hole transfer between the two ethylenic groups in the radical cation of this species should be very fast, to the extent that the hole may well be effectively delocalized over a symmetric (C_{2v}) structure. Indeed, an ESR spectroscopic investigation of the radical cation of **VII** and an isotopically labeled analog has provided evidence that this is the case.[47]

Further evidence for propagation of electronic coupling through solvent molecules trapped in between donor and acceptor groups has been provided through the experimental work of Paddon-Row and coworkers[4h] and by Zimmt and coworkers,[48] who have designed donor-bridge-acceptor systems with a "cleft" between the donor and acceptor groups into which certain size solvent molecules can be introduced. Particularly intriguing is the study of Oliver et al.[4d] who found that the rate of photoinduced ET in a series of DBA systems with *cis* linkages in the bridge displayed much greater sensitivity to the choice of solvent than did the corresponding DBA systems with all-*trans* bridges. This was interpreted in terms of the greater importance of through-solvent coupling in the former case.[4d]

7 $\pi \to \pi^*$ Excited States of Non-conjugated Dienes.

Many of the issues raised in the discussion of the electronic coupling in radical cations and anions are also relevant for the coupling between electronically excited states of remote chromophores. In particular, just as for the ET and HT cases, in the weak coupling limit, one can relate the rate constant for electronic excitation transfer between two chromophores to the square of an electronic coupling. In this case the coupling can be associated with half the splitting between the relevant electronically excited states. Again, it is convenient to use the **II**(n) dienes to illustrate the key issues. The dienes have four singly-excited $\pi \to$

π^* triplet states and four singly-excited $\pi \to \pi^*$ singlet states. For both spin multiplicities, two of the excited states are of A_2 symmetry and two are of B_1 symmetry.

$$A_2: \Psi_1 = \pi_- \to \pi_+^*; \quad \Psi_4 = \pi_+ \to \pi_-^*$$

$$(22)$$

$$B_2: \Psi_2 = \pi_- \to \pi_-^*; \quad \Psi_3 = \pi_+ \to \pi_+^*$$

The Ψ_1 and Ψ_4 configurations mix as do the Ψ_2 and Ψ_3 configurations, with the configuration mixing generally being greater between the latter two configurations. (This follows from consideration of the energy separations.) As a result, it is conceivable that the lowest excited state is actually of B_1 symmetry, rather than of A_2 symmetry (the symmetry of the HOMO-LUMO transition).

An alternative way of interpreting the excited states in $II(4)$ and other dienes is in terms of the localized excitations:

$$\phi_1 = \pi_1 \to \pi_1^* \qquad\qquad \phi_3 = \pi_1 \to \pi_2^*$$

$$(23)$$

$$\phi_2 = \pi_2 \to \pi_2^* \qquad\qquad \phi_4 = \pi_2 \to \pi_1^*$$

ϕ_1 and ϕ_2 denote excitations localized on one of the ethylenic groups, while ϕ_3 and ϕ_4 describe charge-transfer configurations in which an electron is excited from a π orbital on one ethylenic group to a π^* orbital localized on the other. The charge transfer configurations ϕ_3 and ϕ_4 are generally high in energy compared to ϕ_1 and ϕ_2 and can be neglected. For the purpose of understanding EET, it is the coupling between the localized ϕ_1 and ϕ_2 excitations which is of primary importance. For symmetrical molecules such as the $II(n)$ series, ϕ_1 and ϕ_2 can be combined to give the symmetry-adapted functions ϕ_+ and ϕ_-. These mix with excitations involving orbitals localized on the bridge to give $\tilde{\phi}_+$ and $\tilde{\phi}_-$.

In the triplet manifold, the splitting between ϕ_+ and ϕ_- is a result of the TS coupling. However, in the singlet manifold the splitting between ϕ_+ and ϕ_- derives from both TS and dipole-dipole (or Förster) coupling.[49] The difference between the ϕ_+, ϕ_- and $\tilde{\phi}_+, \tilde{\phi}_-$ splittings provides a measure of the TB coupling. The Förster coupling displays a R^{-6} dependence on the separation between the chromophores, whereas, as in the anion and cation states, the TS and TB contributions to the $\tilde{\phi}_+, \tilde{\phi}_-$ splittings depend exponentially on the separation between the chromophores (or on the length of the bridge).

Except for systems with a small separation between the chromophores, the Förster mechanism will dominate the coupling in the singlet manifold. Electron energy loss studies[50] of $\mathbf{II}(4)$ show that the splitting between the $\tilde{\phi}_+$ and $\tilde{\phi}_-$ triplet states in this molecule is less than 0.1 eV, while the splitting between the corresponding singlet states is of the order of 0.6 eV. The greater splitting in the singlet manifold is consistent with the importance of the Förster mechanism.

Calculations of the triplet excitation energies of $\mathbf{II}(n)$, $n = 4 - 10$, reveal that not only are the triplet splittings consistently smaller than the π_+, π_- and π_+^*, π_-^* splittings, they are also attenuated much more rapidly with increasing bridge length.[51] Closs and co-workers[52] have observed that the rates for triplet excitation transfer in DBA systems are approximately proportional to the product of the rates for electron and hole transfer:

$$k^{EET} \propto k^{ET}k^{HT} \tag{24}$$

This implies that

$$\beta^{EET} \approx \beta^{ET} + \beta^{HT} \tag{25}$$

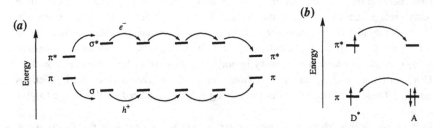

Figure 11. (a) a simplified McConnell-like model for bridge-mediated excitation transfer between two chromophores. h^+ denotes a "hole" or partially filled orbital. Hole transfer actually denotes electron transfer in the opposite direction as illustrated in part (b).

In many systems β^{ET} and β^{HT} are comparable in magnitude, which means that $\beta^{EET} \approx 2\beta^{ET}$, consistent with the much more rapid attenuation of the electronic coupling in the triplet π

$\rightarrow \pi^*$ manifold than in the π or π^* manifolds. Figure 11 illustrates a simple pathway for π $\rightarrow \pi^*$ excitation transfer in a bichromophoric system. If a McConnell-like model were to hold for each of the three electronic couplings, and if electron and hole transfer were "synchronized" as depicted in Figure 11a, Eq. (25) would imply that:

$$\frac{t^{EET}}{\Delta^{EET}} \approx \left(\frac{t^{ET}}{\Delta^{ET}}\right)\left(\frac{t^{HT}}{\Delta^{HT}}\right) \tag{26}$$

where t^{EET} denotes the matrix element giving the coupling between the $\sigma \rightarrow \sigma^*$ excitations on adjacent CC bonds, and Δ^{EET} is the difference in the energies of the localized $\sigma \rightarrow \sigma^*$ and $\pi \rightarrow \pi^*$ excitations. A theoretical justification for Eq. (26) has yet to be provided.

8 Conclusions

Through-bond interactions are very effective at coupling remote chromophores in chromophore-bridge-chromophore molecules. In general, a given bichromophoric system will have associated with it several different TB coupling constants. For the systems considered in the present article, there are separate TB coupling values for the π and π^* manifolds as well as for the triplet and singlet $\pi \rightarrow \pi^*$ manifolds. Although the various TB coupling constants can vary appreciably, there are important general conclusions that hold regardless of the manifold (anions, cations, or electronically excited states). One of the most important general conclusions is that model Hamiltonians retaining nearest-neighbor interactions (with reastic values of the matrix elements) only considerably underestimate the through-bond coupling. In fact, for long all-*trans* alkane bridges the McConnell-type nearest-neighbor pathway contributes a vanishingly small portion of the net coupling. The inclusion of the next nearest-neighbor and second next nearest-neighbor interactions is essential for obtaining a quantitative description of the coupling through alkane bridges. In systems with two or more chains linking the two chromophores there can be considerable destructive interference due to "crosstalk" interactions.

Through-bond interactions can also be propagated via solvent molecules. In general, the net contribution due to through-solvent interactions is likely to be small due to destructive interference between different coupling pathways. However, in systems in which the bridge itself is a relatively poor relay and in which the solvent molecules near the bridge are highly "structured" the net effect of through-solvent coupling may be sizable.

Acknowledgements

This research was supported by the National Science Foundation (KDJ) and the Australian Research Council (MPR). We would like to thank Michael Shephard and Stephen Wong for their contributions to the calculations described in this article.

References

1. (a) M.R.Wasielewski in *Photoinduced Electron Transfer*, M.A. Fox and M. Chanon, Eds., Elsevier: Amsterdam (1988), Part A, 161. (b) G.L. Closs and J.R. Miller, *Science*, 240, 440 (1988). (c) C.A. Stein, N.A. Lewis, and G. Seitz, *J. Am. Chem. Soc.* 104, 2596 (1982).
2. J.F. Smalley, S.W. Feldberg, C.E.D. Chidsey, M.R. Linford, M.D. Newton, and Y.-P. Liu, *J. Phys. Chem.* 99, 13141 (1995), and references therein.
3. *Introduction of Molecular Electronics*, ed. R.R. Birge (ACS, Washington, DC, 1994).
4. (a) N.S. Hush, M.N. Paddon–Row, E. Cotsaris, H. Oevering, J.W. Verhoeven and M. Heppener, *Chem. Phys. Lett.* 117, 8 (1985). (b) K.W. Penfield, J.R. Miller, M.N. Paddon–Row, E. Cotsaris, A.M. Oliver and N.S. Hush, *J. Am. Chem. Soc.* 109, 5061 (1987). (c) H. Oevering, M.N. Paddon–Row, H. Heppener, A.M. Oliver, E. Cotsaris, J.W. Verhoeven and N.S. Hush, *J. Am. Chem. Soc.* 109, 3258 (1987). (d) A.M. Oliver, D.C. Craig, M.N. Paddon–Row, J. Kroon and J.W. Verhoeven, *Chem. Phys. Lett.* 150, 366 (1988). (e) J.M. Lawson, D.C. Craig, M. N. Paddon–Row, J. Kroon and J.W. Verhoeven, *Chem. Phys. Lett.* 164, 120 (1989). (f) M. Antolovich, P.J. Keyte, A.M. Oliver, M.N. Paddon-Row, J. Kroon and J.W. Verhoeven, *Chem. Phys. Lett.* 150, 366 (1988). (g) J. Kroon, J.W. Verhoeven, M.N. Paddon-Row and A.M. Oliver, *Angew. Chem. Int. Ed. Engl.* 30, 1358 (1991). (h) J.M. Lawson, M.N. Paddon-Row, W. Schuddeboom, J.M. Warman, A.H. Clayton and K.P. Ghiggino, *J. Phys. Chem.* 97, 13099 (1993). (i) J.M. Warman, K.J. Smit, S.A. Jonker, J.W. Verhoeven, H. Oevering, J. Kroon, M.N. Paddon-Row and A.M. Oliver, *Chem. Phys.* 170, 369 (1993). (j) M.N. Paddon-Row, *Acc. Chem. Res.* 27, 18 (1994).
5. (a) D.N. Beratan, J.N. Onuchic and J.J. Hopfield, *J. Chem. Phys.* 86, 4488 (1987). (b) D.N. Beratan, J. A. Betts and J.N. Onuchic, *Science* 152, 1285 (1991).
6. K.D. Jordan and M. N. Paddon–Row, *Chem. Rev.* 92, 395 (1992).
7. M.J. Shephard, M.N. Paddon-Row and K.D. Jordan, *Chem. Phys.* 176, 289 (1993).
8. M.N. Paddon–Row and K.D. Jordan, *J. Am. Chem. Soc.* 115 2952 (1993).

9. M.N. Paddon–Row, M.J. Shephard and K.D. Jordan, *J. Phys. Chem.* 97, 1743 (1993).

10. M.N. Paddon–Row, M.J. Shephard and K.D. Jordan, *J. Am. Chem. Soc.* 66, 5328 (1993).

11. (a) C. Liang and M.D. Newton, *J. Phys. Chem.* 96, 2855 (1992). (b) C. Liang and M.D. Newton, *J. Phys. Chem.* 97, 3199 (1993).

12. (a) C.A. Naleway, L.A. Curtiss and J. R. Miller, *J. Phys. Chem.* 95, 8434 (1991). (b) C.A. Naleway, L.A. Curtiss and J.R. Miller, *J. Phys. Chem.* 97, 4050 (1993). (c) L.A. Curtiss, C.A. Naleway and J.R. Miller, *Chem. Phys.* 176, 387 (1993).

13. R.A. Marcus, *J. Chem. Phys.* 24, 979 (1956).

14. N.S. Hush, *Trans. Faraday Soc.* 57, 155 (1961).

15. (a) R. Hoffmann, A. Imamura and W.J. Hehre, *J. Am. Chem. Soc.* 90, 1499 (1968). (b) R. Hoffmann, *Acc. Chem. Res.*, 4, 1 (1971). (c) R. Hoffmann, *J. Am. Chem. Soc.* 97, 4884 (1975).

16. M.N. Paddon–Row, *Acc. Chem. Res.* 15, 245 (1982).

17. H.D. Martin and B. Meyer, *Angew. Chem. Int. Ed. Engl.* 22, 283 (1983).

18. M.N. Paddon–Row and K.D. Jordan in *Modern Models of Bonding and Delocalization*, J.F. Liebman and A. Greenberg, Eds., VCH Publishers, New York, Ch. 3, 115 (1988).

19. T. Koopmans, *Physica* 1, 104 (1934).

20. (a) M.N. Paddon-Row, H.K. Patney, R.S. Brown and K.N. Houk, *J. Am. Chem. Soc.* 103, 5575 (1981). (b) F.S. Jorgensen, M.N. Paddon-Row and H.K. Patney, *J. Chem. Soc., Chem. Commun.* 573 (1983). (c) M.N. Paddon-Row, L.M. Englehardt, B.W. Skelton, A.H. White, F.S. Jorgensen and H.K. Patney, *J. Chem. Soc., Perkin Trans.* 2, 1835 (1987). (d) V. Balaji, L. Ng, H.K. Patney, K. D. Jordan and M.N. Paddon–Row, *J. Am. Chem. Soc.* 104, 6849 (1982).

21. J.S. Binkley, J.A. Pople and W.J. Hehre, *J. Am. Chem. Soc.* 102, 939 (1980).

22. M.N. Paddon–Row and S.S. Wong, *Chem. Phys. Lett.* 167, 4432 (1990).

23. K.D. Jordan and M.N. Paddon–Row, *J. Phys. Chem.* 96, 1188 (1992).

24. M.N. Paddon–Row, S.S. Wong and K.D. Jordan, *J. Chem. Soc. Perkin Trans.* 2, 417 (1990).

25. M.N. Paddon–Row, S.S. Wong and K.D. Jordan, *J. Am. Chem. Soc.* 112, 1710 (1990).

26. M.N. Paddon–Row, S.S. Wong and K.D. Jordan, *J. Chem. Soc., Perkin Trans.* 2, 425 (1990).

27. A.E. Reed, L.A. Curtiss and F. Weinhold, *Chem. Rev.* 88, 899 (1988).

28. (a) E. Heilbronner and A. Schmelzer, *Helv. Chim. Acta*, 58, 936 (1975). (b) R. Gleiter and W. Schaefer, *Acc. Chem. Res.* 23, 369 (1990) (c) A. Imamura and M. Ohsaku, *Tetrahedron* 37, 2191 (1981).

29. H.M. McConnell, *J. Chem. Phys.* 35, 508 (1961).

30. M.D. Newton, *Chem. Rev.* 91, 767 (1991).

31. (a) S. Larsson, *J. Am. Chem. Soc.*, 103, 4034 (1981). (b) S. Larsson, *Soc. Faraday Trans. II* 79, 1375 (1983).

32. J.W. Evenson and M. Karplus, *J. Chem. Phys.* 96, 5272 (1992).

33. J. Reimers and N.S. Hush, *Chem. Phys.* 146, 89 (1990).

34. V. Mujica, M. Kemp and M.A. Ratner, *J. Chem. Phys.* 101, 6849 (1994).

35. J.M. Lopez-Castillo, A. Filali-Mouhim, I.L. Plante and J.P. Gay-Gerin, *J. Phys. Chem.* 99, 6864 (1995).

36. M.J. Shephard and M.N. Paddon-Row, *J. Phys. Chem.* 99, 17497 (1995).

37. S.S. Skourtis and D.N. Beratan, submitted to *Adv. Chem. Phys.* (1996).

38. V.V. Mikkelsen and M.A. Ratner, *Chem. Rev.* 87, 113 (1987).

39. (a) D.M. Pearl, P.D. Burrow, J.J. Nash, H. Morrison and K.D. Jordan, *J. Am. Chem. Soc.* 115, 9876 (1993). (b) D.M. Pearl, P.D. Burrow, J.J. Nash, H. Morrison, D. Nachtigallova and K.D. Jordan, *J. Am. Chem. Soc.* 99, 12379 (1995).

40. J. Nash, D.V. Carlson, K.D. Jordan, A.E. Kaspar, D.E. Love and H. Morrison, *J. Am. Chem. Soc.* 115, 8969 (1993).

41. L.T. Calcaterra, G.L. Closs and J.R. Miller, *J. Am. Chem. Soc.* 105, 670 (1983).

42. A.D. Joran, B.A. Leland, P.M. Felker, A.H. Zewail, J.J. Hopfield and P.B. Dervan, *Nature*, London, 327, 508 (1987).

43. P.O. Löwdin, *J. Chem. Phys.* 18, 365 (1950).

44. J.N. Onuchic, P.C.P. de Andrade, and D. Beratan, *J. Chem. Phys.* 95, 1131 (1991).

45. M.A. Ratner, *J. Phys. Chem.* 94, 4877 (1990).

46. S.S. Skourtis, D.N. Beratan and J.N. Onuchic, *Chem. Phys.* 176 (1993).

47. A.M. Oliver, M.N. Paddon-Row and M.C.R. Symons, *J. Am. Chem. Soc.* 111, 7259 (1989).

48. (a) K. Kumar, Z. Lin, D.H. Waldeck, and M.B. Zimmt, *J. Am. Chem. Soc.* 118, 243 (1996). (b) R.J. Cave, M.D. Newton, K. Kumar, and M.B. Zimmt, *J. Phys. Chem.* 99, 17501 (1995).

49. T. Förster in *Excimers and Exciplexes,* M. Gordon and W.R. Ware, Eds. *The Exciplex,* Academic Press, New York, (1975), 1.

50. D.E. Love, D. Nachtigallova, K.D. Jordan and M.N. Paddon-Row, *J. Am. Chem. Soc.* 118, 1235 (1996).

51. M.N. Paddon-Row and K.D. Jordan, unpublished results.

52. (a) G.L. Closs, P. Piotrowiak, J.M. MacInnis and G.R. Fleming, *J. Am. Chem. Soc.* 110, 2652 (1988). (b) G.L. Closs, M.D. Johnson, J.R. Miller and P. Piotrowiak, *J. Am. Chem. Soc.* 111, 3751 (1989).

THE BREATHING ORBITAL VALENCE BOND METHOD

P.C. HIBERTY

Laboratoire de Chimie Théorique, Université de Paris-Sud,

91405 Orsay, France

A method is proposed for computing compact ab initio valence bond wave functions suitable for diabatic states, or adiabatic states in case the interpretation of the wave function in terms of Lewis structures is needed. The method aims at combining the properties of interpretability and compactness of the classical valence bond method with a reasonable accuracy of the energetics. All Lewis structures relevant to the electronic system are generated, each of them being described by a single valence bond configuration state function. A balanced description of the different Lewis structures is ensured by allowing each configuration to have its specific set of orbitals during the optimization process. In this framework, the dynamical correlation associated to a bond is viewed as the instantaneous adapatation of the orbitals to the electron fluctuation inherent to the bond, hence the "breathing-orbital" denomination. The method is applied to the F_2 and FH molecules and to the F_2^-, $(NH_3)_2^+$, Cl_2^- and $(CH_3)_2^+$ radicals to test its ability to reproduce equilibrium bond lengths and bonding energies for two-electron, three-electron and one-electron bonds. Satisfactory results are obtained in all cases, despite the simplicity of the wave function which is composed of only two or three configurations. The method is subsequently applied to questions of chemical interest. The first application deals with the general problem of symmetry-breaking and with the search for stable three-electron bonded anions of the type RO∴OR'$^-$, the $H_2O_2^-$ potential surface being investigated as a model system. In the second application, the natures of the C-Cl and Si-Cl bonds are compared in terms of interacting purely ionic and purely covalent dissociation curves. The third application deals with the "lone pair bond weakening effect" in X-H bonds (X = C to F). Finally, the last application uses the analysis of dynamical correlation in terms of the breathing-orbital effect to devise an economical procedure, based on Hartree-Fock, to study large-size odd-electron bonded systems.

1 Introduction

Valence bond (VB) theory has an old an intimate relation with structural chemistry and chemical reactivity. Much of our qualitative understanding of electronic structure is couched in the language of local bonds and lone pairs, within the conceptual framework of the VB picture. Indeed, the VB theory is essentially the quantum theoretical formulation of the classical concept of the chemical bond, which is associated in the VB formalism with the spin-pairing of the electrons in the (singly occupied) valence orbitals of the bonded atoms.[1,2] Thus, each term of a VB function can be made to correspond to a chemical structural formula or Lewis structure, which is of great help for the qualitative interpretation of the wave function. Qualitatively, this theory and its simplest variant, resonance theory,[3] have given rise to such fundamental concepts as hybridization, covalency, zwitterionic structures, interplay between resonance structures, resonance stabilization, and often allows the chemists to explain or predict reaction mechanisms or molecular properties by writing VB structures and curved arrows on the back of an envelope.

VB theory however does not reduce to a qualitative model, and a large field of applications is open to quantitative investigations carried out in the VB framework, for the sake of getting some unique chemical insight or specific information that is not available in standard ab initio methods. As a wide-ranging example, VB theory is the choice method to generate diabatic states, whose physical content represents an asymptotic electronic structure that must remain as invariant as possible throughout a reaction coordinate. To mention only a few applications in chemistry, physical chemistry and photochemistry, such calculations apply to: (i) Chemical dynamics, in case the Born-Oppenheimer approximation breaks down; (ii) Chemical reactivity, with the curve-crossing diagrams of Shaik and Pross, in which a reaction barrier is viewed as the avoided crossing of two diabatic curves, one representing the bonding scheme of the reactants and the other that of the products;[4] (iii) Photochemistry, with the harpooning and charge transfer mechanisms;[5] (iv) Fundamental principles of organic chemistry, e.g. the role of electronic delocalization as a stabilizing factor;[6,7,8] (v) Solvation, with theoretical models treating the solvation effects separately on covalent and ionic components of a bond.[9]

For such applications, it is not only important to be able to interpret the wave function in terms of chemical structural formulas (Lewis structures), but also to be able to estimate the energy of each of these individual Lewis structures and their variations along a reaction coordinate before they interact to form the adiabatic states. It is therefore important for a VB method to be useful that its mathematical formulation sticks to the concept of Lewis structure, remains as compact as possible and yield reasonably accurate diabatic/adiabatic energy profiles. It is for gathering these desirable qualities that the breathing orbital VB (BOVB) method has been proposed.[10-12]

2 Historical Background

Historically, the first calculation of the electronic structure of a neutral molecule was carried out by Heitler and London,[13] treating H_2 and using the valence bond (VB) method, in what was to be considered as the greatest single contribution to the clarification of the chemists' conception of valence since Lewis' suggestion in 1916 that the chemical bond between two atoms consists of a pair of electrons held jointly by the two atoms. In this early paper, the molecular wave function for H_2 was considered as purely covalent, and constructed from the atomic orbitals χ_a and χ_b of the separate atoms, as in eq 1 (dropping the normalisation factors):

$$\Psi_{HL} = \chi_a(1)\chi_b(2) - \chi_b(1)\chi_a(2) \tag{1}$$

This simple wave function was able to account for about 66% of the bonding energy of H_2, and performed a little better than the rival Molecular Orbital (MO) method that appeared almost at the same time. It was latter extended to polyatomic molecules,[1,2] thus laying the basis for the classical VB method, a rather primitive form of VB theory in which molecular wave functions are still constructed from orbitals that are optimized for the separate atoms. The underlying philosophy of classical VB was that, as in chemistry molecules are built from atoms, molecular wave functions are to be made from atomic, or atom-like, wave functions. This means that the Schrödinger equation first was solved for the atoms separately and the resulting atomic orbitals subsequently were used to construct molecular wave functions. Much progress has been made since then, as regards accuracy, and in modern VB theory it is now recognized that in a molecule the interatomic interactions cause the original atomic orbitals to be non-optimal, and a variational optimization of the atomic orbitals in the molecular wave function is required in order to obtain accurate results. Meanwhile, classical VB theory played an extremely important part in the early history of molecular quantum mechanics, as representing the first rationalization of empirical chemical-bonding ideas.

Quite different is the basic philosophy of the molecular orbital (MO) theory,[14,15] that conceptually originates in physics and resulted as the straightforward application of the early quantum mechanical methods to the molecular problem. Here the concept of local bond and chemical structural formula is hidden, with the notion that electrons are quasi-independent and located in orbitals delocalized over the whole molecule. This method eventually eclipsed VB theory as a computational method, in spite of its important initial results, and the reason for this has been essentially technical. Indeed, the MO method deals with orbitals which are orthogonal to each others, thus allowing the application of the Slater-Condon rules[16] that considerably simplify the calculation of integrals, a crucial argument in times when computers were still in their early stage of development.

However, despite the dominance of MO ideas for computational purposes, a number of technical papers on VB matrix element evaluation methodology continued to appear.[16-26] Now that modern super computers render the computational difficulties easier to deal with, VB theory as an ab initio computational method is enjoying a clear revival of interest, partly because of its specific possibilities (vide supra) or interpretative power, and partly because of its elegant treatment of electron correlation that allows rather compact wave functions to attain the degree of accuracy of much larger configuration interaction expansions in the framework of MO-CI theory. To better understand this point, let us examine how electron correlation is taken into account in the MO and VB frameworks.

2.1 Left-Right Electron Correlation in MO and VB Theories

The term "electron correlation energy" is usually defined as the difference between the exact nonrelativistic energy and the energy provided by the simplest MO wave function, the mono-determinantal Hartree-Fock wave function. This latter model is based on the "independent particle" approximation, according to which each electron moves in an average potential provided by the other electrons.[27] The Hartree-Fock wave function Ψ_{HF} takes the form of an antisymmetrized product of orbitals, which in the case of H_2 reduces to a Slater determinant involving the spin-up and spin-down counterparts of the bonding orbital σ_g, as in eq 2 (dropping the normalization factors from now on):

$$\Psi_{HF} = |\sigma_g \overline{\sigma}_g| \tag{2}$$

$$\sigma_g = \chi_a + \chi_b$$

The physical meaning of the Hartree-Fock wave function appears most clearly by expanding the MO determinant in eq 2 as a linear combination of determinants constructed from pure atomic orbitals (AOs):

$$|\sigma_g \overline{\sigma}_g| = |\chi_a \overline{\chi}_b| + |\chi_b \overline{\chi}_a| + |\chi_a \overline{\chi}_a| + |\chi_b \overline{\chi}_b| \tag{3}$$

Here the first two determinants are nothing but the determinantal form of the Heitler-London function, and represent a purely covalent interaction between the atoms. The remaining determinants represent zwitterionic structures, $H^- H^+$ and $H^+ H^-$, and contribute 50% to the wave function, whatever the interatomic distance. This is clearly too much at equilibrium distance, according to chemical intuition, and becomes absurd at infinite separation where the ionic component is expected to drop to zero. Qualitatively, this can be corrected by including a second configuration where both electrons occupy the antibonding orbital, σ_u:

$$\Psi_{CI} = C_1 |\sigma_g \overline{\sigma}_g| + C_2 |\sigma_u \overline{\sigma}_u| \tag{4}$$

$$\sigma_u = \chi_a - \chi_b$$

and optimizing the coefficients C_1 and C_2 of the more elaborate wave function, Ψ_{CI} (eq 4). This is the essence of configuration interaction (CI), or multi-configuration SCF (MCSCF) theory if both the coefficients of the configurations and their orbitals are optimized simultaneously in flexible basis sets. As the second configuration also exhibits a 50:50 mixture of covalent and ionic components but coupled with a negative sign, its combination with the first determinant can correct the excess ionic character of Ψ_{HF}, thus leading to a VB expansion (eq 5) that now displays a qualitatively correct behavior at all

$$\Psi_{CI} = \lambda(|\chi_a\overline{\chi}_b| + |\chi_b\overline{\chi}_a|) + \mu(|\chi_a\overline{\chi}_a| + |\chi_b\overline{\chi}_b|) \tag{5}$$

distances, with an optimal covalent/ionic ratio of typically 80:20 at equilibrium distance[28] all the way to 100:0 at infinite separation. This type of electron correlation, by which the two electrons are not allowed to approach each other too often, is called the "left-right" correlation, and constitutes a good deal of the total electron correlation that contributes to chemical bonding.

Totally different is the early VB point of view, since in Heitler-London's approximation of a purely covalent bond the electrons are never allowed to approach each other. Thus, while the Hartree-Fock model underestimates electron correlation, the early VB models overestimate it, so that the true situation is about half-way in-between. In the same way as the Hartree-Fock wave function is improved by CI, the purely covalent VB function can be improved by admixture of ionic structures as in eq 5, in which the coefficients λ and μ would be directly optimized in the VB framework. Both improved models thus lead to wave functions that, although expressed in different languages, are strictly equivalent and qualitatively correct. This statement can be generalized: as both ab initio VB and ab initio MO theories exploit a subspace of the same configuration space, VB and MO wave functions are always interconvertible and become equivalent when both theories are pushed to their highest level of refinement.

A severe inconvenience of describing each bond of a polyatomic molecule by one covalent and two ionic components is that the number of VB structures grows exponentially with the size of the molecule, but Coulson and Fischer[29] proposed a very elegant way to incorporate left-right correlation into a single, formally covalent, VB structure by using deformed, or rather slightly delocalized orbitals as exemplified in eq 6 for H_2. Here each orbital φ_l or φ_r is

$$\Psi_{CF} = |\varphi_l\overline{\varphi}_r| + |\varphi_r\overline{\varphi}_l| \tag{6}$$

$$\varphi_l = \chi_a + \varepsilon\chi_b$$

$$\varphi_r = \chi_b + \varepsilon\chi_a$$

mainly localized on a single center but bears a small component on the other center, so that the expansion of the Coulson-Fischer wave function Ψ_{CF} (eq 7) in AO determinants is in fact equivalent to Ψ_{CI} in eq 5, provided the coefficient ε is properly optimized.

$$\Psi_{CF} = (1 + \varepsilon^2)(|\chi_a \overline{\chi}_b| + |\chi_b \overline{\chi}_a|) + \varepsilon(|\chi_a \overline{\chi}_a| + |\chi_b \overline{\chi}_b|) \tag{7}$$

The Coulson-Fischer proposal gave rise to the "separated electron pair theory" which was initiated by Hurley, Lennard-Jones and Pople[30] and later developed extensively as the "Generalized Valence Bond" (GVB) method by Goddard.[31] In the latter method, each bond in a polyatomic molecule is considered as a pair of spin-coupled orbitals which are non-orthogonal to each other, but can be constrained to be orthogonal to the other GVB pairs without much loss in numerical accuracy. Each GVB orbital is centred on one atom and can be hybridized, but have tails on the neighboring atoms, so that a VB structure that formally displays purely covalent bonds implicitly contains some hidden ionic structures, necessary for a reasonable description of the bonds. In the most popular version of GVB theory, that involves the so-called "perfect pairing" approximation, only one spin-coupling is considered, for example that linking together each sp^3 hybrid of a carbon atom to its nearest neigboring hydrogen in methane.

As an extension of the GVB method, the Spin-coupled VB (SCVB) method of Gerrat and Cooper[32] removes any orthogonality restrictions and consider all possible spin-couplings between the singly occupied orbitals. Note that the shape of the orbitals (e.g. sp^3-like in the carbon atom of methane) and their degrees of delocalization are not a priori imposed, but naturally arise from the optimization of the orbitals for self-consistency. The lone pairs can be treated either as doubly occupied localized orbitals, or as pairs of strongly overlapping singly occupied orbitals.

The GVB and SCVB methods take care of all the left-right correlation in molecules, but do not include the totality of the "non-dynamical" correlation which must be retrieved through a "Complete Active Space" MCSCF calculation (CASSCF) involving all possible configurations that can be constructed within the space of valence orbitals, the latter being defined as the set of lone pairs, bonding combinations and lowest antibonding combinations of orbitals.

2.2 Dynamical Electron Correlation

The importance of electron correlation for the description of the bond is best appreciated in the case of the F_2 molecule, for which the Hartree-Fock energy at the experimental bonding distance is actually *higher*, by ca 36 kcal/mol,[33] than twice the energy of a fluorine radical, at variance with a fairly large experimental bonding energy of 38 kcal/mol. Matters are much improved, of course, at the GVB or CASSCF levels (see Tables 1 and 2 below) which are nearly equivalent for this molecule, yet the calculated bonding energy is still disappointingly small, reaching only half of the full CI estimation in the same basis set. What misses is the so-called *dynamical* electron correlation, which is retrieved in MO-CI theory through rather large CI expansions including configurations constructed from non-valence orbitals. However the qualitative defect of the GVB or CASSCF wave function of F_2 appears most clearly once the wave function is expanded in terms of covalent and ionic VB structures with strictly localized AOs, in a manner similar to eq 7 and as pictorially represented in eq 8.

$$\Psi_{GVB} = \quad \text{} \quad (8)$$

$$F \cdot \text{---} \cdot F \qquad\qquad F^- F^+ \qquad\qquad F^+ F^-$$

What we are dealing with is a wave function in which the orbitals and coefficients of the covalent and ionic structures are optimized simultaneously, but careful scrutiny indicates a subtle inadequacy. Indeed, for lack of degrees of freedom is the orbital optimization, the VB expansion of the GVB function in eq 8 deals with atomic orbitals which are nearly identical for the covalent and ionic structures, while common sense suggests that, for instance, the left fluorine fragment should exhibit orbitals of different sizes and shapes according as it is neutral as in $F \cdot \text{---} \cdot F$, anionic as in $F^- F^+$ or cationic as in $F^+ F^-$. In fact, all the orbitals are optimized for an average neutral situation, which is about correct for the covalent structure but disfavors the ionic ones. One can therefore anticipate that this mean-field constraint underestimates the weight of the ionic structures, leading to a poor description of the bond.

Relaxing this constraint during the orbital optimization would allow each VB structure to have its own specific set of orbitals, different from one structure to the other, and should improve the description of the bond without increasing the number of VB structures. In such a wave function, the orbitals can be viewed as instantaneously following the charge fluctuation by rearranging in size and shape, hence the name "breathing-orbital valence bond" (BOVB) that has been attached to the corresponding method. Our working hypothesis is that the qualitative improvement brought by this breathing-orbital effect closely corresponds to the contribution of dynamical correlation to the formation of the bond.

3 The Breathing Orbital Valence Bond Method

The idea of using different orbitals for different VB structures is not new, and has been successfully applied to molecules qualitatively represented as a pair of resonating degenerate Lewis structures.[34-36] What is advocated here is just the systematic application of this principle to the description of the chemical bonds in reacting systems, with the aim of defining a VB method gathering the following features: (i) unambiguous interpretability of the wave function in terms of Lewis structures; (ii) compactness of the wave function; (iii) ability to calculate diabatic as well as adiabatic states; (iv) reasonable accuracy (say a few kcal/mol) of the calculated energetics; (v) consistency of the accuracy at all points of the calculated surfaces. The two latter points require the method to be able to accurately describe the elementary events of a reaction, i.e. bond-breaking or bond-forming. Thus, a crucial test for the method will be its ability to reproduce dissociation curves, for two-electron as well as odd-electron bonds.

3.1 General Principles

The basic principle underlying this method is very simple and just consists of generating *all* the Lewis structures necessary to describe a reacting system in VB terms, and providing the corresponding VB structures with the best possible orbitals to minimize the energy of the final multi-structure state. This kind of "absolute" optimization of the orbitals is attained by getting rid of the above discussed mean-field constraint, i.e. by allowing different orbitals for different VB structures, so as to remove a source of imbalance in the description of the different structures. The general philosophy is that the representation of an electronic state in terms of Lewis structures is not just a model but reflects the true nature of the chemical interactions, and only needs a rigorous quantum mechanical formulation to become a quantitative computational method. The method is thus grounded on the basic postulate that *if all relevant Lewis structures of an electronic state are generated and if these are described in a balanced way by a wave function, then this wave function should accurately reproduce the energetics of this electronic state throughout a reaction coordinate.*

The requirement that all Lewis structures are generated implies that both covalent and ionic components of the chemical bonds must be considered. As the number of VB structures grows exponentially with the number of electrons, it is already apparent that the BOVB method will not be applied to large systems of electrons, but rather to that small part of a molecular system that effectively takes part in a reaction, the rest of the electrons being considered as spectators and treated at the MO level.

Choice of an Active Sub-system. Consider a typical S_N2 reaction as an example (eq 9). The reaction consists of the breaking of an F-C bond

$$Cl^- + CH_3F \longrightarrow [Cl...CH_3...F]^- \longrightarrow ClCH_3 + F^- \qquad (9)$$

followed by the formation of a new C-Cl bond. Three lone pairs of fluorine, three other lone pairs of chlorine and three C-H bonds of carbon will keep their status unchanged during the reaction and will form the "spectator" or "inactive" system. On the other hand, the remaining four electrons and three orbitals involved in the C-F bond and in the attacking lone pair of Cl^- will constitute the heart of the reaction, and will form the "active" system. More generally, the active system will be composed of those orbitals and electrons that undergo bond-breaking or bond-forming in a reaction.

While the inactive system will be treated at the simple MO level, i.e. the corresponding lone pairs or bonds will be described as localized doubly occupied MOs, the active system will on the contrary be subject to a detailed VB treatment involving the complete set of chemically relevant Lewis structures. In the above example, this would mean consideration of the full set of the six VB structures (1-6) that one can possibly imagine for a system of four electrons in three orbitals. The active electrons are thus explicitly correlated,

Cl⁻ C••F	Cl••C F⁻	Cl⁻ C⁺ F⁻	Cl⁻ C⁻ F⁺	Cl⁺ C⁻ F⁻	Cl• C⁻ F•
∕\	∕\	∕\	∕\	∕\	∕\
1	**2**	**3**	**4**	**5**	**6**

while the inactive electrons are not, but one expects that this latter lack of correlation results in an error that carries over through the whole potential surface and therefore just uniformly shifts the calculated energies relative to fully correlated surfaces. Note that in this model the inactive electrons are not considered as unaffected by the reaction, since their orbitals rearrange at all points of the reaction coordinate. It is simply their mutual correlation that is considered as constant. From this picture it emerges that: (i) Most of the total electron correlation is neglected, as the method aims at reducing the computational (and conceptual) effort to treating that part of the correlation that varies along a reaction. (ii) Only one part of the molecule is examined from the VB point of view, hence the small number of VB structures.

The above definitions of active/inactive subsystems is of course not restricted to the study of reactions but can be generalized to all static systems whose qualitative description can be made in terms of resonating Lewis structures, like conjugated molecules, mixed valence compounds, etc..

VB Formulation of the Lewis Structures. Next step after the choice of the relevant Lewis structures is their quantum mechanical formulation. Each Lewis structure corresponds to a set of atomic orbitals which are singly or doubly occupied, as illustrated in **7-9** for difluorine, and is represented by a sin-

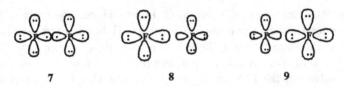

7 **8** **9**

gle VB spin-eigenfunction Ψ_7-Ψ_9 more simply called "VB structure" in what follows. Such VB structures are linear combinations of Slater determinants involving the same occupied AOs as the corresponding Lewis structures, as in eqs 10-12 where ϕ_i, ϕ_i' and ϕ_i" represent the set of inactive orbitals for each VB structure, L and R are the active orbitals of the left and right fragments, respectively, and the subscripts n and a stand for neutral and anionic fragments, respectively.

$$\Psi_7 \;=\; |...\phi_i...L_n\overline{R}_n| + |...\phi_i...R_n\overline{L}_n| \tag{10}$$

$$\Psi_8 \;=\; |...\phi_i'...L_a\overline{L}_a| \tag{11}$$

$$\Psi_9 \;=\; |...\phi_i"...R_a\overline{R}_a| \tag{12}$$

Note that the inactive orbitals ϕ_i, ϕ_i' and ϕ_i" of Ψ_7-Ψ_9 are all different from each other, as are the active orbitals L_n, L_a, or R_n, R_a, as pictorially represented in **7-9** where the orbitals are drawn with different sizes.

As an important feature of the BOVB method, the active orbitals are chosen to be strictly localized on a single atom or fragment, without any delocalization tails. If this were not the case, a so-called "covalent" structure, defined with more or less delocalized orbitals like, e.g., Coulson-Fischer orbitals, would implicitly contain some ionic contributions which would make the interpretation of the wave function questionable.[37] The use of pure AOs is thus a way to ensure an unambiguous correspondence between the concept of Lewis structural scheme and its mathematical formulation. Another reason for this choice is that for the breathing orbital effect to be effective and to lead to a significant stabilization of the VB wave function, the charge fluctuation has to be truly reflected in the VB structures, meaning that the ionic structures are really ionic and the covalent ones really covalent. This cannot be the case if the orbitals are not local, as formally ionic structures are in fact contaminated by covalent ones and can at best reflect some damped charge fluctuation. Finally, it is to be noted that, since the whole set of Lewis structures is generated, the choice of purely localized active orbitals is *not* a restriction in the orbital optimization. Indeed, as noted above, letting the spin-coupled orbitals of a covalent structure be delocalized is equivalent to adding ionic structures, so that this extra degree of freedom together with the generation of a complete set of Lewis structures would lead to a redundant set of parameters in the orbital optimization.

On the other hand, letting the inactive orbitals be delocalized does not matter much from a conceptual point of view. Taking for example the local π_x lone pairs of difluorine or their doubly occupied bonding and antibonding combinations does not change the physical picture of this four-electron interaction which in both cases represents two lone pairs facing each other. However the delocalized representation has slightly more degrees of freedom in flexible basis sets, in that the AOs that compose both delocalized combinations can differ, leading to a slightly better description of the inactive interactions. Therefore, the delocalization of inactive orbitals will be used as a possible option in the BOVB method.

Various Possible levels. Several theoretical levels are conceivable within the BOVB framework. First, (vide supra), the inactive orbitals may or may not be allowed to delocalize over the whole molecule. To distinguish the two options, a calculation with *localized* inactive orbitals will be labeled "L", as opposed to the label "D" that will characterize *delocalized* inactive orbitals. The usefulness and physical meaning of this option will be discussed below on particular cases.

Another optional improvement concerns the description of the ionic VB structures. At the simplest level, the active ionic orbital is just a unique doubly occupied orbital as in **8** or **9**. However this description can be improved by taking care of the radial correlation (also called "in-out" correlation) of the two active electrons, and this can be done most simply by splitting the active orbital in two singly occupied orbitals, much as in GVB theory, as pictorially represented in **10** and **11** which represent improved descriptions of **8** and **9**. This improved level will be referred to as "S" (for "split") while the simpler

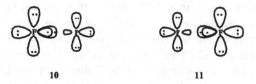

10   **11**

level will carry no special label. Combining the two optional improvements, the BOVB calculations can be performed at the L, SL, D or SD levels.[a] These will be tested below on bonding energies and/or dissociation curves of classical test cases representative of two-electron and odd-electron bonds.

3.2 Dissociation of Two-electron Bonds

The Difluorine Molecule. The dissociation of difluorine is a demanding test case that is traditionally used to appreciate the qualities of new computational methods, and the complete failure of the Hartree-Fock method to account for the bonding in F_2 has already been mentioned. Table 1 displays the calculated energies of F_2 at a fixed distance of 1.43 Å, relative to the separated atoms, in 6-31G* basis set. Note that at infinite distance, the ionic structures disappear, so that one is left with a pair of singlet-coupled neutral atoms which just corresponds to the Hartree-Fock description of the separated atoms.

Table 1: L-BOVB calculation on the F_2 molecule at a fixed distance of 1.43Å. The 6-31G* basis set has been used. See Ref. 10 for more details.

Iteration	Energy(au)	De(kcal/mol)	Coefficients(Weights)	
			Covalent	Ionic
0	-198.71314	-4.6	0.840(0.813)	0.194(0.094)
1	-198.75952	24.6	0.772(0.731)	0.249(0.134)
2	-198.76494	27.9	0.754(0.712)	0.258(0.144)
3	-198.76572	28.4	0.751(0.709)	0.260(0.146)
4	-198.76600	28.5	0.752(0.710)	0.259(0.145)
5	-198.76608	28.6	0.750(0.707)	0.261(0.146)
Projected GVB[a]	-198.74554	15.7	– (0.768)	– (0.116)

[a] The VB weights are calculated after projecting the GVB wavefunction onto a basis of pure VB functions defined with strictly localized AOs.

As quite large basis sets are required for this molecule, the reference bonding energy for calculations in 6-31G* basis set is best taken as the full CI value in the same basis set, about 30 kcal/mol, rather than the experimental bonding energy. The classical VB level, referred to in the Table as iteration 0, is a simple non-orthogonal CI between one covalent and two ionic structures, the orbitals being the pure atomic orbitals of fluorine as optimized in the free atoms. As can be seen, the bonding energy at this latter level is extremely

---

[a]The L, SL and SD levels were referred to as levels I, II and III in Ref. 11.

poor (though better than Hartree-Fock) and does not even have the right sign. The GVB level, which nearly corresponds to the same VB calculations but with optimized orbitals (all VB structures sharing the same set of orbitals), is much better but still far from quantitative. However, as soon as the orbitals are allowed to be different from one structure to the other (iterations 1-5) the bonding energy is considerably increased and converges rapidly to a value close to the full CI estimation. Thus, *the breathing orbital effect just corresponds to that part of the dynamical electron correlation that vanishes as the bond is broken.* This provides a clear picture for the physical meaning of the dynamical correlation associated to the single bond, which is nothing but the instantaneous adaptation of the orbitals to the charge fluctuation experienced by the two bonding electrons.

Table 1 also displays the weights of the covalent and ionic structures, as calculated by means of the popular Chirgwin-Coulson formula,[b] thus emphasizing the imbalanced ionic/covalent ratio that characterize low levels of calculation. The classical VB calculation, with its orbitals optimized on the free atoms, severely disfavors the ionic structures with a weight being much too small as compared with the best calculation, iteration 5. The GVB wave function (projected on a basis of VB functions defined with pure atomic orbitals), with its orbitals optimized for the bonded molecule, is a little better in that respect but still suffers from the mean-field constraint. Now when full freedom is given to the ionic structures to have their orbitals different from the covalent ones, the ionic weights gradually increase after each iteration. This clearly supports the above stated intuitive proposal that the lack of dynamical correlation, that characterizes the classical VB, GVB, SCVB or valence-CASSCF levels, results in an imbalance in the treatment of covalent vs ionic situations, the latter being disfavored.

The above single point calculation[10] corresponds to the simplest level of the BOVB method, referred to as L-BOVB. All orbitals, active and inactive, are strictly local, and the ionic structures are of closed-shell type, as represented in **8** and **9**. However the theory can be pushed a bit further, and the performances of the various levels, in terms of optimized interatomic distances and bonding energies, are displayed in Table 2, using two different basis sets involving both diffuse and polarization functions. It appears that the L-BOVB level in 6-31+G* basis set, although providing a fair bonding energy, yields an equilibrium distance that is rather too long as compared to sophisticated estimations in the region of 1.44-1.45 Å. This is the sign of an incomplete description of the bond, and indeed this simple level does not fully take into

---

[b]The weight W_n of a VB structure V_n is calculated as: $W_n = \sum_m C_n C_m S_{nm}$, where C_n and C_m are the coefficients of V_n and V_m in the wave function and S_{nm} is their overlap.

account the correlation of the active electrons, which are located in doubly occupied orbitals in the closed-shell ionic structures **8** and **9**. Splitting the active orbitals of the ionic structures as in **10** and **11** remedies for this deficiency, and indeed the corresponding SL-BOVB level displays an increased bonding energy and a shortened bond length as compared to L-BOVB in Table 2.

Table 2: Dissociation energies and optimized equilibrium bond lengths for the F_2 molecule

Method	$Re(\text{Å})$	$De(\text{kcal/mol})$	Ref.
6-31+G* Basis Set			
GVB(1/2)	1.506	14.0	11
CASSCF	1.495	16.4	11
L-BOVB	1.485	27.9	11
SL-BOVB	1.473	31.4	11
SD-BOVB	1.449	33.9	11
Estimated full CI		<33	33
Dunning-Huzinaga Basis Set [a]			
SD-BOVB	1.443	31.6	11
Estimated full CI	1.44 ± 0.005	28 - 31	39,40
Experimental	1.412	38.3	41

[a] A modified Dunning-Huzinaga basis set used by Laidig, Saxe and Bartlett [39]. The normal (4,1) p contraction is extended to (3,1,1) and a set of six d functions of exponent 1.58 is added.

The optimized equilibrium distance is still too large however, and now the interatomic interactions between inactive electrons have to be considered. In the F_2 case, they are constituted of three lone pairs on each atom, facing each other. Their local AO or delocalized MO descriptions would be strictly equivalent in a minimal basis set, but it can be shown than in more flexible basis sets the delocalized MO description implicitly allows some charge transfers from one lone pair of an atom to some out-valence orbitals of the other atom.[38] Most of this charge transfer corresponds to some back-donation in the ionic structures, i.e. the fragment F^- that has an excess of electrons in its σ orbitals back-donates some charge to the F^+ fragment through its π orbitals. Thus, allowing the π lone pairs to delocalize (SD-BOVB entries in Table 2) results in a significantly shortened calculated bond length which is now in the expected range (Table 2).

For the sake of comparison, Table 2 also displays some full CI estimations by Laidig, Saxe and Bartlett[39] (LSB), along with SD-BOVB calculations using the same basis set. The BOVB bonding distance appears as perfectly correct, while the bonding energy seems a bit too large, however still remaining within an acceptable error margin. Trying to interpret the remaining error would probably be meaningless, having regard to the simplicity of the three-configuration BOVB wave function and to the large part of the total electron correlation that is ignored in the calculation.

Why not pursuing the delocalization further, and delocalizing *all* orbitals? As has been mentioned above, this would break the interpretability of the wave function but this is not the only problem. Let us take an extreme case, in which the active orbitals would be delocalized in the Coulson-Fischer way, as in eq 6. In such a case it is quite conceivable that the wave function would most efficiently lower its energy by using the same active orbitals in all VB structures, but different inactive orbitals. This way, some left-right correlation would be taken into account in the active space, because of the delocalized nature of the active orbitals, but some correlation would also be introduced in the inactive space for the inactive orbitals being different from one structure to the other. The inactive electrons would therefore be correlated in the bonded molecule but uncorrelated in the separated fragments. Obviously, the description of the dissociation energy profile would be much imbalanced in this extreme case and would result in too large a bonding energy. Even without falling in such an obvious artefact, it is clear that the delocalization of the active orbitals would be a source of imbalance in the description of the bond in the BOVB framework. It is therefore the rule to *keep the active orbitals strictly localized at any level of the BOVB method*, not only for interpretative purposes, but also for a correlation-consistent description of an electronic system throughout a potential surface.

The hydrogen fluoride molecule. Hydrogen fluoride is another classical test case, representative of a typical polar bond between two atoms of very different electronegativities. Owing to this peculiarity, the molecule is expected to display one ionic structure, F^-H^+ (**13**) nearly as important as the covalent one (**12**), so that any deficiency in the description of ionic structures should result in significant errors in the bonding energy and dissociation curve. Another distinctive feature of the FH bond is its very high experimental bonding energy of 141 kcal/mol. The recombination of the two fragments is therefore an extremely exothermic reaction, and one may wonder if under such conditions the inactive electrons may still keep their identity, as assumed as a basic hypothesis of the BOVB method. For these two reasons, hydrogen fluorine is a challenging case, all the more interesting as some benchmark full CI calculations are available for the bonding energy and dissociation curve.

As usual, the single bond is described by three VB structures, **12-14**, among which the last one, F^+H^- (**14**), is expected to be very minor but is

| **12** | **13** | **14** |

nevertheless added for completeness. Table 3 displays the optimal bond lengths and bonding energies calculated at various theoretical levels, in 6-31+G** basis set and in another basis set of comparable quality used by Bauschlicher and Taylor.[42] Dynamic electron correlation effects appear once again to be

Table 3: Dissociation energies and optimized equilibrium bond lengths for the FH molecule

Method	$Re(\mathring{A})$	De(kcal/mol)	Ref.
6-31+G Basis Set**			
GVB(1/2)	0.920	113.4	11
L-BOVB	0.918	121.4	11
SL-BOVB	0.911	133.5	11
SD-BOVB	0.906	136.3	11
Extended SD-BOVB	0.916	137.4	11
BT Basis Set [a]			
SD-BOVB	0.906	136.5	11
Extended SD-BOVB	0.912	138.2	11
Full CI [b]	0.921	136.3	42
Experimental	0.917	141.1	41

[a] A double-zeta + polarization + diffuse basis set used by Bauschlicher and Taylor [42].
[b] The 2s orbitals are not included in the CI.

an important component of the bonding energy, since the GVB calculation in 6-31+G** basis set yields a value of only 113 kcal/mol, quite far from the experimental value. However the simple L-BOVB level also proves to be quite insufficient, with a bonding energy being still much too small. This was expected (vide supra), owing to the importance of the F^-H^+ ionic structure **13** that is rather poorly described as a closed-shell configuration at this level. Splitting the active orbital of this structure, as in **15**, leads to a spectacular improvement of the bonding energy, by ca 12 kcal/mol (SL-BOVB/6-31+G** entry in Table 3), putting it already close to a reasonable value. As in the

15

F_2 case, further improvement is gained by delocalizing the π inactive orbitals to reach the SD level that yields a bonding energy of 136.3 kcal/mol, in very reasonable agreement with the experimental value having regard to the lack of high momentum polarization functions in the basis set.

Due to its polar nature, hydrogen fluoride is a severe test for the key approximation that the correlation of the inactive electrons remains nearly constant throughout the dissociation process. Indeed, as the inactive electrons of F^- in the F^-H^+ structure feel a different electric field than those of the neutral F fragment at infinite separation, one might expect the intra-pair correlation energy of the active electrons to vary with the interatomic distance, owing to the importance of the ionic structure at the equilibrium geometry. To probe this possible source of error, we have pushed the BOVB calculation to a further level, splitting *all* lone pairs, active and inactive, in the two main VB structures **12** and **13**, leading to **16** and **17** in which the inactive electrons are provided some radial correlation. This test calculation, referred to as "Ex-

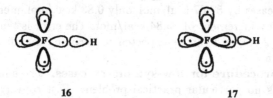

16   **17**

tended SD-BOVB" in Table 3, results in an improvement of only 1.1 kcal/mol of the bonding energy relative to the standard SD level, thus confirming the near-constancy of the correlation within inactive electrons, and the futility of further rising the level of complexity.

Table 3 also displays a comparison of a full CI calculation by Bauschlicher and Taylor with the best BOVB levels using a common basis set involving diffuse and polarization functions. Once again very little difference is found in the optimal distances and bonding energies yielded by the SD-BOVB level and its extension that splits inactive orbitals, and both levels are in very satisfying agreement with full CI results.

By nature, the BOVB method should describe the two-electron interaction equally well at any interatomic distance from equilibrium to large separation. Therefore, one can expect dissociation curves to be faithfully reproduced. As a basis for comparison, Bauschlicher et al.[43] have performed a full CI calculation of three points of the FH dissociation curve, corresponding to different multiples of the experimental bond length Re. Table 4 displays a comparison of this latter calculation with the extended SD-BOVB level using the same

Table 4: Energies of the FH molecule at three interatomic distances. Re is the experimental equilibrium distance of 0.917 Å . Both the full CI[43] and the BOVB calcul ations use the same basis set of double-zeta + polarization type[43]

| Distance | Full CI | | Extended SD -BOVB | |
	E(Hartrees)	ΔE(kcal/mol)	E(Hartrees)	ΔE(kcal/mol)
Re	-100.250969		-100.129176	
		56.84		57.67
$1.5Re$	-100.160393		-100.037277	
		49.75		49.83
$2Re$	-100.081108		-99.957863	

basis set and the same interatomic distances, and shows very little difference between the BOVB and exact energy profiles. Indeed, from Re to 1.5 Re, the VB energy increases by 57.67 kcal/mol, only 0.83 kcal/mol in error relative to the full CI energy increment of 56.84 kcal/mol. The error is ten times smaller from 1.5 Re to 2 Re, and can be expected to become insignificant from 2 Re to infinite distance.

General procedure for low-symmetry cases. The simplest level, L-BOVB, presents no particular practical problem. Fast convergence is generally obtained by using well adapted guess orbitals, that can be chosen as the Hartree-Fock orbitals of the isolated fragments with the appropriate electronic charge. Thus, the guess orbitals for the covalent structure are those of the isolated radicals, while the guess orbitals for the ionic structures can be chosen as those of the isolated anions and cations.

Going to the more accurate SL-BOVB level just requires making sure that the orbital that is split in an ionic structure is an active orbital, and that the corresponding pair of singly occupied orbitals does not eventually belong to the inactive space after the optimization process. While this condition is generally met by choosing an appropriate guess function in high symmetry cases (F_2 or FH above), in the general case nothing a priori guarantees that such an exchange between active and inactive spaces will not take place, leading for instance to **18** instead of the correct structure **19** in the H_2N-NH_2 example. To circumvent this difficulty, a general procedure consists of localizing the

18   **19**

orbitals by any standard method after the L-BOVB step,[c] then splitting the active orbital while freezing the inactive ones during the optimization process.

Delocalization of the inactive orbitals (D-BOVB or SD-BOVB) is important for getting accurate energetics, especially in cases when the inactive orbitals are not lone pairs but local bonding orbitals (e.g. C-H bond in H_3C-F). This is because this additional degree of freedom allows for some charge transfer to take place between inactive orbitals, and low-lying valence antibonding orbitals are better suited than high-lying out-valence lone pair orbitals to accept an extra electron from the neighboring atom[44] (see Section 5 below). Now it is important to make sure that the orbitals that are delocalized are indeed the inactive ones, while the active set remains purely localized as assumed as a basic principle of the method, as otherwise any artefactual solution might be found. In the general case when the active and inactive spaces cannot be distinguished by symmetry, once again any spurious exchange between the active and inactive spaces must be prevented during the orbital optimization process. Practically, this can be done by starting from an L-BOVB or D-BOVB wave function, then allowing delocalization of the inactive orbitals while freezing, this time, the *active* orbitals during the further optimization process that leads to the D-BOVB or SD-BOVB levels.

3.3 Dissociation of Odd-e Bonds

Theoretical models for odd-electron bonding. Two-electron bonds are not the only type of interactions that can strongly hold two fragments together. Odd-electron bonds also play an important role in chemistry, and, as such, constitute a compulsory test case for a computational method. First described by Pauling[45] in 1931, odd-electron bonded systems are represented as two resonating Lewis structures that are mutually related by charge transfer, as shown in eq 13 for two-center, one-electron (2c,1e) bonds and in eqs 14 and 15

---

[c]This requires prior orthogonalization of the orbitals within each fragment.

for typical two-center, three-electron (2c,3e) bonds.

$$(A \cdot B)^+ = A \cdot B^+ \leftrightarrow A^+ \cdot B \tag{13}$$

$$(A \because B)^+ = A^+ \because B \leftrightarrow A \because B^+ \tag{14}$$

$$(A \because B)^- = A \cdot \because B^- \leftrightarrow A \because^- \cdot B \tag{15}$$

According to qualitative VB theory, such bonds owe their strength to the resonance energy associated with the mixing of the two limiting structures. As a significant resonance energy requires a similar stability of the two resonating structures, one expects many of the observed odd-electron bonds to be homonuclear (A = B), in agreement with experimental facts.

In MO theory the stability of these bonds is readily interpreted by inspection of orbital interaction diagrams **20** and **21**, where σ and σ^* are bonding and antibonding combinations of active orbitals, respectively. Both diagrams dis-

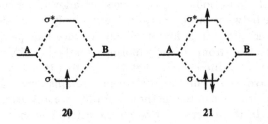

20   **21**

play one net bonding electron, hence a bond order of 1/2 and the qualification of "hemibond" that is often attached to one-electron and three-electron interactions. Both VB and MO qualitative models[46] agree on the bonding energy expressions (16) and (17) for (2c,1e) and (2c,3e) interactions, respectively,

$$De(\text{1-e}) = \frac{\beta}{1 + S} \tag{16}$$

$$De(\text{3-e}) = \frac{\beta(1 - 3S)}{1 - S^2} \tag{17}$$

where S is the overlap between the atomic orbitals that are involved in the bond and β is the usual resonance integral.

Diagrams **20** and **21** can be further considered from the point of view of left-right electron correlation. In **20**, the active space reduces to a unique electron, which eliminates the need for electron correlation within this space. On the other hand the active space of **21** involves three electrons, however the only configuration one might add, within this space, to the simple Hartree-Fock wave function would be a single excitation $\sigma^1 \sigma^{*2}$ which by virtue of Brillouin's theorem does not mix with the lowest configuration. It follows that the concept of left-right correlation is meaningless in such systems, and that the description of both one-electron and three-electron bonds is already qualitatively correct at the Hartree-Fock level, contrary to two-electron bonds.

In view of the preceding analysis, the complete failure of Hartree-Fock ab initio calculations to reproduce three-electron bonding energies is paradoxical. As shown by Clark[47] and Radom,[48] who carried out systematic calculations on series of cation radicals involving three-electron bonds between atoms of the first and second rows of the periodic table, the Hartree-Fock error is always large, sometimes of the same order of magnitude as the bonding energy itself. Thus, F_2^- that is experimentally bound by 30 kcal/mol is found to be unbound at the Restricted-Open-Shell Hartree-Fock (ROHF) level.[33] The error is much smaller in the case of one-electron bonds,[47] yet it may be as large as 13 kcal/mol in the $(H_3C.CH_3)^+$ cation. Interestingly, the Hartree-Fock error is not constant, but gradually increases as the bonded atoms are taken from left to right or from bottom to top of the periodic table.

Focusing on the three-electron case, that displays by far the largest electron correlation effect, one may analyze the amazing Hartree-Fock deficiency by expanding the corresponding wave function in terms of VB structures, as has been done in the two-electron case (vide supra). Taking the F_2^- case as an example, the Hartree-Fock wave function $\Psi_{HF}(3\text{-e})$ reads:

$$\Psi_{HF}(3\text{-e}) = |...\phi_i...\sigma_g\overline{\sigma}_g\sigma_u| \tag{18}$$

where ϕ_i represent the set of inactive orbitals, and the active orbitals σ_g and σ_u are defined as in eqs 2 and 4. Expanding σ_g and σ_u leads, after elimination of nil determinants, to eq 19:

$$\Psi_{HF}(3\text{-e}) = |...\phi_i...\chi_a\overline{\chi}_a\chi_b| + |...\phi_i...\chi_a\overline{\chi}_b\chi_b| \tag{19}$$

where it is clear that the Hartree-Fock wave function is equivalent to a two-configuration VB wave function that directly arises from Pauling's description of three-electron bonds, and displays a resonance between structures **22** and **23**:

22	**23**

However physically correct, the latter wave function suffers from the same defect as the GVB wave function for two-electron bonds, i.e. the unique set of

AOs for the two structures, which makes the active orbitals unadapted to their instantaneous occupancy and the inactive ones unadapted to the instantaneous charges of the atoms. Once again, a remedy to this defect consists of allowing for the breathing orbital effect by letting the VB wave function have different orbitals for different structures, as in eq 20, where the orbitals are defined in the same way as in eqs 10-12.

$$\Psi_{BOVB}(3\text{-e}) = C_1|...\phi_i...L_a\overline{L}_aR_r| + C_2|...\phi_i'...L_r\overline{R}_aR_a| \tag{20}$$

Whether or not this simple effect can retrieve the considerable contribution of dynamical electron correlation to three-electron bonding energies will be tested below on representative examples.

The F_2^- radical anion. As its neutral homologue, the difluorine radical anion is a difficult test case for the calculation of its bonding energy. As the computational studies of Clark[47] and Radom[48] have shown that the importance of electron correlation effects increases with the elecronegativity of the three-electron bonded atoms, it is not surprising that the Hartree-Fock error on the bonding energy reaches its maximum with F_2^-. Besides, the otherwise most versatile and successful Density Functional Theory (DFT), in its B3LYP popular version, also fails to reproduce a realistic bonding energy for this compound (Table 5).

The equilibrium distance and bonding energy of F_2^- are displayed in Table 5, as calculated at various levels of theory, in 6-31+G* basis set. Let us consider first some VB calculations in which all orbitals are purely local and all lone pairs have a closed shell form as in **22** and **23**, and let us introduce the breathing orbital effect by steps, in order to understand its nature in more details. In a first step, no orbitals are included in the breathing set, so that both determinants in eq 20 share the same unique set of orbitals. At this level, the VB wave function is nearly equivalent to the ROHF wave function, and displays the full Hartree-Fock error. In a second step, the active orbitals (L_a, R_r, L_r, R_a), and only them, are included in the set of breathing orbitals, while the inactive orbitals are optimized but constrained to remain identical in both determinants:

$$L_a \neq L_r; \quad R_r \neq R_a; \quad \phi_i = \phi_i' \tag{21}$$

This already results in a very significant improvement of the dissociation energy profile, which now displays an attractive potential well, with a positive dissociation energy of ca 13 kcal/mol, to be compared with the ROHF value -4 kcal/mol.[33] However, as sizeable as it can be, this improvement is far from yielding a realistic bonding energy, indicating that the inactive orbitals are not

Table 5: Calculated equilibrium distances and dissociation energies for the F_2^- radical anion, in 6-31+G* basis set. BOVB calculations are performed with all valence orbitals being included in the set of breathing orbitals (fully-breathing option) unless otherwise specified (entries 2 and 3)

Method	$Re(\text{Å})$	$De(\text{kcal/mol})$	Ref.
ROHF		-4	33
L-BOVB (active set only)	1.954	13.3	12
L-BOVB	1.964	29.7	" "
L-BOVB (HF orbitals) [a]	2.050	24.9	" "
D-BOVB	1.954	30.1	" "
SD-BOVB	1.975	28.0	" "
SD-BOVB (4-structure)	1.976	28.0	" "
MP2	1.916	26.2	" "
PMP2	1.935	29.5	" "
MP4	1.931	25.8	49
DFT (B3LYP)	2.000	41.6	50
Experiment		30.2	41

[a] The atomic orbitals come from Hartree-Fock and ROHF calculations on the isolated fragments F^- and F, and are not further optimized.

to be neglected. Therefore, as a third step, all the orbitals, active and inactive (with the exception of the frozen core), are included in the breathing set:

$$L_a \neq L_r; \quad R_r \neq R_a; \quad \phi_i \neq \phi_i' \tag{22}$$

The resulting fully-breathing wave function, corresponding to the standard L-BOVB level, can be represented as in **24, 25** and yields a bonding energy of 29.7 kcal/mol (Table 5, entry 3), in amazing agreement with the experimental bonding energy of 30.2 kcal/mol.[41]

24   25

The Hartree-Fock error is thus completely corrected by the BO effect, which can be attributed for 17.3 kcal/mol to the active orbitals, and for 16.8

kcal/mol to the inactive ones. On a per orbital basis, each active AO contributes for 8.6 kcal/mol to the overall BO stabilization, while the inactive lone pairs have a lesser influence, about 2.8 kcal/mol each. This quantitative difference easily finds a qualitative interpretation: in the absence of BO effect, as in **22**, **23**, all orbitals are unadapted to the instantaneous charge of the fragment to which they are linked, but the active orbitals have the extra defect to be unadapted to their instantaneous occupancy, that can be single or double. It is therefore expected that the latter orbitals will be more sensitive to the charge fluctuation than the inactive ones and will exhibit a more stabilizing BO effect.

Another question of interest is the nature of the orbitals in each determinant of the BOVB wave function, and one may wonder whether they bear any relationship to the orbitals of F^- and F isolated fragments. This can be checked by putting the latter orbitals, respectively calculated at the RHF and ROHF Hartree-Fock levels, into eq 20 and calculating the energy of Ψ_{BOVB}(3-e) without any further optimization. The resulting bonding energy and equilibrium distance (Table 5, entry 4) are amazingly good, relative to the performances of the fully optimized L-BOVB wave function, indicating that the F^- and F fragments keep their identity in **24** and **25**. This also shows that the BO effect in F_2^- corresponds to an orbital change in size rather than a polarization effect, since the latter is ineffective in isolated fragments.

As has been done for the two-electron bonds, the calculation can be further sophisticated by allowing the π inactive orbitals to delocalize over the whole molecule, leading to the D-BOVB level. As a result, the equilibrium bond length is shortened by 0.01 Å, and the bonding energy is increased by 0.4 kcal/mol relative to the L-BOVB level (Table 5). These rather weak consequences of increasing the level of theory indicate that the fully localized atomic orbitals are, right at the outset, well adapted to the description of the three-electron interaction, contrary to what is generally observed in two-electron bonds. This difference may be due in part to the long equilibrium distance that characterize three-electron bonds, which results in weak interatomic repulsions between inactive lone pairs. Another reason for the ineffectiveness of π delocalization is that neither of **22** or **23** VB structures display a polar σ bond that would need be counterpolarized by some π back-donation as in two-electron bonds.

Somewhat more significant is the effect of splitting the active orbitals, leading to structures **26** and **27** at the SD-BOVB level, where the local singlet spin-couplings are indicated by curved lines. However this improvement does not lead to an increase, but rather to a small decrease (2.1 kcal/mol) of the bonding energy. This is because the effect of splitting the active orbitals stabilizes both the separated fragments and the bonded molecules, so that both stabilizations nearly compensate one another and may lead to a small correction of any sign in the bonding energy. By contrast, in the case of two-electron bonds, splitting the doubly occupied active orbitals may only benefit to the ionic structures that are present at equilibrium distance but vanish at infinite separation, which can only result in an increased bonding energy.

Up to now we have dealt with the two Lewis structures that a chemist might think to generate for the three-electron interaction. However, mathematically speaking, there are two ways of spin-coupling three electrons in three orbitals so as to generate a doublet spineigenfunction, so that one might think of further rising the level of complexity of the SD-BOVB wave function by adding structures 28 and 29 that exhibit the same orbital occupancy as 26 and 27 but different couplings. This latter calculation, referred to as "SD-BOVB (4-structure)" in Table 5, yields a coefficient of only 0.0008 for structures 28 and 29, with a bonding energy and a equilibrium distance practically unchanged relative to the standard SD-BOVB level, thus fully confirming the validity of Pauling's simple model based on chemical intuition.

The performances of the various BOVB levels can be compared to those of Möller-Plesset perturbation theory, displayed in Table 5, yet with some caution as the various MP orders do no not converge well. This is due to a rather large spin contamination in F_2^- at the unrestricted MP2 level, with an expectation value of 0.78 for the spin-squared operator, which pleads in favor of the spin-projected (PMP2) value. Keeping in mind that the breathing orbitals of F_2^- are not much polarized[12] (vide supra), the bonding energy is not expected to be much basis set dependent, so that the SD-BOVB value of 28.0 kcal/mol is entirely reasonable relative to the experimental value of 30.2 kcal/mol. On the other hand, the BOVB calculated equilibrium bond lengths are rather long relative to the values calculated at the various MP levels (no experimental value is available), and both sets of values display significant variations from one level to the other. This inaccuracy is however normal, owing to the extreme flatness of the potential surface near the energy minimum: at the MP4 level, the force constant is only 0.55 mdyn/Å, which means that stretching the bond by 0.02 Å away from equilibrium results in an energy rise of only 0.03 kcal/mol.

The Cl_2^- radical anion. By nature, even with similar exponents, 3s and 3p orbitals are much more spread in space than 2s and 2p orbitals, owing to their different radial components. As in addition the optimized exponents in chlorine are smaller than those of fluorine (2.03 in Cl^- vs 2.40 in F^-), the valence orbitals of Cl_2^- are much larger than those of F_2^-, which should a priori have some consequences on the BO effect. Two electrons occupying the same orbital should be more confined if this orbital is small, and consequently more sensitive to the active breathing orbital effect that adapt orbitals to their occupancy. Perhaps to a lesser extent, small inactive lone pairs should be more sensitive than large ones to the charge fluctuation within the active space. Thus, one can anticipate the breathing orbital effect to be generally weaker in Cl_2^- than in F_2^-, within both active and inactive spaces, and more so within the active space.

Table 6 displays some bonding energies for Cl_2^-, as calculated in 6-31+G* basis set at the simple D-BOVB level and other theoretical levels including Hartree-Fock and Möller-Plesset perturbation theory. Unlike the F_2^- case, the Möller-Plesset series converges well around the values 24-25 kcal/mol which can be taken as references for the bonding energy in this basis set. The fully breathing D-BOVB result is once again satisfying, being in good agreement with the various Möller-Plesset values and with the POL-CI calculation of Wadt and Hay[51] in similar basis set. The BO effect among inactive orbitals, estimated by comparing the bonding energies in entries 1 and 2 in Table 6, only amounts to 7.1 kcal/mol, thus significantly smaller than the corresponding value of 16.4 kcal/mol in F_2^-. Moreover, the BO effect among the active orbitals alone (compare entries 2 and 3) only brings a stabilization of 4.5 kcal/mol in Cl_2^- relative to the Hartree-Fock level, to be compared to the value 17.3 in F_2^-, thus fully supporting the qualitative predictions based on orbital size. As a result, the Hartree-Fock error, though still quite significant, is much smaller than in the preceding case, as Cl_2^- is found slightly bound by 11.0 kcal/mol at the ROHF level.

Comparison of BOVB-calculated bonding energies with experiment is difficult, owing to considerable experimental uncertainty, but some very accurate calculations by Curtiss et al.[52], using the G2 method, and by Roos et al.[53], using the Coupled-Pair Functional approach, are available. Both calculations

Table 6: Calculated dissociation energies for the Cl_2^- radical anion, in 6-31+G* basis set, with an MP2-optimized bond length of 2.653 Å, except otherwise specified

	De(kcal/mol)	Ref.
D-BOVB		
fully-breathing	22.6	12
active set only	15.5	12
ROHF	11.0	12
MP2	24.7	12
PMP2	25.5	12
MP4	24.4	12
POL-CI [a]	24.0	51
G2 [b]	27.5	52
CPF/6s5p4d3f2g [c]	27	53

[a] Optimized bond length of 2.69 Å. [b] Optimized within the G2 procedure. [c] Optimized bond length of 2.59 Å.

use some basis sets close to complete, so that their reported bonding energy of ca 27 kcal/mol for F_2^- is in satisfactory agreement with the D-BOVB result, having regard to our rather modest 6-31+G* basis set. It must be added that the bonding energy of another three-electron bonded diatomic of the second row, Ar_2^+, has been calculated by Archirel,[54] using the D-BOVB method, this time in an appropriate basis set of 4s4p2d1f type. His value of 30.5 kcal/mol is in good agreement with the experimental value of 30.7 kcal/mol, or 32.0 kcal/mol after removing the estimated spin-orbit interaction.[55]

The $(NH_3)_2^+$ radical cation. The atomic orbitals of nitrogen are, as those of chlorine, smaller than those of fluorine, with optimized exponents of 1.96 for NH_3 vs 2.40 for F^-, in minimal basis set. Therefore, one can once again predict a smaller BO effect for both the active and inactive orbitals of $(H_3N \therefore NH_3)^+$, relative to F_2^-. However an additional factor can be anticipated to play a role, referring to the nature of the inactive orbitals that represent single N-H bonds in one system and lone pairs in the other. It is clear that the inactive electrons are closer to the nuclei in the second case than in the first, and therefore more sensitive to the charge fluctuation of the active space. This together with the orbital size effect leads to the prediction that the BO effect should be definitely weaker in $(H_3N \therefore NH_3)^+$ than in F_2^-, and probably more so in the inactive space than in the active one.

The bonding energies of the $(H_3N\dot{\cdot}NH_3)^+$ cation, as calculated in 6-31G* basis set at the D-BOVB level and at various levels of MO theory, are reported in Table 7 and support the above qualitative deductions. The total BO effect only amounts to 17.6 kcal/mol, that can be decomposed in 7.1 kcal/mol for the inactive orbitals and 10.5 kcal/mol for the active ones, much reduced relative to F_2^-, especially in the inactive space. As in the preceding case, the BOVB-calculated bonding energy of $(H_3N\dot{\cdot}NH_3)^+$ is in satisfying agreement with the results of the Möller-Plesset series which is rather well converged and probably reflects the basis set limit.

Table 7: Calculated dissociation energies for the $(H_3N\dot{\cdot}NH_3)^+$ radical cation in its D_{3h} conformation, in 6-31G* basis set. MP2-optimized geometries are used everywhere

	De(kcal/mol)	Ref.
D-BOVB		
fully-breathing	37.9	12
active set only	30.8	" "
ROHF	20.3	" "
MP2	40.0	" "
PMP2	41.4	" "
MP4	38.9	" "

One-electron bonds. Contrary to three-electron bonds, one-electron bonds are already rather well described at the simple Hartree-Fock level, as shown by Clark[47] in a comprehensive computational study of $(H_nX\dot{\cdot}XH_n)^+$ radical cations (X = Li to C, Na to Si). This is because the active system contains a single electron, so that the BO effect is ineffective in the active system in which each orbital is either empty or singly occupied as illustrated in **30** $\leftrightarrow$ **31** for the C.C bond. Therefore, the BO effect is restricted to the

30   **31**

inactive space and, in accord, the Hartree-Fock error is nearly proportional to the number of inactive orbitals and gradually increases in the series (X = Li to C), to reach 13 kcal/mol in $(H_3C\dot{\cdot}CH_3)^+$, for a total bonding energy of 51.0

kcal/mol,[56] as calculated at the MP4 level in 6-31G* basis set. In accord with the above qualitative analysis of the BO effect in terms of orbital size, compounds of the second row (X = Na to Si) exhibit less correlation effects than analogs of the first row of the periodic table.

The $(H_3C.CH_3)^+$ radical cation has been selected, owing to its largest correlation effect in the series, to test the ability of the BOVB method to describe one-electron bonds. Satisfactorily, the bonding energy, as calculated at the D-BOVB level, amounts to 48.7 kcal/mol, in fair agreement with the MP4 value.

3.4 Summary of the Computational Tests.

The two-electron as well as odd-electron interactions appear to be well described with the simple BOVB wave functions as can be judged from the dissociation energies which are in good agreement with accurate calculations using identical or comparable basis sets. The lack of internal electron correlation within the inactive space has no significant consequences on the relative energies, meaning that the inactive electrons, although their orbital undergo some important changes in size, polarization or hybridization, have some nearly constant correlation energy. In fact, just the minimum electron correlation is taken into account since the method becomes equivalent to a Hartree-Fock calculation of the separated fragments at the dissociation limit. Thus, the method only calculates the differential electron correlation, that involves the left-right electron correlation of the active electrons, and the dynamical correlation associated to the formation of the bond. As the latter term is nothing but an instantaneous adaptation of the orbitals to the charge fluctuation of the active electrons, dynamical correlation effects are particularly important in three-electron bonds, because in such systems the stabilizing interaction is 100% made of charge fluctuation.

While all levels provide nearly equally good bonding energies for the three-electron bonds, the same does not hold true for two-electron bonds which require the best levels for being accurately described. Splitting the active orbitals in the ionic structures is particularly important when the bond is very polar. Moreover, the interatomic interactions between inactive orbitals are important in two-electron bonds, owing to their short equilibrium bond lengths. Such interactions, that involve some charge transfer terms,[d] are particularly important when the inactive orbitals are bonding orbitals rather than lone pairs

---

[d]It has been checked by counterpoise calculations that the stabilization due to the delocalization of the inactive orbitals is much larger than the spurious basis set superposition effect.[11,12]

and are adequately accounted for by allowing delocalization of the inactive orbitals.

3.5   Diabatic vs Adiabatic States

The computational tests have shown that the BOVB method is able to account, with a reasonable accuracy, for the elementary event of a chemical reaction: bond-breaking or bond-forming, be the bond of the 1-e, 2-e or 3-e type. The method is therefore suitable for generating reaction profiles or potential surfaces for adiabatic states as well as diabatic states, as will be done in Sections 4, 5, 6 and 7 below.

While the definition of an adiabatic state is straightforward, as an eigenfunction of the hamiltonian within the complete basis of VB structures, the concept of diabatic state is less clear-cut and accepts different definitions. Strictly speaking, a basis of diabatic states (ϑ, ϑ',...) should be such that eq 23 is satisfied for any variation δQ of the geometrical coordinates.

$$< \vartheta \mid \frac{\delta}{\delta Q} \mid \vartheta' >= 0 \qquad (23)$$

However this condition is impossible to fulfill in the general case with more than one geometrical degree of freedom, so that one has to search for a compromise in the form of a function whose physical meaning remains as constant as possible along a reaction coordinate. Clearly, a single VB structure, that keeps the same bonding scheme whatever the geometry of the system, is the choice definition for a diabatic state in the general case. For example, if we consider the A-B molecule in the BOVB framework, the ground state (made of three VB structures) will be adiabatic, while the three single VB structures, respectively $A \cdot - \cdot B$, A^+B^- and A^-B^+, will be the diabatic states.

Diabatic states of this type will be exemplified in Section 5 below, where the BOVB method will be applied to the study of the H_3C-Cl and H_3Si-Cl bonds. Note however that a diabatic state can be possibly made of more than one formal VB structure. For instance, in the S_N2 reaction (9), one diabatic state could be the bonding scheme of the reactants, $Cl^- + H_3C$-F, while the other would represent the products, Cl-$CH_3 + F^-$. In this case, each diabatic state would be made of three VB structures, respectively **1, 3, 4** and **2, 3, 5**, so as to accurately describe the Cl-C and F-C bonds. As an applicatory example, such diabatic states constitute the crossing curves of the VB correlation diagrams of Shaik and Pross.[4]

Now that the diabatic states have been defined, there remains to specify which way they are calculated, in other words which orbitals they are made of.

One first possibility is to keep for the diabatic states the orbitals that have been optimized for the ground state. This has the advantage of simplicity. Practically, once the orbitals have been determined at the end of the BOVB orbital optimization process, the hamiltonian matrix is constructed in the space of the VB structures and the adiabatic energies are calculated by diagonalization of the hamiltonian matrix while the energies of the diabatic states are just the respective diagonal matrix elements. This technique is suitable as a practical mean to diabatize potential surfaces for computations in chemical dynamics, as for such applications the adiabatic states should result from simple CI in the space of the diabatic states, without any further orbital optimization.

One inconvenience of this practical procedure is that it does not guarantee the best possible orbitals for the diabatic states. Indeed, the BOVB orbitals are optimized so as to minimize the energy of the multi-structure ground state and are therefore the best compromise to lower the energy of the individual VB structures and to maximize the resonance energy between these VB structures. This latter requirement implies that the final orbitals are not the best possible orbitals to minimize each of the individual VB structure taken separately. It follows that the diabatic states calculated this way are not the best possible diabatic states to represent the respective bonding schemes, and in practical calculations they may appear surprisingly high in energy. For instance, the purely covalent $H_3C \cdot - \cdot Cl$ bond appears as repulsive, if calculated this way, which is rather unexpected.

Another logic consists of optimizing each diabatic state separately, in an independent calculation involving this diabatic state alone. As a result, the orbitals of the diabatic states come out different from those of the adiabatic states, and we now get for each diabatic state its best possible set of orbitals. The diabatic energies are obviously lower as calculated by this latter procedure than by the previous one; still taking the H_3C-Cl bond as an example, the latter procedure now yields a bonding energy profile for the purely covalent structure, with a bonding energy of 34 kcal/mol, more in agreement with common sense than a repulsive covalent interaction. We therefore believe that the separate calculations of the diabatic states yields the best possible results in terms of chemical interpretation, and this is the procedure that we use in all applications and in particular in Sections 5 and 6 below.

320

4 Application to the Problem of Symmetry-breaking. The $H_2O_2{}^-$ Potential Surface

Given that all dihalogen radical anions $X \colon\! X^-$ are fairly stable relative to three-electron bond breaking, with a dissociation energy in the region of 30 kcal/mol, one may expect isoelectronic species of the type $ROOR^-$ to be similarly stable in symmetrical conformations in which two equivalent OR fragments are linked by a three-electron $O \colon\! O$ bond. The $HOOH^-$ complex is a suitable model for such peroxide radical anions, and has the further interest to be subject to the doublet instability, by which various SCF or MCSCF levels lead to wave functions whose symmetry is lower than that of the molecular conformation.

The symmetry breaking problem is very general and exists in an enormous variety of open-shell electronic states that can be represented by more than one low-energy VB structure, as for example allyl radicals[57,58] or radicals of allylic type,[34,59-61] core-ionized diatoms,[35a] n-π^* excited molecules containing two equivalent carbonyl groups,[35b] n-ionized molecules having two equivalent remote lone pairs,[62] charged clusters,[63,64] etc.. As discussed by McLean et al.,[60] the root cause behind this well documented artefact lies in a poor description of conformations that lend themselves to resonance between several VB structures, relative to other portions of the potential surface. It originates in a conflict between two stabilizing factors, the energy of each individual VB structure and the resonance energy, that are each favored by a different set of orbitals. Let us consider for example the $HOOH^-$ complex in a three-electron bonded conformation, represented by the resonating VB structures **32** and **33**. Although the nuclear frame is symmetrical, it is for sure that

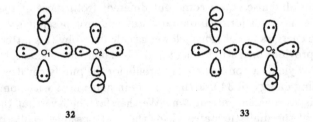

32   **33**

each VB structure is better stabilized with a set of specific, symmetry-broken, set of orbitals, as has been found above for F_2^- and as represented in **24** and **25**. If a unique set of orbitals is used, following the logic of MO-CI calculations, a set of symmetry-broken orbitals will favor one VB structure to the detriment of the other. On the other hand such orbitals will disfavor the resonance stabilization since the latter is maximized when both VB structures are of equal energies. In cases where the resonance factor is predominant, as is found in F_2^- and in most three-electron bonded radical cations, the calculation yields symmetry-adapted orbitals, at least as far as the bonding distance is shorter than some critical value. However for larger distances, or even at equilibrium distance as in the case of HOOH$^-$ (vide infra), the resonance effect becomes weak enough so that the electronic system is better stabilized by sacrifying the resonance energy to the benefit of one particular VB structure that imposes its optimal set of symmetry-broken orbitals, *despite the symmetrical confor-mation of the nuclear frame.* That is, using McLean's terminology,[60] the orbital size effect prevails over the resonance effect. The symmetry dilemma is thus a generalization of the problem we have encountered in the Hartree-Fock description of odd-electron bonds.

In cases the symmetry of the nuclear frame is kept in all investigated conformations, as for example in diatomics or if we are studying the mere dissociation of a symmetrical complex, the problem is easily dealt with at the MO-CI level: it suffices to impose the symmetry of the electronic wave function and to correct for the lack of breathing orbital effect by introducing electron correlation at rather low levels.

The problems gets more difficult, however, in the general case where no obvious conflict exists between the symmetry of the nuclear frame and that of the electronic wave function. The same type of artefact has, indeed, all chances to occur in quasi-symmetrical conformations like regions of the potential sur-face close to a symmetrical conformer, or in derivatives of open-shell complexes bearing non-equivalent substituents, etc. In such cases there is no way to im-pose a set of orbitals that respects the quasi-symmetry of the molecule, and one must deal with a set of more or less unphysical orbitals that most probably favors one VB structure with respect to the other(s). Moreover, there is no way to estimate the seriousness of the artefact. This may have several consequences: (i) At the Hartree-Fock level, unsymmetrical conformations are preferred over symmetrical ones, so that a symmetrical minimum may be missed. (ii) At the correlated level, electronic configurations displaying symmetry-broken orbitals are a bad starting point for CI or perturbation. As a consequence, unsymmet-rical conformations are disfavored relative to symmetrical ones if symmetry-adapted orbitals are imposed in the latter. Relaxing any symmetry constraint on the wave function is not a solution to the problem; in such a case, the symmetry-broken orbitals are more unphysical in symmetrical conformations than in quasi-symmetrical ones, and the imbalance remains.

The HOOH$^-$ potential surface is an illustration of the above dilemma. By analogy with dihalogen anion radicals, one expects a symmetrical three-electron bonded conformation of the type **34** to be much more stable than the dissociation products OH + OH⁻. However one also expects hydrogen-bonded forms of the type **35** or **36** to be low-lying[65] and presumably much more stable than **34**. The hydrogen-bonded forms are so much unsymmetrical that

they cannot be affected by the symmetry-breaking artefact. On the other hand **34** is symmetrical and obviously affected, while the transition state connecting **34** to hydrogen-bonded forms is an intermediate case and must be affected to some extent. As the relative energy of this latter transition state relative to the other conformers is of crucial importance as regards the stability of **34**, the question arises of which level of theory is required to know if HOOH⁻ exists as a metastable three-electron bonded conformer of if the latter collapses to hydrogen-bonded forms by a barrierless process.

4.1 Experimental and Theoretical Background.

The experimental background around the $H_2O_2^-$ complex is summarized in a recent theoretical paper by Hrusak et al..[66] This complex has been postulated to be an intermediate in the reaction of O⁻ with H_2O, leading to the dissociation products OH⁻ and OH.[65,67] The same species was also observed by Bowers et al.,[68] in the photodissociation of the $(CO_3^-).H_2O$ cluster, and by Buntine et al.[69] in the photodissociation of the $(O_2^-).H_2O$ complex. Two conformations, respectively of the types [HOH...O⁻] (**35**) and [HO⁻...HO] (**36**) were first proposed by Lifshitz[65] and suggested to be easily interconverting low-lying conformers.

   Numerous experimental studies were complemented by rather scarce theoretical investigations aiming at characterizing the various minima on the $H_2O_2^-$ potential surface. A minimum of type **35** was characterized by Benassi and Taddei,[70] and Bowers et al.[68] found an additional low-lying bifurcated complex **37**. Just recently, a symmetrical complex **38** (in fact not fundamentally different from the hypothetical conformer **34** above) was further reported, almost simultaneously, by Hrusak et al.[66], using a number of DFT methods together with standard high-level methods as MP4 or CCSD(T), and by Humbel et al.[71], using BOVB in addition to standard methods. It is this latter work that is summarized in the present section. Rather than an

37   38

exhaustive study of the $H_2O_2^-$ potential surface, the scope of the following theoretical study is a search for a balanced description of regions of the potential surface displaying very different propensities to the symmetry-breaking artefact. In that spirit, the study is restricted to the symmetrical three-electron bonded species **34** or **38**, the unsymmetrical hydrogen-bonded conformations **35** and/or **36**, the reaction pathway connecting these two types of conformers and the dissociation products OH^- + OH. Except otherwise specified, the 6-31+G* basis set has been used.

4.2   Results of Hartree-Fock Calculations.

In order to search for a possible symmetrical minimum of type **34** or **38**, the left-right symmetry has been imposed in a first step by constraining both OH bond lengths and both HOO angles to be equivalent. While a conformation of type **34** might intuitively be expected for a three-electron bonded species, the UHF optimization leads to a stable conformation that displays very small HOO angles (68°), a geometry rather corresponding to conformation **38**, an indication of the strong electrostatic attraction between the hydrogen of one fragment and the oxygen of the other. This conformation nevertheless corresponds to a true three-electron bond, as indicated by the doubly occupied σ(O-O) orbital and the singly occupied σ^*(O-O) antibonding one. The rather long O-O bond distance, 2.29 Å, and the negative charge equally shared by the two OH fragments are also consistent with three-electron bonding. The geometry of **38**, displayed in Table 8, is therefore a compromise between the requirements of three-electron bonding and those of hydrogen-bonding.

As expected in view of the above noted deficiency of the Hartree-Fock method, the bonding energy of **38** relative to the dissociation products OH^- + OH is found to be much too small (4.3 kcal/mol) at the UHF level, while a realistic value should be in the region of 25 kcal/mol by comparison with the isoelectronic species F_2^- (30 kcal/mol) and $HSSH^-$ (19 kcal/mol). By contrast, the UHF values for the cohesion energy and the geometry of the hydrogen-bonded complex **35** (Table 8) are in good agreement with higher levels of theory.[70]

Relaxing the symmetry constraints leads to the disappearance of any extremum, on the UHF potential surface, corresponding to **38**. The latter conformation collapses to **35** without energy barrier. Even in its fixed symmetrical geometry, **38** displays some symmetry-broken orbitals, leading to an artefactual energy lowering of 2.3 kcal/mol relative to the symmetry-adapted UHF solution. Another sign of the extent of the symmetry-breaking artefact in **38** is the huge imbalance of the spin densities on the oxygen atoms (respectively 0.1 and 0.9) in the symmetry-broken electronic wave function.

4.3 CASSCF Calculations.

In view of the inadequacy of the Hartree-Fock method for a potential surface like that of $H_2O_2^-$, higher levels of theory are required. Following Benassi and Taddei,[70] we have pursued the investigations at the CASSCF level, including the complete set of the 3300 configurations that belong to the valence space. As a result, the most obvious manifestations of the symmetry-breaking artefact no longer appear, as the CASSCF electronic wave function of **38** now has the symmetry of the nuclear frame. Noteworthily, **38** is not a minimum but a transition state, and its bonding energy relative to the dissociation products is still surprisingly small (11.1 kcal/mol), a sign that the valence CASSCF level is not much more correct than Hartree-Fock for three-electron bonded conformations. This is not surprising, if one remembers that the Hartree-Fock defect is fundamentally a lack of breathing orbital effect, which can be retrieved through proper orbital optimization.

This optimization, in the BOVB framework, can be done with the help of the super-CI technique,[72] that consists of performing a CI in a space involving all the single excitations out of the two basic configurations (**32, 33** above or $\sigma^2\sigma^{*1}$ and $\sigma^1\sigma^{*2}$ in the MO description), then transforming the orbitals and iterating. To first order, this rigorous process can be approximated by a simple CI or perturbation calculation involving the same CI space,[56] and this is why MP2 provides realistic three-electron bonding energies. The defect of the valence-CASSCF is that does not involve *all* such single excitations, and in particular not the most useful ones. This can be clarified by reasoning in terms of localized orbitals. Some optimization of the O-H bonding orbitals, for instance, can be simulated by a CI including single excitations to the corresponding virtual antibonding orbitals. However the virtual orbitals that should be used to optimize the lone pairs do not belong to the valence space, so that a valence-CASSCF completely misses the breathing orbital effect of the lone pairs, hence the poor three-electron bonding energy which is very little improved relative to UHF.

Table 8: Optimized geometries and energies of some conformers on the $(H_2O_2)^-$ potential surface, as calculated at various computational levels, in 6-31+G* basis set. Bond lengths in Å, angles in deg and energies in kcal/mol. The system of coordinates is that of **39**

Conformation	UHF	MP2	CASSCF	CCSD(T)	BOVB
Separate fragments					
$(H_1-O_1)^-$	0.953	0.978	0.980	0.981	0.952
H_2-O_2	0.959	0.981	0.983	0.987	0.958
O_1-O_2	∞	∞	∞	∞	∞
E	0.0	0.0	0.0	0.0	0.0
3-e bonded 38					
H_1-O_1	0.950	0.977	0.992	0.981	0.954
H_2-O_2	0.950	0.977	0.992	0.981	0.954
O_1-O_2	2.286	2.248	2.324	2.329	2.345
$H_1-O_1-O_2$	68.2	64.1	50.3	60.8	65.8
$H_2-O_2-O_1$	68.2	64.1	50.3	60.8	65.8
$H_1-O_1-O_2-H_2$	180 [a]	149.0	159.5	147.9	180 [a]
E	-4.3	-24.3	-11.1	-25.7	-26.6
H-bonded 35					
H_1-O_1	1.690	1.595	1.667		
H_2-O_2	0.947	0.970	0.971		
O_1-O_2	2.679	2.640	2.678		
$H_1-O_1-O_2$	5.4	8.8	3.9		
$H_2-O_2-O_1$	97.8	87.5	93.2		
$H_1-O_1-O_2-H_2$	180 [a]	180 [a]	180 [a]		
E	-35.1	-35.3	-38.0		
H-bonded 36					
H_1-O_1			1.045		
H_2-O_2			0.976		
O_1-O_2			2.555		
$H_1-O_1-O_2$			0.5		
$H_2-O_2-O_1$			108.0		
$H_1-O_1-O_2-H_2$			180 [a]		
E			-34.8		

[a] Planar constrained.

Table 8 (continued)

Conformation	UHF	MP2	CASSCF	CCSD(T)	BOVB
Transition state 39					
H_1-O_1		0.978		0.984	0.964
H_2-O_2		0.973		0.980	0.964 [b]
O_1-O_2		2.219		2.320	2.605
H_1-O_1-O_2		57.3		55.5	26.9
H_2-O_2-O_1		80.1		68.1	90.6
H_1-O_1-O_2-H_2		180 [a]		180 [a]	180 [a]
E		-17.2		-25.3	-24.9

[a] Planar constrained. [b] Both HO bond lengths constrained to be identical.

No such problems are encountered in the hydrogen-bonded regions of the potential surface, and a minimum of type **35** is found with a geometry and a cohesion energy in agreement with UHF and higher levels (Table 8). Note-worthily, a second hydrogen-bonded minimum, of the type **36**, is found at the CASSCF level. This conformation correspond to the [HO⁻..HO] complex postulated by Lifshitz.[65]

4.4 Möller-Plesset Perturbation Calculations.

All the calculations using Möller-Plesset perturbation theory are performed within the unrestricted formalism. As argued above, MP2 calculations are expected to yield realistic three-electron bonding energies, provided the symmetry-breaking artefact is avoided by imposing the symmetry of the wave function. Under such a restriction, a minimum of type **38** is indeed found at the MP2 level, with a geometry not much different from that found at the SCF level, and a bonding energy (24.3 kcal/mol) now in the expected range relative to the dissociation fragments OH and OH⁻.

Relaxing the symmetry constraint on the wave function has the spectacular effect of raising the energy of **38**, by 7.7 kcal/mol, contrary to what was observed at the UHF level. This is because the symmetry-broken orbitals, while leading to a lower UHF solution than symmetry-adapted orbitals, are a bad starting point for CI or perturbation calculations that eventually tend to restore the symmetry of the wave function. High levels of electron correlation would of course correct this inadequacy (at the full CI limit the energy becomes independent of the set of orbitals), however it is clear that the MP2 level is far from being sufficient. As a further indication, the MP2-calculated spin densities on each oxygen atoms amount to 0.3 and 0.7 in **38**, when the symmetry is relaxed, which is in-between the symmetry-broken UHF values and the expected densities of 0.5 on each oxygen.

A transition state on the pathway connecting **38** to **35** is easily located at the MP2 level, with a conformation of type **39**, and lying 7.1 kcal/mol above **38** (Table 8). While this result might be used as an argument in favor of the

$$\mathrm{O_1} \overset{\mathrm{H_1}}{\underset{\mathrm{H_2}}{\cdots}} \mathrm{O_2} \quad \ominus$$

39

stability of **38**, one must keep in mind that all the region of the potential surface that is situated in the vicinity of **38** is likely to be subject to some symmetry-breaking artefact although symmetry is not formally present. Therefore, the MP2 relative energy of **39** relative to **38**, the latter being calculated with symmetry-adapted orbitals, is likely to be overestimated. The magnitude of this overestimation is a priori unknown, as there is no way to argue if the symmetry-breaking artefact will raise the energy by a quantity close to 7.7 kcal/mol, as in **38**, or by much less as in geometries close to **35**. As this latter uncertainty on the height of the transition state is slightly larger than the calculated stability of **38**, the MP2 level is insufficient to decide whether the three-electron bonded conformer is slightly stable or definitely unstable.

Pushing the calculation to the MP4 level does not help much. While the three-electron bonding energy of **38** is almost unchanged relative to MP2 (24.2 vs 24.3 kcal/mol), the transition state is found to lie 4.1 kcal/mol above the symmetrical conformation **38**, the latter being calculated with symmetry-adapted orbitals. However the symmetry-breaking artefact persists and leads to an energy rising of 4.4 kcal/mol in **38** once the symmetry of the wave function is relaxed. It follows that neither MP2 nor MP4 levels are conclusive on the stability of **38**. On the other hand, both levels agree to find this intermediate (if stable) to lie some 11 kcal/mol above the hydrogen-bonded species **35**, and to have a significant three-electron bonding energy in the expected range as compared to isoelectronic species.

4.5 Coupled-Cluster Calculations.

Since Coupled-Cluster (CC) methods are now recognized to yield geometries and relative energies close to full CI, provided the Hartree-Fock configuration is the dominant component of the correlated wave function,[73] this theory has been used to get rid of the symmetry-breaking problem so as to conclude on the stability of **38** in a meaningful way. A very simple test for checking the adequacy of CC computational levels for the problem in hand consists of comparing the energies of **38** as calculated with and without symmetry constraint on the electronic wave function. The CCSD level, that includes all the single and double excitations relative to the ground configuration, proves to be much better than MP2 and MP4 in that respect, but still not quite sufficient as the calculation with symmetry-broken orbitals yields an energy 0.8 kcal/mol higher than with symmetry-adapted orbitals. This difference drops to 0.2 kcal/mol at the higher CCSD(T) level, that includes triple excitations in a perturbative way. This latter level has therefore been finally chosen, as displaying a practically negligible symmetry-breaking artefact.

The CCSD(T) results are displayed in Table 8. For convergence difficulties, the search has been restricted to planar conformations, a restriction that was estimated not to affect relative energies by more than 0.1 kcal/mol.[71] A minimum **38** and a transition state **39** are indeed characterized at the CCSD(T)level, displaying geometries in agreement with the MP2 calculations. However both conformers are almost degenerate in energy, so that the barrier to rearrangement of the three-electron bonded conformer **38** to hydrogen-bonded conformers is extremely small, 0.4 kcal/mol, much reduced as compared to MP2 and MP4 levels, showing that the symmetry-breaking artefact is at work at these latter levels even in somewhat unsymmetrical regions of the potential surface.

Improving the basis set by adding polarization functions does not change the barrier, which still amounts to 0.4 kcal/mol at the CCSD(T)/6-31+G** level.

4.6 BOVB Calculations.

The BOVB method is by nature free from the symmetry-breaking artefact. Because the orbitals of each VB structure are independent, both the size effect and the resonance effect can be made optimal without conflicting with each other. As a result, the BOVB wave function of **38** is a superposition of two symmetry-broken determinants, thus fully respecting the symmetry of the nuclear framework and gathering the stabilizing features of both symmetry-broken and symmetry-adapted orbitals. This sound description of the two-structure electronic system smoothly extends to unsymmetrical conformations, in which the two VB structures have unequal weights, all the way to hydrogen-bonded conformations, so that the whole potential surface is expected to be described in a balanced way. This latter prediction can be tested by calculating the barrier to rearrangement of **38** towards **35**, a difficult test in view of the poor performances of the UHF, valence-CASSCF, MP2 and MP4 methods in that respect.

The relative energies of **38** and **39** have been calculated, still in the 6-31+G* basis set, at the basic BOVB level with all orbitals being localized and all lone pairs having a closed-shell structure (L-BOVB). In a first step, the CCSD(T) geometries have been used. As a result, the transition state **39** is found to lie very slightly above **38**, by 0.1 kcal/mol, to be compared to the CCSD(T) value of 0.4 kcal/mol. The bonding energy of **38**, 25.1 kcal/mol, is also in good agreement with the CCSD(T) calculation. In a second step, the geometries of **38** and **39** have been optimized (in a planar-constrained conformation) at the L-BOVB level, still leading to relative energies close to the CCSD(T) values. While the geometry of **38** is not much changed with respect to CCSD(T), that of **39** is shifted towards the hydrogen-bonded conformer on the **38**→**35** pathway. This however is not surprising in view of the extreme flatness of this region of the potential surface, as already noted by Hrusak et al. who found the calculated geometry of **38** to be strongly method-dependent.[66]

4.7 Conclusion.

It is clear that the hydrogen peroxide anion cannot be observed under the form of a three-electron bonded conformer, even at low temperature. Its barrier to rearrangement to hydrogen-bonded forms is of the order of one kcal/mol and would not anyway persist after zero-point vibrational energy correction. On the other hand, **35** may serve as a model for other three-electron bonded anions of the type ROOR'$^-$ that do not lend themselves to easy rearrangements to hydrogen-bonded forms. The present calculations suggest that the O∴O three-electron bond is quite stabilizing in this type of anions which should be experimentally observable with a dissociation energy of the order of 20 kcal/mol.

The symmetry-breaking artefact appears as a difficult obstacle to the study of the $H_2O_2^-$ potential surface, leading either to discontinuities or to imbalance in the description of various conformers. In the MO framework, rather large levels of computation are required to get a balanced description of the potential surface. Interestingly, the various DFT options perform rather poorly in that respect, finding the symmetrical conformer **38** very close in energy to **35** and sometimes even lower, while accurate calculations put **38** some 10 kcal/mol above **35**.[66]

By contrast, the BOVB method offers of simple remedy to the above artefact by cutting off its root cause right at the outset. By its very nature indeed, the latter method does not privilege one structure over the other, nor does it exhibit any conflict between the resonance energy and the stabilization of each Lewis structure. It therefore appears as an appropriate method for the large class of electronic systems that exhibit more or less apparent manifestations of the symmetry-breaking problem.

5 The Compared Natures of the C-Cl and Si-Cl Bonds

Bonding in first vs higher row atoms poses a number of interesting problems, among which the curious reluctance of R_3Si-X bonds to heterolyze[74-78] in solution, in comparison with the ease of heterolysis in R_3C-X compounds. So rare are the $R_3Si^+X^-$ species in condensed phase that a compound like Ph_3SiClO_4 that appears initially as an excellent candidate for an ionic bond was found to be a covalent solid exhibiting a Si-O bond.[79] In contrast, the carbon analog is definitely ionic, $Ph_3C^+ClO_4^-$.[80] It appears that Si has a very strong affinity for covalent interactions; much stronger indeed than carbon, and that it takes counterions such as hexabromocarborane to approach, albeit not completely, an ion pair; $R_3Si^+X^-$.[81] This difference between bonding at silicon and carbon cannot be explained by electronegativity considerations since silicon is much more electropositive than carbon and might as such be expected to be more prone to form free cations.

Such intriguing experimental facts raise fundamental questions that require understanding, which can only be gained through a detailed investigation of the covalent and ionic interactions and their interplay in Si-X vs C-X bonds. The BOVB method is suitable to this purpose, and has been used to generate the purely ionic and purely covalent dissociation curves of the model systems H_3Si-Cl and H_3C-Cl.[82] The interplay between the ionic and covalent components of the bond has also been investigated, by calculating the resonance integral that couples the ionic and covalent structures, as well as the resonance energy arising from their mixing, and the variations of these quantities as a function of the bonding distance.

5.1 Technical Details.

As only a qualitative understanding of bonding rather than accurate dissociation energies was sought for, the simple 6-31G* basis set has been used throughout. On the other hand, to ensure a balanced treatment of covalent vs ionic VB structures, the BOVB method has been used within its most accurate level, SD-BOVB as defined in Section 3. The bonding energies of H_3Si-Cl and

H_3C-Cl have also been calculated at the lower SL level, to appreciate the effect of delocalizing the inactive orbitals of π symmetry in this type of compounds.

The VB potential energy curves (Figures 1 and 2) were plotted by performing the SD-BOVB calculations of the individual configurations and of the final adiabatic state on geometries determined at the GVB(1/2) level, i.e., with local correlation of the electron pair of the M-Cl bond in the framework of the Generalized Valence Bond method.[31] These geometries correspond to a preoptimization of all the geometric parameters of H_3M-Cl (M = C, Si) at different R_{MCl} distances taken as the dissociation coordinate in Figures 1 and 2.

The importance of each of the three relevant VB structures, namely $H_3M \cdot - \cdot Cl$ (covalent) and $H_3M^+Cl^-$ and $H_3M^-Cl^+$ (ionic) as a contributor to the ground state has been appreciated by calculating their weights by means of the Chirgwin-Coulson formula (vide supra).

5.2 Results.

Table 9 shows the adiabatic bond energies calculated at the SL- and SD-BOVB levels, as compared to the experimental and G1 and G2 energies. It is seen

Table 9: Bond Energies De (kcal/mol) for H_3C-Cl and H_3Si-Cl at various computational levels

species	exptl[a]	SL-BOVB[b]	SD-BOVB[b]	G1[c]	G2[c]
H_3C-Cl	87.3	72.8	79.9	88.9	88.3
H_3Si-Cl	110.7	91.2	101.7	111.9	110.7

[a] De obtained from experimental D_0 values quoted in ref 83b and corrected by a calculated ΔZPE (ref 86a). [b] Optimized geometric values (GVB(1/2)/6-31G*) are, $R_{CCl} = 1.815$ Å; $R_{CH} = 1.078$ Å; θ(HCCl) = 108.1°; $R_{SiCl} = 2.086$ Å; $R_{SiH} = 1.468$ Å; θ(HSiCl) = 108.3°. [c] G1 and G2 values from ref 83.

that the difference between the SL and SD levels is considerable, 7.1 and 10.5 kcal for the bonding energies of the C-Cl and Si-Cl bonds, respectively, much larger than was found for the F-F or F-H bonds (vide supra). This is because the delocalization of the π inactive orbitals, at the SD level, allows for some electron donation from the π lone pairs of chlorine to the symmetry-adapted combinations of σ^*_{MH} antibonding orbitals, as shown in 40. Such orbitals are low-lying and more prone to accept some electrons than the virtual orbitals of F_2 or HF. A detailed analysis of the effect in the individual VB structures shows that the ionic structure M^+Cl^- exhibits the largest effect (9.5 and 14.9

40

kcal for M = C and M = Si, respectively). The effect is smaller in the covalent structure M·— ·Cl and about the same for the C-Cl and and Si-Cl bonds, respectively 6.4 and 7.3 kcal/mol.

The SD-BOVB bonding energies can be compared to the results obtained by the G1 and G2 methods and to the experimental data. The underestimation of the BOVB-estimated bonding energies relative to experiment is severe, 7.4 and 9.0 kcal/mol respectively for the C-Cl and Si-Cl bond, but in the expected range in view of the paucity of the basis set used in this semi-quantitative study, much smaller than the basis set used at the G2 level.[83]

The dissociation curves representing the ground states, the purely covalent and lowest purely ionic curves of the two molecules are displayed in Figures 1 and 2, respectively, as calculated at the SD-BOVB level. It is seen that in both molecules, the covalent structure is bonded relative to the separate fragments by a significant amount. This bonding energy is the covalent contribution that arises solely from the spin-pairing of the two electrons which are, in turn, localized on their respective fragments. Interestingly, these covalent interaction energies are almost the same for both systems, 34 kcal/mol for H_3CCl and 37 kcal/mol for H_3SiCl. On the other hand the optimal distance for the purely covalent bond is, expectedly, longer in the Si-Cl than in the C-Cl bond.

Quite more surprising are the compared features of the ionic curves. While the minimum of the $H_3C^+Cl^-$ curve is located at a bonding distance of 2.452 Å, that of $H_3Si^+Cl^-$ is found at a significantly shorter distance, 2.159 Å, making the effective radius of the H_3Si^+ cation *smaller* than that of H_3C^+, in opposite order to the sizes of the silicon and carbon atoms. Moreover, the ionic potential well of $H_3Si^+Cl^-$ is much deeper than that of $H_3C^+Cl^-$, with an ionic bonding energy of 139 vs 89 kcal/mol, relative to the asymptotic values calculated at a separation of 10 Å.

These different features of the $H_3Si^+Cl^-$ and $H_3C^+Cl^-$ bond pairs can be understood by considering the detailed net charges on both cations, as calculated by means of a natural bond orbital (NBO) analysis. It appears that while the positive charge in the CH_3^+ cation is dispersed on all atoms, it is on the contrary quite concentrated on the silicon atom in SiH_3^+. Quantitatively, the calculated net charges amount to only +0.284 on carbon vs +1.464 on silicon in the two respective cations, as calculated at the same geometries they

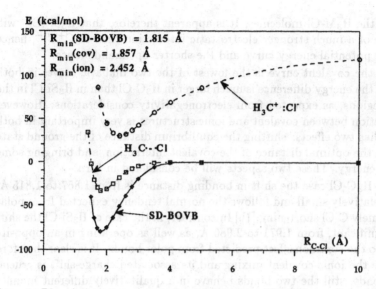

Figure 1: Dissociation energy curves for the pure ionic ($H_3C^+Cl^-$) and the pure covalent ($H_3C\cdot-\cdot Cl$) structures, and the SD-BOVB ground state for the H_3C-Cl bond.

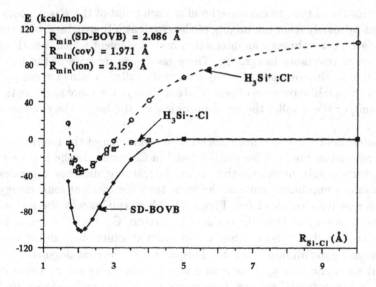

Figure 2: Dissociation energy curves for the pure ionic ($H_3Si^+Cl^-$) and the pure covalent ($H_3Si\cdot-\cdot Cl$) structures, and the SD-BOVB ground state for the H_3Si-Cl bond.

possess in the R_3M-Cl molecules. It is apparent therefore that $H_3Si^+Cl^-$ will be the site of a much stronger electrostatic interaction than $H_3C^+Cl^-$, hence the deeper potential energy curve and the shorter optimal distance.

While the covalent curve is the lowest of the two diabatic curves in both molecules, the energy difference is much larger in H_3C-Cl than in H_3Si-Cl in the bonding regions, as expected from electronegativity considerations. However the interaction between covalent and ionic structures is very important in both cases and has two effects: shifting the equilibrium distance of the ground state away from the optimal distance of the covalent interaction, and bringing some resonance energy. These two aspects will be considered in turn.

In the H_3C-Cl case the shift in bonding distance is from 1.857 to 1.815 Å, which is relatively small and follows the normal tendency expected from polar bonds, namely C-Cl shortening.[1b] In contrast, in the case of H_3Si-Cl the shift is both significant, from 1.971 to 2.086 Å, as well as operating in an opposite direction to the common effect predicted from polar bonds. It is clear therefore that, while the ionic-covalent mixing and its associated charge-shift is crucial in both bonds, still the two bonds behave in a qualitatively different manner in their response to the charge fluctuation. To understand this contrasted behavior, let us examine in details the mechanism of the ionic-covalent mixing in both systems.

The formation of the ground state bond at each point of the $R_{M...Cl}$ coordinate arises primarily from the mixing of the covalent VB structure with the lowest ionic structure, through an interaction matrix element which is nothing but the classical resonance integral β. There results an increase of the bond energy, relative to the covalent curve, by a quantity called "resonance energy" which obeys the qualitative rules of perturbation theory: the more negative the β integral and/or the smaller the covalent-ionic gap, the larger the resonance energy.

The variation of the resonance integral β as a function of the interatomic distance is plotted in Figure 3 for the C-Cl and Si-Cl bonds. In the first case, β becomes increasingly more negative as the interatomic distance decreases from the covalent minimum, while at the same time the covalent-ionic energy gap remains relatively constant (see Figure 1). This explains why the ground state minimum is shorter than the covalent minimum. On the other hand, in the Si-Cl bond the β integral reaches a maximum absolute value in-between the covalent and ionic minima (ca 2 Å). This, added to the near degeneracy of the covalent and ionic curves, is the reason why the resonance energy peaks in this region of the potential surface, thus imposing a final bond distance that is an average between the minima of the covalent and ionic VB structures.

Let us now turn our attention to the nature of the M-Cl bond at its

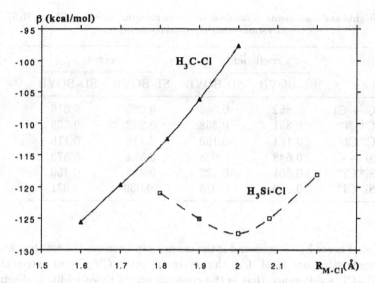

Figure 3: Variation of the β resonance integral in the H_3C-Cl and H_3Si-Cl cases, as a function of the M-Cl distance (M = C, Si).

equilibrium geometry. The coefficients of the covalent and ionic structures in the ground states of H_3SiCl and H_3CCl are displayed in Table 10, along with their calculated weights. At the SD-BOVB level, it appears that the C-Cl bond is mostly covalent, with a weight of ca 62%, as compared with a weight of 27% for the lowest ionic structure, C^+Cl^-, while the other ionic structure is marginal. On the other hand, the Si-Cl bond has rather similar covalent and ionic weights (respectively 57% and 46%), in agreement with the near degeneracy of the corresponding VB structures, at equilibrium bonding distance. The highest ionic VB structure, Si^-Cl^+, is totally negligible and even comes out with a negative weight, which in the Chirgwin-Coulson definition is interpreted as close to zero.

As may be seen from Figures 1 and 2, the resonance energy is strikingly large, ca 46 kcal/mol for H_3CCl and ca 66 kcal/mol for H_3SiCl. In fact, in both cases the major bonding interaction that glues the two fragments is the resonance energy, and for the Si-Cl bond this contribution is truly dominant being about 65% of the total bond energy. Two factors join to make the resonance energy more dominant in the H_3SiCl case in comparison with H_3CCl. The first is the effective interaction matrix element β that is slightly larger (ca 0.125 vs 0.117 kcal/mol) in the Si-Cl than in the C-Cl bond at equilibrium

Table 10: Weights and coefficients of covalent and ionic structures for H_3C-Cl and H_3Si-Cl at various computational levels

	coefficients		weights	
	SL-BOVB	SD-BOVB	SL-BOVB	SD-BOVB
$H_3C\cdot-\cdot Cl$	0.652	0.646	0.622	0.616
$H_3C^+Cl^-$	0.351	0.358	0.262	0.269
$H_3C^-Cl^+$	0.191	0.190	0.116	0.115
$H_3Si\cdot-\cdot Cl$	0.648	0.628	0.594	0.572
$H_3Si^+Cl^-$	0.501	0.522	0.436	0.459
$H_3Si^-Cl^+$	0.076	0.075	-0.030	-0.031

distances (see Figure 3). The second is the energy gap between the covalent and the major ionic structure $H_3M^+Cl^-$ that is significant in CH_3Cl but extremely small in SiH_3Cl, a difference that is the consequence of three additive effects: (i) the lower electron affinity of SiH_3^+ relative to CH_3^+, that globally lowers the $H_3Si^+Cl^-$ curve relative to $H_3C^+Cl^-$; (ii) the deeper ionic potential well of $H_3Si^+Cl^-$ relative to $H_3C^+Cl^-$ (vide supra); (iii) the coincidence of the ionic and covalent minima on the dissociation coordinate in the SiH_3Cl case, in contrast with the different ionic and covalent optimal distances in CH_3Cl.

How would these bonding features be modified in a polar solvent? Because of its higher dipole moment, the ionic component of an heterolytic bond is thought to be much more stabilized than the covalent component by its interaction with the solvent, so that the bond is essentially ionic in solution. As in addition the effect of the solvent is to flatten the ionic dissociation curve, an easy heterolytic cleavages ensues. Why does this mechanism, that is well accepted for the C-X dissociation in R_3C-X, seem to be inefficient in the R_3Si-X case? While this problem must await a proper solvation treatment, the above results have shed enough light on the compared natures of the C-Cl and Si-Cl bonds to suggest some working hypotheses.

One first observation is that the large resonance energy due to covalent-ionic mixing in the Si-Cl bond might well persist in solution. Indeed, the net dipole of the $R_3Si^+Cl^-$ ionic structure should be quite small at short Si-Cl, due to the Si→R polarization of each Si-R bond,[82] so that solvation is not expected to drastically lift the near degeneracy of the covalent and ionic structures in the vicinity of the equilibrium distance. The result will be again a large resonance energy at short distance (vs a smaller resonance energy at long distance, because the β integral is distance-dependent), which will oppose the heterolytic bond-breaking. Another contributing reason for the rarity of free R_3Si^+ cations in solution might lie in the structure of the cation itself. On the one hand, its reduced effective size (vide supra) along the missing coordination site will make it an extremely efficient coordinating species for an electron-rich ligand X, making the breaking of the Si^+X^- interaction hardly compensated by solvent interactions. This is all the more true as the R_3Si^+ cation takes the form of a highly positive Si atom surrounded by negatively charged alkyl groups,[82] which inhibits the solvation of a separated $R_3Si^+X^-$ ion pair.

5.3 Conclusion.

The pure covalent and pure ionic dissociation energy curves, as well as the state curves resulting from the ionic-covalent mixing have been computed for the H_3C-Cl and H_3Si-Cl bonds. In both systems, the purely covalent interactions contribute to the bonding energy approximately the same quantity, 34-37 kcal/mol. In contrast, the ionic interactions are distinctly different, and the difference is rooted in the properties of the R_3Si^+ and R_3C^+ species. The R_3C^+ ion has a delocalized charge distribution, and hence its effective size is large; offering no preferred direction of approach to the counterion Cl^-. The result is a diminished electrostatic interaction in the $R_3C^+Cl^-$ structure and a relatively long C^+Cl^- distance. The mixing of this structure into the covalent structure shortens slightly the R_{CCl} distance, and remains of secondary influence, eventhough the resonance energy due to mixing is significant. In contrast, in R_3Si^+ the positive charge is localized on the silicon atom, which thereby acquires a diminished effective size along the missing coordination site. This allows a close approach of the Cl^- counterion to R_3Si^+ in the ionic structure and therefore a very large electrostatic interaction ensues. The ionic energy curve $R_3Si^+Cl^-$ approaches therefore the covalent curve to a near degeneracy, leading thereby to a large covalent-ionic resonance energy which is responsible for more than half of the Si-Cl bond energy.

Solvent effects are not expected to change drastically the H_3Si-Cl bonding picture, which can be generalized to R_3Si-Cl bonds. The reason why this latter system does not appear to solvolyze via free ions may be attributed, on the one hand, to the loss of the large ionic-covalent resonance energy which necessarily attends a heterolytic process, and on the other hand to the lack of available direction of approach for the solvent to stabilize the $R_3Si^+X^-$ separated ion-pair.

6 The Lone Pair Bond Weakening Effect in H_nX-H Bonds

The lone pair bond weakening effect (LPBWE) has long been recognized in homonuclear bonds.[84-88] As the effective nuclear charge experienced by the valence electrons in the series (X = C, N, O, F) gradually increases, one might logically anticipate X-X single bonds of H_nX-XH_n (n = 3 to 0) molecules to get stronger and stronger in the same series, while on the contrary the X-X bonding energies gradually decrease from H_3C-CH_3 to F_2. This bond weakening, which is also effective if X belongs to lower rows of the periodic table, has generally been ascribed to the effect of lone pairs adjacent to the X-X bond.

Some authors[84,85] have interpreted the LPBWE as arising from inter-atomic mutual repulsions between lone pairs, but Sanderson, in his book *Polar Covalence*,[89] rather viewed the effect as being intra-atomic and intrinsic to the X atom, and predicted that the LPBWE should also be effective in X-H bonds. As can be noted, this effect is much less apparent, if it exists at all, in X-H bonds than in homonuclear X-X bonds, since the increasing number of lone pairs on the N, O and F atoms does not prevent the X-H bonds to get stronger and stronger in the series H_3C-H, H_2N-H, HO-H and F-H, while the corresponding bond lengths get shorter and shorter.

An intriguing clue exists, however. Pauling has predicted that the bonding energy of single bonds should be an increasing function of the electronegativity difference between the two atoms being bonded (see more details below). If the X-H experimental bonding energy is plotted against Pauling's electronegativity difference between X and H (Figure 4), a linear correlation is indeed found in the series H_2N-H, HO-H, FH, while the H_3C-H bonding energy departs from the straight line, as if this latter bond were reinforced or, equivalently, as if the X-H bonds were weakened as soon as the X atom bears one or several lone pairs.

This break in the correlation curve, which is also observed with analogous hydrides of the second row of the periodic table and is independent of the electronegativity scale,[90] might be indicative of an effective LPBWE in X-H bonds, as suggested by Sanderson. In this context, the nature of the X-H bond and its dissociation process in a series of H_nX-H molecules (X = C, N, O, F; n = 3 to 0) have been investigated in detail,[90] in order to answer a number of questions that naturally arise from Sanderson's statement, Pauling's electronegativity scale, and the tendency displayed in Figure 4 : (i) Does the LPBWE extend to X-H bonds and can it be related to any significant energy quantity? (ii) If it does, what is the physical effect that connects the presence of a lone pair on the X atom to the weakening of the X-H bond? (iii) If the latter effect is removed, do the so-calculated unweakened bond energies exhibit

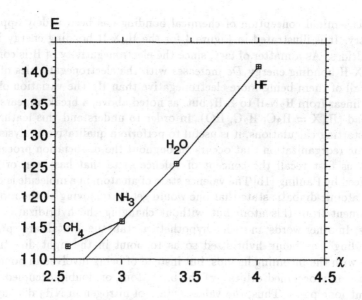

Figure 4: Experimental H_nX-H bond energy (De), in kcal/mol, as a function of the electronegativity of X.

smooth correlation with the X-H electronegativity difference, in accord with Pauling's basic idea? With this purpose, the BOVB method has been used for its unique ability to calculate a quantity that will be at the center of the debate: the energy of a radical H_nX in its "valence state",[1b] a diabatic state in which the orbitals are constrained to keep the same hybridization as in the molecule H_nX-H.

6.1 Qualitative Analysis.

It is a fact that strong bonds are generally heteropolar. For instance, the sum of the bonding energies of H_2 and F_2 is smaller than twice the bonding energy of FH. Pauling[1b] has given a simple explanation resting on the VB description of the A-B bond in terms of the three VB structures A·—·B (covalent), A^+B^- and A^-B^+ (ionic). If A and B have very different electronegativities (say $\chi_a \gg \chi_b$), then one ionic structure, A^-B^+, is low in energy and may lie at the same level as the covalent structure, hence a large resonance energy which results in a strengthening of the A-B bond. The electronegativity scale proposed by Pauling is based on this relationship: the stronger the A-B bond, the larger the electronegativity difference $(\chi_a - \chi_b)$.

This seminal conception of chemical bonding has been widely applied in chemistry. It is illustrated in Figure 4 for the H_nX-H bonding energy in first row hydrides. As a matter of fact, since the electronegativity of H is constant, the H_nX-H bonding energy De increases with the electronegativities of the X atoms, all of them being more electronegative than H. The variation of De is almost linear from H_2N-H to F-H, but, as noted above, a break appears in the first triad ($H_nX = H_3C$, H_2C, HO). In order to understand this feature, and before starting calculations, it is useful to perform a qualitative analysis of the electronic reorganization that occurs throughout the dissociation process.

Let us first recall the concept of valence state that has been originally introduced by Pauling.[1b] The valence state of an atom in a molecule is defined as that atomic diabatic state that one would get by removing all the molecular environment from this atom but without changing the hybridization of its orbitals. In other words, in such a hypothetical state the orbitals are "prepared for bonding", as being hybridized so as to point in the right directions to overlap with the incoming ligands, but instead of being involved in bonds they are just singly occupied orbitals of the bare atom, or doubly occupied if they represent lone pairs. Thus, the valence state of nitrogen in NH_3 displays four sp^3 hybrids among which three are singly occupied and one is doubly occupied. The valence state of carbon in CH_4 has four singly occupied sp^3 hybrids, etc..

In this framework, the atomization energy of a molecule XH_n is viewed as the sum of two terms: the promotion energy that is required to bring the atom X from its ground state to its valence state, and the stabilization that is gained by bonding each hybrid to its respective hydrogen atom. There follows the general principle that all other things being equal, the lower the valence state, the more stable the molecule. In connection with this, one important factor of stability for the valence state is its average s vs p character: since s orbitals are lower than p orbitals, the larger the average s character of the electrons, the lower the valence state. The s or p character of a given atomic valence state is readily calculated from the occupancy of it hybrids, whose s/p ratio is given by its very denomination, e.g. 25% s and 75% p character for an sp^3 hybrid, and so on. For instance, the nitrogen atom in ammonia has five electrons distributed in four sp^3 hybrids, which corresponds to an average occupancy of 1.25 electrons for the s orbital and 3.75 electrons for the p orbitals, or $s^{1.25}p^{3.75}$ in a condensed notation.

Let us now consider the N-H bond dissociation in NH_3, reaction (24).

$$NH_3 \longrightarrow NH_2 + H \tag{24}$$

An important feature of this reaction is that the dissociation product, the NH_2 radical, is sp^2 hybridized, unlike the starting ammonia molecule that is sp^3

hybridized. As a result, the ground state of NH_2, a π radical ($^2B_1$, **41**), has its unpaired electron located in a pure p orbital, while the two N-H bonds and the lone pair are built with the remaining three sp^2 orbitals of the nitrogen atom. Note that an alternative electron distribution is possible ($^2A_1$, **42**) in which the unpaired electron lies in a sp^2 hybrid while the pure p orbital becomes

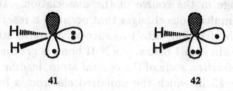

41   **42**

the lone pair: this σ radical is an excited state, for its non-bonding electrons have, in average, less s character and more p character than they have in **41**. Therefore, reaction (24) is the site of two simultaneous phenomena: (i) the breaking of the bond between an sp^3 hybrid orbital and an hydrogen atom, and (ii) a nitrogen rehybridization ($sp^3 \rightarrow sp^2$) in the departing NH_2 fragment.

This latter electronic reorganization has some important consequences as it affects the s/p ratio and therefore the energy of the nitrogen's valence state during the course of the reaction. Indeed, in the π NH_2 radical **41** the valence state of the nitrogen atom has one electron in a pure p orbital and four electrons shared by the three sp^2 hybrids, leading to an average occupancy of $s^{1.33}p^{3.67}$, to be compared with the occupancy $s^{1.25}p^{3.75}$ in ammonia. By contrast, no such increase in s character appears in reaction (25), the dissociation of a C-H

$$CH_4 \longrightarrow CH_3 + H \tag{25}$$

bond in methane, if the same analysis is applied. Indeed, the carbon atom in CH_4 has a valence state displaying four singly occupied sp^3 hybrids, leading to the average orbital occupancy s^1p^3. On the product side, the CH_3 radical is planar and corresponds to a different hybridization, with one pure p orbital and three sp^2 hybrids, however leading to the same valence state average occupancy, s^1p^3, as in the case of methane.

There lies a qualitative difference between reactions (24) and (25). Although the same rehybridization ($sp^3 \rightarrow sp^2$) occurs in both reactions, *the valence state of the carbon atom remains constant in (25)* while that of nitrogen increases its s character in (24). More specifically, throughout the H_2N-H dissociation the nitrogen's valence state evolves so that a *singly* occupied hybrid, that involved in the bond being broken, gradually ceases its s character to a *doubly* occupied hybrid, the lone pair. The same qualitative changes of the central atom's valence state can be anticipated in the hydrogen extrusions from

OH_2 and FH. There results a stabilization of the dissociation product in the latter reactions and, therefore, a weakening of the bond linking the departing hydrogen to the central atom.

In order to relate the above effect to quantitative energies, it is interesting to estimate what the bonding energies would be if the central atom's valence state did not change in the course of the dissociation. To this aim, let us separate the two simultaneous changes that occur in a reaction, namely bond-breaking and rehybridization, into two successive steps, still taking reaction (24) as an example. In step 1 below, the N-H bond is gradually broken *without changing the hybridization state* of the central atom, leading to an NH_2 radical in a diabatic state **43** in which the unpaired electron is located in a hybrid

(grossly sp^3) orbital. The bonding energy corresponding to this step, De^*, has not been weakened by the valence state stabilization that is associated to the presence of a lone pair and will therefore be called "unweakened bonding energy", following Sanderson.[89]

The second step consists of the electronic relaxation of **43** to the ground state π radical **41**, by a rehybridization of the central atom from sp^3 to sp^2. The energy difference between NH_3 and the ground state dissociation products (**41** + a hydrogen atom) is the actual bonding energy De, so that the difference $(De^* - De)$ can be taken as a quantitative measure of the LPBWE (Figure 5). Step 2 can also be viewed as a configuration interaction within the space of the two diabatic configurations **43** and **44**, the latter being deduced from **43** by inverting the occupancies of the two hybrid lone pairs (see Figure 6). The resonance energy E_R resulting from this non-orthogonal configuration mixing is another way of estimating the LPBWE and should not differ much from the $(De^* - De)$ if our qualitative analysis is correct. In this scheme, the bond weakening effect, that is directly related to the presence of a lone pair on the central atom, can be expected to be efficient as well in the H-OH and F-H bonds and should gradually increase with the central atom s-p gap in the series H_2N-H, HO-H, F-H. In H_3C-H, on the other hand, the resonance energy E_R is zero by definition since there is no lone pair on the CH_3 fragment, so that the H_3C-H bond is not weakened.

Without claiming to have fully described the root cause for the LPBWE, what we have identified in the above qualitative analysis is an energetic quan-

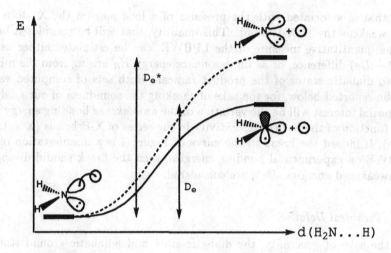

Figure 5: Adiabatic (solid line) and diabatic (dotted line) dissociation energy curves for the NH₃ molecule dissociating to **41** + H and **43** + H, respectively.

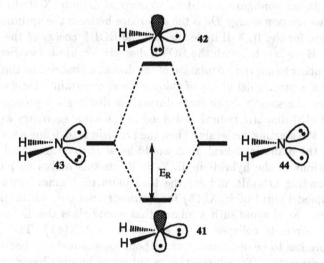

Figure 6: Nonorthogonal configuration interaction between the two diabatic radicals **43** and **44** leading to the ²B₁ (**41**) and ²A₁ (**42**) adiabatic states.

tity, that is associated with the presence of a lone pair on the X atom and that weakens the H_nX-H bond. This quantity, that will be considered below as the quantitative measure of the LPBWE, can be estimated either as the $(De^* - De)$ difference, or as the resonance energy E_R arising from the mixing of two diabatic states of the product radical. Both sets of computed values will be reported below, for the sake of checking the soundness of our analysis. Of special interest will be the variation of the *unweakened* bonding energy De^* as a function of the X electronegativity in the series of X-H bonds (X = C, N, O, F). If indeed the break in the curve of Figure 4 is a manifestation of the LPBWE on experimental bonding energies, then the break should disappear if unweakened energies, De^*, are considered.

6.2 Technical Details.

For the sake of generality, the diabatic state and adiabatic ground state of the H_nX radical will be referred to as $H_nX(43)$ and $H_nX(41)$, respectively, by analogy with drawings **43** and **41** which represent the H_2N particular case.

Following the basic principles of the method, the BOVB wave function for H_nX-H involves one covalent and two ionic configurations. As the coefficients of the two latter configurations drop to zero at infinite X-H distance, the adiabatic dissociation energy De is the difference between the optimized BOVB wave function for the H_nX-H molecule and the ROHF energy of the products, $H_nX(41)$ + H, in keeping with the BOVB dissociation limits (see Section 3.4).

On the other hand, the calculation of the diabatic dissociation limit $H_nX(43)$ + H requires a precise definition of the technical constraint that forces the X atom to keep the same valence state during the dissociation process. First of all the $H_nX(43)$ diabatic radical is defined in the same geometry as the H_nX fragment in the starting molecule. Then the problem is to define a set of hybrid orbitals for the diabatic radical, that would be at once optimal and matching the orientations of the hybrids in H_nX-H. While this makes no problem for the X-H bonding orbitals and for the lone pairs that cannot mix with the singly occupied hybrid of $H_nX(43)$ for symmetry reasons, for the remaining lone pair one has to avoid such a mixing that would allow the diabatic radical to rehybridize and to collapse to the ground state $H_nX(41)$. The latter lone pair therefore has to be optimized, yet without loosing its hybridization nor its directional character. To achieve this, a subset of hybrids having essentially the same s/p ratio and orientation as the relevant lone pair of H_nX-H has to be defined, so as to form an orbital space within which the lone pair optimization in $H_nX(43)$ can be performed. Details of the procedure that has been employed can be found in the original report.[90]

The (2X2) mixing of $H_nX(\mathbf{43})$ and $H_nX(\mathbf{44})$ (Figure 6) is performed with the VBSCF technique[91] of nonorthogonal MCSCF type. Note that in this latter calculation, the non-bonding orbital whose hybridization is frozen is constrained, as in $H_nX(\mathbf{43})$ above, to optimize itself in a restricted subset of hybrid orbitals.

The 6-31G** standard basis set has been used throughout. The BOVB calculations are restricted to the SL level, for lack of a suitable strategy to meaningfully delocalize the inactive orbitals in low-symmetry cases (vide supra) by the time this investigation has been originally carried out.[90] Thus, our results should be regarded as semi-quantitative in nature.

6.3 Results.

Table 11 displays the calculated dissociation energies De, the unweakened bonding energies De^* and the resonance energies E_R associated with the mixing of diabatic configurations $H_nX(\mathbf{43})$ and $H_nX(\mathbf{44})$ above.

As anticipated in view of the semi-quantitative character of our calculations, our calculated De values are only in moderate agreement with experimental values, the error reaching 8.6% of the potential well in the F-H case. This is mainly due to the deficiencies of the rather modest 6-31G** basis set, which lacks diffuse functions as well as high momentum polarization functions. This basis set is therefore all the more inappropriate as the molecule contains electronegative atoms, and indeed the error gradually increases in the (H_3C-H, H_2N-H, HO-H, F-H) series. However what matters for the present purpose is that the calculated tendency for the dissociation energies, including the break in the first triad, is correctly reflected.

Table 11: Adiabatic (De) and diabatic (De^*) bonding energies and resonance energies (E_R) for the H_nX-H dissociation, as functions of the electronegativity (χ) of the X atom. All energies are in kcal/mol

H_nX-H	χ^a	$De(\exp)^b$	De	De^*	$E_R{}^c$	De^*-De
H_3C-H	2.6	112.0	114.5	114.5	0.0	0.0
H_2N-H	3.0	116.0	114.4	135.3	20.0	20.9
HO-H	3.4	125.4	119.1	153.0	33.3	33.9
F-H	4.0	141.1	129.0	183.8	54.6	54.8

^{a} According to Pauling's electronegativity scale. ^{b} For details see ref 90. ^{c} Calculated by 2×2 mixing of the diabatic states $H_nX(\mathbf{43})$ and $H_nX(\mathbf{44})$.

The two possible ways of calculating the LPBWE, as the ($De-De^*$) energy difference or as the resonance energy E_R, yield very similar values, differing from each other by less than 1 kcal/mol in all cases. This excellent agreement nicely confirms that the rehybridization undergone by the H_nX fragment can be adequately modelized by a simple mixing of two diabatic states as in Figure 6. It also shows that the hybrid orbitals, optimized for these diabatic states by the procedure described above, are realistic and form a convenient basis of orbitals to generate the ground state H_nX $^2B_1$ radical.

The resonance energies display a regular tendency, gradually increasing from 0 in H_3C (by definition), to 20, 33 and 55 kcal/mol in H_2N, HO and F, respectively. This tendency is in agreement with the variation of the s-p orbital energy gap in the series (N, O, F). As the gap increases, the σ radical $H_nX(42)$ is increasingly less stable than the π radical $H_nX(41)$, meaning that the rearrangement from $H_nX(43)$ to $H_nX(41)$ is increasingly stabilizing. As a consequence, the LPBWE turns out to be roughly proportional to the number of lone pairs on the X atom, as proposed by Sanderson.[89]

It is interesting to compare the above LPBWE values with the empirical estimations of Sanderson[89] for the LPBWE in the same series: 0, 28, 35 and 36 kcal/mol, respectively. One may note in passing that Sanderson's value for the F atom does not follow the monotonous tendency of the other members of the series. This can be explained by the rather inaccurate bond energies that were used for these estimations based on homonuclear bonds: the F-F bond energy was overestimated, and the O-O one was taken as an averaged value from a series of peroxides. Using accurate homonuclear bond energies of 65.8, 51.0 and 38.3 kcal/mol for the H_nX-XH_n dimers (X = N to F),[92] associated with Sanderson's estimated unweakened bond energies, leads to a more coherent set (0, 15, 27 and 37 kcal/mol) of empirical LPBWE values. These are consistently smaller but of the same orders of magnitude as our computed resonance energies.

In view of the above discussed Pauling's relationship and of the break in the De variation in Figure 4, the plot of the unweakened bonding energy De^* against the electronegativity of the central X atom, as displayed in Figure 7, exhibits a rather striking result: the variation is practically linear. While our definition of the unweakened bonding energy is not claimed to be universal, this smooth variation of the unweakened bonding energies as X is taken from left to right of the periodic table is an indication that our calculated De^* values are realistic, since they vary like the effective nuclear charge experienced by the bonding electrons in harmony with first principles of chemical bonding. It also supports Pauling's idea that the strength of the A-B bond is a simple function of the electronegativity difference between A and B, all other things being equal.[1b]

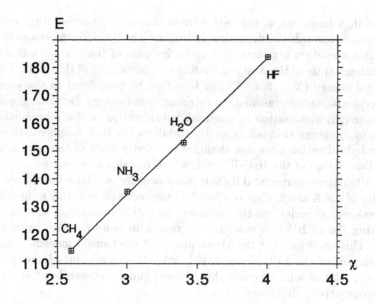

Figure 7: Calculated unweakened H_nX-H bond energy (De^*), in kcal/mol, as a function of the electronegativity of X.

6.4 Conclusion.

The present work is not claimed to provide some accurate values for the LPBWE, which would probably be a meaningless endeavour. What has been evidenced here is an energetic quantity, the stabilization of the central X atom's valence state in the course of the X-H dissociation, that is attached to the presence of a lone pair and that weakens the X-H bond. However not universal, this definition of the LPBWE is realistic and yields a set of unweakened bonding energies, after Sanderson's concept, that are numerically coherent with first principles of chemical bonding. The LPBWE calculated values are quite significant and support Sanderson's proposal that the weakening effect is not restricted to homonuclear X-X bonds but indeed exists in X-H bonds as well. This indicates that the general bond weakening effect exerted by a lone pair does not solely arise from interatomic repulsions between lone pairs, but also has an important component that is intra-atomic and intrinsic to the X atom, whatever the nature of the ligand.

Physically, the breaking of an H_nX-H bond is accompanied with a gradual change in the hybridization of the X atom. In the series (X = N to F), this rehybridization has the effect of increasing the s character of the lone pair(s) of the H_nX fragment, to the detriment of the singly occupied fragment orbital that points towards the departing hydrogen. The net outcome is a stabilization of the dissociation products, due to an increase of the average s character in the valence state of the X atom, leading to a weakening of the bond. The H_nX-H bond energy (X = N to F) can therefore be considered as the sum of two components: an unweakened, or intrinsic, bond energy De^* that corresponds to a thought dissociation at constant hybridization, and a reorganization (or resonance) energy that subsequently stabilizes the H_nX fragment. In contrast, the rehybridization does not change the valence state of the central atom in the dissociation of the H_3C-H bond, which remains unweakened.

While the experimental H_nX-H bond energies, as plotted vs the electronegativity of the X atom, display a break in the series (X = C to F), the calculated unweakened energies, on the contrary, smoothly increase in the same series, showing the LPBWE is indeed non-existent in bond energies calculated this way. This confirms that the above proposed mechanism indeed comprehends the essence of the LPBWE in X-H bonds. It also supports Pauling's idea of a direct relationship, all other things being equal, between bond strength and electronegativity difference.

7   The Uniform Mean-Field Hartree-Fock Procedure

The present application does not in fact report actual BOVB computations, but uses the qualitative aspects of the BOVB method, and in particular the analysis of the Hartree-Fock error in odd-electron bonded systems (see Section 3.3), to devise an economical ab initio procedure suitable for such systems.

Experimentally, one-electron and especially three-electron bonds are abundant and well characterized.[93] Numerous $(R_2S\therefore SR_2)^+$ radical cations, $(RS\therefore SR)^-$ radical anions and $R_2S\therefore SR$ neutral radicals have been identified, as well as $N\therefore N$, $P\therefore P$, $As\therefore As$, $Se\therefore Se$, $I\therefore I$ and more generally all kinds of $X\therefore Y$ (X, Y = N, S, P, halogen, etc.) two-center-three-electron (2c,3e) bonds. Despite the ample observations of these species, very little experimental data exist for the strengths of their odd-electron bonds, and thermodynamic data are remarkably sparse, so that most of the information about odd-electron bonding in chemistry rather comes from theoretical studies. This situation makes ab initio calculations particularly useful, however these are limited by two requirements which are not easy to satisfy simultaneously: (i) the accuracy is rather sensitive to the quality of the basis set,[48] and (ii) the electron correlation is essential in such calculations (vide supra). As a consequence, most of the calculations are reduced to model systems, which is all the more unfortunate as substituent effects seem to be important but largely unknown.[56]

Were the economical Hartree-Fock (HF) method reliable for this type of system, it is clear that ab initio investigations using good basis sets could be performed for large systems in real size. We know that it is not the case, yet the HF framework still deserves some further consideration, for two reasons: (i) It provides a wave function that is qualitatively correct in terms of resonating Lewis structures (see Section 3.3), and (ii) the origin of the quantitative HF error is well understood in terms of the breathing-orbital effect. Using this knowledge, we have devised a non-empirical remedy to the latter error, in the form of a HF-based procedure, dubbed Uniform Mean-Field Hartree-Fock (UMHF), aimed at calculating accurate odd-electron bonding energies without resort to electron correlation. Such a remedy needs not necessarily involve an improvement of the *absolute* energies. But in keeping with the nature of the problem, the procedure should lead to a consistent description of both the molecule and the separated fragments. In this sense, the consistency of the dissociation energy profile for an odd-electron bond rests on two conditions: (i) a consistent description of the individual resonance structures at all distances, and (ii) a satisfactory estimate of the mixing resonance energy of these resonance structures. These two aspects are considered below in turn.

7.1 Technical Details.

All the computational results reported in the Tables, including those quoted from other sources, were obtained with the standard 6-31G* basis set, with the exception of the He_2^+ cation radical that was calculated with the 6-31G** basis set that contains p-type polarization functions for the helium atom.

Although the procedure that is proposed here could use unrestricted (UHF) as well as restricted (ROHF) Hartree-Fock calculations, the former type was chosen because of its greater flexibility. Alike, the Möller-Plesset (MP) perturbation theory was used in its unrestricted formalism throughout this study. The MP dissociation energies were calculated at the MP2//MP2 and MP4//MP2 level, i.e. the geometries of the molecules and separated fragments have been optimized at the second order of perturbation (MP2), while the final energies were calculated at the second (MP2) and fourth (MP4) orders; the frozen-core approximation is applied in this latter case. The MP4//MP2 and MP2//MP2 results are quoted from ref. 48 for the three-electron bonds, while those for the one-electron bonds were calculated in this work (Table 12). The MP2-optimized geometries of the one-electron bonded species and their separate fragments are displayed in the original paper referring to this work.[56]

For the sake of consistency, the geometries employed to calculate odd-electron UMHF bonding energies were those optimized at the UHF level, and quoted from refs 47 and 48, since the UMHF and UHF methods are consistent in the vicinity of the local minima of the odd-electron bonded species.

Some test ab initio valence bond calculations on the F_2^- radical anion are reported in the text and have been performed in 6-31+G* basis set , derived from the 6-31G* by adding an s,p set of diffuse functions, of exponent 0.1076, on each fluorine atom.

7.2 A Consistent Description of the Individual Resonance Structures.

The reason for the Hartree-Fock deficiency in odd-electron bonded systems has been illustrated above (see Section 3.3) on the example on F_2^-. At short distances, the bonding interaction is nothing but a fluctuation between two limiting situations that each represent an anion flanked by a radical. On the other hand, the UHF orbitals are optimized within a mean-field environment and therefore come out equivalent for the F^- and F fragments of the molecule, as in 22, 23. As a consequence, the UHF orbitals are not adapted to the instantaneous electron density which is either one or the other limiting situation but not the average.

Now it should be noted that the Hartree-Fock defect does not carry over to all distances, and this is where the real problem lies. Indeed, at infinite separation distance, the charge does not fluctuate any more between both fragments but localizes on one of them. Thus, the Hartree-Fock description of the supersystem consists of one anion and one radical, each one having its specific set of orbitals optimized so as to fit the particular neutral or ionic situation. It follows that the orbital inadequacy that is present at bonding distances vanishes at large distance, so that the HF level describes the separated fragments better than the bonded molecule, hence the underestimation of bonding energies.

This imbalanced description of the dissociation process is illustrated in Figure 8 (dotted line), for a three-electron bonded system of the type X_2^+. At the UHF level, the wave function is symmetric at bonding distances but symmetry-broken at large distances. Somewhere in-between these two distances, the dissociation curve displays a singularity due to sudden symmetry breaking. The BOVB method, with its charge-fluctuation-adapted orbitals, cures the defect by improving the description at short distance, as shown in Figure 8 (lower curve, full line). Thus, while the BOVB and UHF curves approximately coincide [e] at large distance, both curves gradually deviate from each other as the distance is shortened and the BOVB curve smoothly

---

[e] They coincide only approximately, because the BOVB dissociation limit corresponds to the ROHF energies of the separate fragments, and therefore does not exactly match the UHF limit.

descends towards a physically realistic situation that is a superposition of two symmetry-broken VB structures with adequate orbitals. However, as the VB technology is known to be CPU consuming, this method is not the answer to our search for a procedure suitable for large systems.

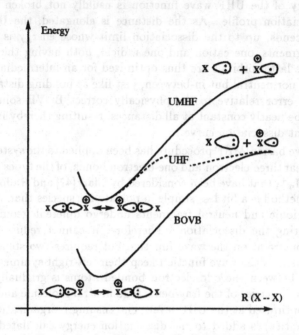

Figure 8: Schematic dissociation profiles as calculated by the BOVB, UHF, and UMHF methods. The R coordinate is the interatomic distance between the odd-electron bonded atoms. The UHF curve, dotted line, merges into the UMHF one at short distances and into the BOVB one at infinite distance. In-between, it displays a discontinuity due to sudden symmetry-breaking.

An alternative economical solution consists of removing the imbalance by *extending the Hartree-Fock mean-field defect to any distances up to that of separated fragments*. This can be done by optimizing the orbitals of each fragment with a frozen mean occupancy, that is kept constant throughout the dissociation coordinate. In the case of symmetrical diatomic molecules, the procedure is particularly simple and consists of maintaining the symmetry of the wave function all the way to the limit of separated fragments which are considered as a symmetrical supersystem in a mean field. This is the main principle of the Uniform Mean-Field Hartree-Fock procedure (UMHF), whose aim is not to improve the Hartree-Fock wave function but rather to generate a dissociation curve being uniformly shifted relative to the reference BOVB curve.

Figure 8 displays the expected qualitative profile of the UMHF dissociation curve. At bonding distances, it exactly coincides with the UHF curve since the symmetry of the UHF wave function is usually not broken in this part of the dissociation profile. As the distance is elongated, the UMHF curve smoothly ascends, up to the dissociation limit whose energy is that of two separate fragments, one cation and one radical, both having the same set of orbitals. The latter orbitals are thus optimized for an intermediate situation, neither ionic nor neutral but in-between, just like at bonding distance, so that the absolute error relative to the physically correct BOVB solution can be expected to be nearly constant at all distances, resulting thereby in a balanced and consistent dissociation curve.

The above basic UMHF procedure has been applied to the systematic series of homonuclear three-electron and one-electron bonds, of the types $H_n X \therefore X H_n{}^+$ and $H_n X \cdot X H_n{}^+$, that have been considered by Clark[47] and Radom.[48] Practically, the method is a bit less simple in polyatomic species than in diatomics, because the ionic and neutral fragments undergo different geometry reorganizations during the dissociation. Therefore, it cannot reduce to a simple symmetry constraint on the wave function but requires two steps. First, the molecule as well as the wave function keep their left-right symmetry, and only the distance between the odd-electron bonded atoms is gradually elongated. Second, the geometries of the fragments (one neutral molecule and one radical cation) are optimized at the UHF level. The ensuing reorganization energy (a negative quantity) is added to the dissociation energy calculated in the first step, to get the final dissociation energy.

The first part of Table 12 displays the three-electron bonding energies for the $H_n X \therefore X H_n{}^+$ species, as calculated at the UHF, MP2 and MP4 levels, and by the above UMHF procedure in its primary form, referred to as "uncorrected". It first appears that the Hartree-Fock error is considerable relative to the MP4 level, ranging from 15 to 28 kcal/mol for X atoms of the first row, and a bit smaller for atoms of the second row. On the other hand, the uncorrected UMHF results are consistently much better. In all cases the orders of magnitude are correct, and the errors relative to MP4 never exceed 7 kcal/mol. Moreover, we have checked, on the Ne_2^+ example, that the UMHF method still behaves well when the basis set is increased; in the 6-31G(2d,f) basis set, the MP4 and uncorrected UMHF methods provide bonding distances of 1.698 and 1.718 Å, respectively, and bonding energies of 38.2 and 41.5 kcal/mol.

The calculated bonding energies for the one-electron bonded species are displayed in the second part of Table 12. Here the Hartree-Fock error is much

Table 12: Calculated dissociation energies, in kcal/mol, of some odd-electron bonded radical cations[a]

	UHF//UHF[b]	UMHF[b] (uncorrected)	β	S	UMHF[b,c] (corrected)	MP2//MP2[d]	MP4//MP4[d]
Three-electron bonds							
HeHe+ ($D_{\infty h}$)	43.2	58.1	.901	.973	49.6	49.9	53.2
NeNe+ ($D_{\infty h}$)	9.1	40.7	.997	.971	39.4	38.8	37.0
ArAr+ ($D_{\infty h}$)	11.2	26.2	.988	.969	25.0	24.2	23.7
HFFH+ (C_{2h})	19.8	51.9	.985	.952	48.5	48.2	45.4
HClClH+ (C_2)	17.4	31.9	.978	.960	29.9	29.5	28.9
H2OOH2+ (C_{2h})	22.9	51.1	.975	.946	47.0	46.5	44.0
H2SSH2+ (C_{2h})	19.8	32.6	.961	.956	30.0	30.1	29.6
H3NNH3+ (D_{3d})	24.1	46.2	.948	.942	40.2	40.3	39.2
H3PPH3+ (C_{2h})	19.4	30.1	.990	.937	27.3	27.3	26.8
One-electron bonds							
HBeBeH+ ($D_{\infty h}$)	48.1	51.3	1.	.991	50.4	49.4	49.7
HMgMgH+ ($D_{\infty h}$)	31.4	33.7	1.	.989	32.9	31.8	32.1
H2BBH2+ (D_{2d})	45.5	54.4	1.	.982	52.1	54.6	56.0
H2BBH2+ (D_{2h})	39.3	48.2	1.	.982	46.1	46.1	47.1
H2AlAlH2+ (D_{2d})	28.5	32.9	1.	.987	31.7	31.3	32.0
H2AlAlH2+ (D_{2h})	26.7	31.0	1.	.987	29.8	29.4	30.1
H3CCH3+ (D_{3d})	38.2	54.4	1.	.977	51.7	51.4	51.0
H3CCH3+ (D_{3h})	36.7	52.8	1.	.977	50.1	49.6	49.2
H3SiSiH3+ (D_{3d})	30.4	37.2	1.	.985	35.8	36.0	36.7
H3SiSiH3+ (D_{3h})	29.5	36.3	1.	.985	35.0	35.0	35.6

[a] Except for He2+, all calculations use the 6-31G* basis set. The 6-31G** basis set has been used for He2+. [b] UHF-optimized structures from refs 47 and 48 except for He2+ (this work, 1.078 Å). [c] Dissociation energy corrected for resonance energy. [d] MP2-optimized structures from ref 48, except for He2+ (this work, 1.0815 Å).

354

smaller than in the preceding case, for the reasons already discussed in Section 3.3 : the breathing orbital effect, the lack of which is the source of the Hartree-Fock error, is non-existent in the active system of one-electron bonds. Still, the UHF bonding energy is 13 kcal/mol too small in the $H_3C \cdot CH_3^+$ case. The uncorrected UMHF results are once again improved relative to UHF, with a maximum error of 3.6 kcal/mol.

It is noteworthy that our procedure, as used at the above simple stage (second column of Table 12), yields dissociation energies which are systematically overestimated relative to MP2 and MP4 results. This suggests that the remaining error might be due to a well-defined physical reason rather than to mere inaccuracy. As the energies of the individual VB structures are in principle calculated in a balanced way throughout the dissociation curve, a possible source of error lies in the calculation of the resonance energy (RE), i.e. the stabilization associated with the mixing of the two VB structures. Is this RE correctly estimated by a method of Hartree-Fock type (UHF or UMHF) that does not take the breathing orbital effect into account? And if not, how can we correct the UMHF procedure to remedy this defect? These questions have to be considered in order to give the UMHF procedure its definitive form.

7.3 The Influence of Breathing Orbitals on the Resonance Energy.

A good indication that the resonance energy might be overestimated in a calculation of Hartree-Fock type (in both UHF and UMHF) is provided by a computational experiment on the F_2^- molecule using calculations of VB type. Thus, a simple L-BOVB calculation, in 6-31+G* basis set, puts the individual VB structures, at the equilibrium geometry, 10.1 kcal/mol above the separated fragments, and yields an RE of 39.8 kcal/mol, thus leading to the value 29.7 kcal/mol for the dissociation energy. On the other hand, a VB calculation using averaged orbitals, while putting the individual VB structures much too high, yields a significantly *larger* resonance energy, 44.8 kcal/mol. As this latter calculation is nearly similar to a Hartree-Fock calculation at bonding distance, one may reasonably conclude that the Hartree-Fock method overestimates the resonance term in the interaction energy of F_2^- with respect to the reference BOVB calculation with charge-fluctuation adapted orbitals. This tentative statement can be generalized and understood with the help of qualitative VB theory, and a detailed derivation is given in the original investigation.[56] Here, only the leading ideas will be summarized, for the sake of conciseness.

The resonance energy arising from the mixing of two Lewis structures is proportional to a quantity βS, where S is the overlap integral between the two VB structures (e.g. **22** and **23** above) and β is the resonance integral, itself nearly proportional to S. A rough estimation of this latter term can be made by ranking the spinorbitals of each VB structure so as to put them in maximum correspondence, then computing the product of all the 2×2 overlaps between spinorbitals of the same rank. If the two VB structures share the same set of orbitals, as is the case for **22** and **23**, most of the orbital overlap terms are unity and there only remains the overlap between the two active orbitals that directly interact. If on the other hand the VB structures have breathing orbitals as in **24** and **25**, then the overlaps between orbitals of the same rank are all smaller than unity, and their product contributes to diminish the final value of S. The same reasoning holds for the β integral. In other words, the two VB structures are more different if they have breathing orbitals than if they share the same set of orbitals, and thereby their overlap and resonance integrals S and β are larger in the latter case, hence the overestimation of the resonance term at the Hartree-Fock level.

It follows from the above reasoning that the Hartree-Fock overestimation of the resonance term in an odd-electron interaction is only a matter of orbital size. As it has been shown that the breathing orbitals in an odd-electron bonded species are roughly similar to the respective orbitals of the isolated ion and radical, an estimation of this aspect of the Hartree-Fock error should be possibly gained from the orbitals of the ionic and neutral fragments, without resort to BOVB calculations. Indeed, a simple non-empirical formula can be derived to correct the UMHF bonding energies.[56] For three-electron bonded diatomics, the correction takes the form of a product of two multiplicative factors, β and S, that have the effect of scaling the uncorrected UMHF values as in eq (26). The estimation of the β and S corrective factors only necessitate simple calculations of UHF type.[56]

$$De(RE{-}corrected) = \beta S\, De(uncorrected) \qquad (26)$$

For one-electron bonded diatomics, the correction is even simpler since there is no breathing orbital effect among the atomic orbitals that directly interact, so that eq 26 can be applied by setting β to unity in this type of system. The method can be used as well in polyatomic systems, by applying the correction to the intermediate dissociation energy that is calculated in the first step of the procedure, before relaxation of the geometries of the fragments.

Table 12 displays the calculated corrective β and S factors, as well as the corrected UMHF bonding energies for the three-electron and one-electron bonded systems. One first result is that, in accord with our qualitative analysis, all the calculated β and S factors are consistently smaller than unity, and therefore the βS term lowers the uncorrected UMHF bonding energies, as expected. As a result, the final UMHF three-electron bonding energies, displayed in the sixth column of Table 12, are extremely close, within less than 1 kcal/mol, to the MP2//MP2 values themselves in good agreement with MP4 calculations. Moreover, the errors are not systematic but can be positive or negative, an indication that they do not reflect a physically based inadequacy of the method.

The corrected UMHF dissociation energies for the one-electron bonded species are also in excellent agreement with the MP2 and MP4 levels, with a maximum deviation of nearly 1 kcal/mol. The only exception is $B_2H_4^+$ in its D_{2d} conformation: here the UMHF bonding energy amounts to 52.1 kcal/mol, vs 54.6 and 56.0 at the MP2 and MP4 levels, respectively. However the error does not lie in the description of the odd-electron bond, as indicated by the good UMHF value for the D_{2h} conformation. Rather, it lies in the incorrect description, at the Hartree-Fock and MP2 levels, of the interaction between the empty orbital of one boron atom and the hydrogens of the other, in the twisted D_{2d} conformation, a hyperconjugative phenomenon that falls outside the scope of the present method.

Another remarkable result of Table 12 is the near-constancy of the calculated β and S factors from one system to the other. However economical in CPU, the rigorous calculation of these factors requires some tedious handworking that could be saved by adopting some average scaling factors for all systems. To this aim, we have proposed the following two standard scaling factors respectively suitable for three-electron and one-electron bonds, in the following form:

$$De(RE-corrected) = \eta De(uncorrected) \qquad (27)$$
$$\eta(3-e) = 0.926$$
$$\eta(1-e) = 0.985$$

These average scaling factors lead to energies in excellent agreement with accurate values in Table 12. They have been used throughout the UMHF calculations of substituent effects in Table 13 (vide infra).

7.4 Application: Substituent Effects on Three-electron Bonding Energies.

As noted above, the substituent effects on three-electron bonding energies can be important, and rather unpredictable. As an example, some recent results of Illies et al. clearly show that the $R_2S \therefore SR_2^+$ bond strength is almost constant in the series with R = H, Me, and Et.[94] In contrast, the CH_3 substituent exerts a strong weakening effect on the $O \therefore O$ bond, which is found to be 15 kcal/mol weaker in $Me_2O \therefore OMe_2^+$ than in $H_2O \therefore OH_2^+$, at the MP2 level.[56]

In order to check the validity of the UMHF procedure employed with the above average scaling factors (eq 27), we have used this latter method, together with the standard MP2 method, to calculate the bonding energies of a number of symmetrical three-electron bonded systems corresponding to the substituted analogs of the species displayed in Table 12. Two kinds of substituents have been employed, F and SiH_3, chosen for their very differenciated inductive character, the former being a strong σ attractor and the latter being a σ donor. For each radical, the average value $< S^2 >$ of the spin-squared operator has been calculated and compared to the value of 0.75 required for a pure doublet, to make sure the spin contamination was not excessive. As UHF geometries and energies are known to be unreliable in cases of important spin contamination, we have not considered radicals in which the UHF $< S^2 >$ value was larger than a threshold of 0.79. The results, displayed in Table 13, first show quite a sizable substituent effect, if the bonding energies are compared to those of the unsubstituted species in Table 12. They also display some regular tendencies, whose interpretation passes the scope of this section and is detailed elsewhere.[97] More important for the question in hand is the comparison between UMHF and MP2 calculated bonding energies. Although a bit less good than when the corrective factor is rigorously calculated as in Table 12, the agreement between MP2 energies and UMHF energies, as calculated with the standard scaling factor of eq 27, remains quite satisfying and confirms the validity of the method as a computational procedure suitable for large systems.

7.5 Extension to Unsymmetrical Species.

Most odd-electron bonds are observed between identical fragments, because the strength of the odd-electron bond falls off exponentially with the increase of the electronegativity difference between the fragments.[47] However some heteronuclear odd-electron bonds are still observed, and it may be useful to generalize the UMHF procedure to cases that do not possess left-right symmetry. This generalization requires the use of an open-shell Fock operator accepting non-integer electron occupancy in the Hartree-Fock orbital optimization process.

Whether any left-right symmetry is present or not, the basic principle of the UMHF procedure is to optimize the orbitals of the separate fragments with the same average occupancy, possibly fractional, as the bonded molecule. Consider one fragment in an odd-electron bonded species, involving some permanently occupied orbitals φ_j and a highest spin orbital φ_a which is alternatively occupied or empty, following the charge fluctuation. At equilibrium distance,

Table 13: Calculated dissociation energies, in kcal/mol, of some three-electron bonded radical cations with fluorine or silyl substituents[a]

	$< S^2 >$[b]	corrected UMHF[c]	MP2//MP2
$FH_2N \therefore NH_2F^+$	.777	31.7	28.0
$F_2HN \therefore NHF_2^+$	.775	24.6	22.6
$F_3N \therefore NF_3^+$	.772	18.3	16.2
$FH_2P \therefore PH_2F^+$	.784	29.5	32.4
$F_2HP \therefore PHF_2^+$	.781	27.9	27.2
$F_3P \therefore PF_3^+$	.773	18.9	16.8
$FHS \therefore SHF^+$	.788	24.2	23.1
$FCl \therefore ClF^+$	.771	19.4	18.2
$(SiH_3)H_2N \therefore NH_2(SiH_3)^+$	.782	29.6	28.0
$(SiH_3)_2HN \therefore NH(SiH_3)_2^+$	.788	17.8	20.1
$(SiH_3)HO \therefore OH(SiH_3)^+$	.780	31.4	30.5
$(SiH_3)_2O \therefore O(SiH_3)_2^+$	.786	13.7	16.2
$(SiH_3)H_2P \therefore PH_2(SiH_3)^+$	.775	26.7	27.3
$(SiH_3)_2HP \therefore PH(SiH_3)_2^+$	.768	22.3	23.4
$(SiH_3)_3P \therefore P(SiH_3)_3^+$	.766	20.0	22.9
$(SiH_3)HS \therefore SH(SiH_3)^+$	.770	26.5	26.4
$(SiH_3)_2S \therefore S(SiH_3)_2^+$	.769	24.0	25.6
$(SiH_3)Cl \therefore Cl(SiH_3)^+$	.765	26.7	26.5

[a] The 6-31G* basis set is used for fluorine-substituted species, while the CEP-31G* basis set with effective core potentials is used for all atoms of the silyl-substituted species. [b] Average value of the spin-squared operator. [c] The average scaling factor of 0.926 (eq 26) is used in the first step of the procedure.

this fragment has a probability P_1 to bear the fluctuating electron in φ_a, and a probability P_2 to have ceased this electron to the other fragment. In accord with the above principle, the orbitals that we search for the same fragment at infinite separation should minimize the following weighted sum of determinants Ψ (eq 28), in which the squares of the coefficients C_1 and C_2 match the density probabilities P_1 and P_2. Thus, the functional Ψ represents a pseudo-

$$\Psi = C_1 |...\varphi_j...\varphi_a| + C_2 |...\varphi_j...| \qquad (28)$$
$$C_1{}^2 = P_1$$
$$C_2{}^2 = P_2$$
$$C_1{}^2 + C_2{}^2 = 1$$

electronic state of the fragment under scrutiny, considered as isolated but having the same average orbital occupancy as in the bonded species.

Applying the traditional Hartree-Fock equations leads to the result that the functional Ψ is minimal in energy if the following F operator is diagonal in the basis of the optimized orbitals:

$$F = h + \sum_{j}^{occ}(J_j - K_j) + C_1{}^2(J_a - K_a) \tag{29}$$

where the h, J and K operators have their usual meaning. The orbitals are then determined in the usual way by iteratively diagonalizing the open-shell Fock operator displayed in eq 29. Note that this operator is of UHF type, but that an ROHF Fock operator with arbitrary fractional occupancies might also be used. The advantage of the second type is that it is already implemented in several standard ab initio Hartree-Fock programs, however the UHF type may be recommended for its greater flexibility.

There remains to determine the C_1 and C_2 coefficients which must reflect the average electron distribution at the equilibrium geometry. This can be done by means of a simple Mulliken or NBO population analysis of the bonded molecule, which yields the orbital population $C_1{}^2$ for φ_a. Note by the way that the inaccuracy which is usually attached to the Mulliken population analysis does not apply here. Indeed, the Mulliken population of an orbital is inaccurate only when this orbital strongly overlaps with other ones, but in the odd-electron species the overlap between the two interacting atomic orbitals is always rather weak (about 0.17 as a standard value for three-electron bonds), as has been shown by Radom.[48]

7.6 On Hartree-Fock Optimized Geometries.

One of the most striking paradoxes arising from Clark's and Radom's systematic theoretical studies[47,48] is the apparent ability of the UHF method to provide nearly correct optimized geometries for odd-electron bonded systems, contrasting with the complete failure of the same method to account for bonding energies. Based on the above understanding on the Hartree-Fock error, we can now understand this surprising and rather fortunate finding, and show that it is general. The reasoning will be illustrated on the particular case of symmetrical species, but could be extended to unsymmetrical cases as well.

In the vicinity of the local minimum the left-right symmetry is generally not still broken at the UHF level. This means that the UHF and UMHF methods, the latter having being shown to give satisfactory relative energies at all distances, do not fundamentally differ in this portion of the potential surface, except that the UHF potential well should be a little sharper. It follows that the UHF potential surface has *locally* the right shape — although much too high in energy relative to other conformations — and therefore yields optimized geometries in good agreement with those calculated at the MP2 level. In some cases however,[48] the odd-electron bonded species is found, at the UHF level, to be a transition state which by breaking its left-right symmetry rolls down to low energy hydrogen-bonded species. Clearly, if the geometry optimization were to be carried out within the UMHF framework, the symmetry-breaking mode would have been frozen, resulting *precisely in the UHF geometry,* but now as a real minimum (no imaginary frequency).

The reasoning readily extends to the case of unsymmetrical species, provided the latter do not exhibit any Hartree-Fock instability due to quasi-symmetry-breaking, i.e. the same type of instability as the symmetry-breaking artefact but in a molecule that does not display formal symmetry (vide supra). This particular case would not however lead to erroneous geometries, as in fact no odd-electron bonded extremum would be found at the UHF level. It follows that all the stationary points calculated at the UHF level should display reasonable geometries, even if they are not found to be of the right nature. As note above, the latter statement is of course limited to cases not displaying an excessive UHF spin contamination, a condition that applies, at any rate, to any kind of spin-unrestricted calculations.

7.7 Conclusion.

The Hartree-Fock failure to describe odd-electron bonding is characterized by two deficiencies. The main one is the poor description of the individual resonance structures at bonding distances, due to an inherent constraint which freezes the orbitals at average shapes and sizes and prevent their adaptation to the instantaneous charge fluctuation. The second one is an overestimation of the resonance energy arising from the mixing of the two resonance structures. The computational defect typifies energies, but does not affect optimized geometries which are found close to being correct at the UHF level; albeit at the same time the UHF stationary point is found not to be a minimum but a saddle point. The UMHF procedure corrects for these two deficiencies, and yields odd-electron dissociation energies in satisfactory agreement with accurate calculations performed in the same basis set. It is based on the simple UHF method and has the same computational requirements. Further, it can be used by means of standard ab initio programs, without any extra programming effort. The procedure may therefore be used to study large size species and thereby broaden the range of odd-electron bonded species that can be investigated theoretically.

8   Summary and Conclusion

The BOVB method is a procedure of ab initio valence bond type that aims at reconciling the properties of interpretability, compactness and reasonable accuracy of the energetics. The interpretability is the same as the classical VB method, which ensures the best correspondence between the mathematical formulation and the concept of Lewis structure displaying a specific bonding scheme. The accuracy is ensured by orbital optimization, but with the additional improvement with respect to current VB methods that the orbitals are optimized independently for each VB structure, which is thereby allowed to have its specific set of orbitals, different from one structure to the other.

This specificity allows an accurate description of bond-breaking and bond-forming, by taking care of the associated dynamical electron correlation which is considered in this framework as the instantaneous adaptation of all orbitals to the charge fluctuation inherent to the bond, and therefore directly calculated as such. The computational tests that have been performed in notoriously difficult test cases show that the description is reasonably accurate, in view of the compacity of the wavefunctions that only involve three configurations for the two-electron bonds, and two for the odd-electron ones. In all cases, the bonding energy and/or dissociation energy profile are close to the full CI result in the same basis set, within a few kcal/mol. Besides, the BOVB method is free, by nature, from troublesome artefacts like basis-set-superposition error or symmetry-beaking.

Practically, only one small part, called the active part, of a molecular system is treated in the VB framework, while the rest is treated at the ordinary MO level. The active part includes these orbitals and electrons that undergo effective changes throughout a potential surface, like bond-breaking or bond-forming. The inactive part undergoes orbital optimization to follow the changes of the active part, but its electrons are not explicitly correlated, in keeping with the assumption that the absolute error so introduced is quasi-constant throughout a potential surface.

The VB structures that are generated are chosen so as to represent the complete set of Lewis structures that are relevant for the description of the active electronic system. This and the use of breathing orbitals ensures that the active electrons are explicitly correlated. The VB structures, whose coefficients and orbitals are optimized simultaneously, can be defined different ways according to the various levels of accuracy, but all levels agree on the principle that the active orbitals should be strictly localized on their specific atom or fragment, and not allowed to delocalize in the course of the orbital optimization process. This latter condition is important for keeping the interpretability of the wavefunction in terms of Lewis structures, but also for a correlation-consistent description of the system throughout a potential surface.

The BOVB method is of course not aimed at competing with the standard ab initio methods in the general case, but has its specific domain. It has been mainly devised as a tool for computing diabatic states, with applications to chemical dynamics, chemical reactivity with the VB correlation diagrams, photochemistry, resonance concepts in organic chemistry, reaction mechanisms, and more generally all cases where a valence bond reading of the wave function or the properties of one particular VB structure are wished in order to better understand the nature of an electronic state. The method has passed its first tests of credibility and is now ready for routine applications.

Acknowledgments

I am most grateful to the members of the Utrecht group, C.P. Byrman, J.H. Langenberg, J.H. van Lenthe and J. Verbeek for their collaboration to the development of the BOVB method and for their VBSCF program TURTLE,[96] with which most of the valence bond calculations displayed here have been performed. Thanks are also due to the members of the Jerusalem group, D. Danovich, S. Shaik and A. Shurki, for a long-term collaboration in this field, and to the other workers who contributed to the research presented here: P. Archirel, N. Berthe-Gaujac, I. Demachy, S. Humbel, D. Lauvergnat, P. Maitre M. Menou and F. Volatron.

References

1. (a) L. Pauling, *J. Amer. Chem. Soc.* **53**, 1367 (1931); (b) L. Pauling, J. Amer. Chem. Soc. **54**, 998, 3570 (1932); (c) L. Pauling, in *The Nature of the Chemical Bond*, 3rd edn. (Cornell University Press, Ithaca, NY, 1960).
2. (a) J.C. Slater, *Phys. Rev.* **35**, 509 (1930); (b) J.C. Slater, *Phys. Rev.* **38**, 1109 (1931); (c) J. C. Slater, *Phys. Rev.* **41**, 255 (1932).
3. G.W. Wheland, in *Resonance in Organic Chemistry* (John Wiley, New York, 1955).
4. S.S. Shaik, *J. Amer. Chem. Soc.* **103**, 3692 (1981); (b) A. Pross and S.S. Shaik, *Acc. Chem. Res.* **16**, 363 (1983); (c) A. Pross, *Adv. Phys. Org. Chem.* **21**, 99 (1985); (d) A. Pross, *Acc. Chem. Res.* **18**, 212 (1985); (e) S.S. Shaik, in *New Concepts for Understanding Organic Reactions*, eds.

J. Bertran and I.G. Csizmadia (Kluwer, Dordrecht, 1989, NATO ASI Series Vol. C267); (f) S.S. Shaik and P.C. Hiberty, in *Theoretical Models of the Chemical Bonding, Part 4*, ed. Z.B. Maksic (Springer, Heidelberg, 1991), pp 269-322.

5. (a) A. Sevin, P.C. Hiberty and J.-M. Lefour, *J. Am. Chem. Soc.* **109**, 1845 (1987); (b) A. Sevin, P. Chaquin, L. Hamon and P.C. Hiberty, *J. Am. Chem. Soc.* **110** 5681 (1988); (c) A. Sevin, C. Giessner-Prettre, P.C. Hiberty and E. Noizet, *J. Phys. Chem.* **95**, 8580 (1991).

6. (a) S.S. Shaik, P.C. Hiberty, G. Ohanessian and J.M. Lefour, *J. Phys. Chem.* **92**, 5086 (1988); (b) S. Shaik, A. Shurki, D. Danovich and P.C. Hiberty, *J. Amer. Chem. Soc.* **118**, 666 (1996); (c) P.C. Hiberty, D. Danovich, A. Shurki and S. Shaik, *J. Amer. Chem. Soc.* **117**, 7760 (1995) an references therin.

7. P.C. Hiberty and C.P. Byrman, *J. Amer. Chem. Soc.* **117**, 9875 (1995).

8. J.D. da Motta Neto and M.A.C. Nascimento, *J. Phys. Chem.* **100**, 15105 (1996).

9. J.-K. Hwang, G. King, S. Creighton and A. Warshel, *J. Amer. Chem. Soc.* **110** 5297 (1988).

10. P.C. Hiberty, J.P. Flament and E. Noizet, *Chem. Phys. Lett.* **189**, 259 (1992).

11. P.C. Hiberty, S. Humbel, C.P. Byrman and J.H. van Lenthe, *J. Chem. Phys.* **101**, 5969 (1994).

12. P.C. Hiberty, S. Humbel and P. Archirel, *J. Phys. Chem.* **98**, 11697 (1994).

13. H. Heitler, F. London, *Z. Phys.* **44**, 455 (1927).

14. F. Hund, *Z. Phys.* **40**, 742 (1927); **42**, 93 (1927); **51**, 759 (1928).

15. R.S. Mulliken, *Phys. Rev.* **32**, 186, 761 (1928); **33**, 730 (1929); **40**, 55 (1932); **41**, **49**, 751 (1932).

16. R. McWeeny, *Proc. Roy. Soc. (London)* **A223**, 63, 306 (1954).

17. P.O. Lowdin, *Phys. Rev.* **97**, 1474 (1955).

18. F.A. Matsen, *Adv. Quant. Chem.* **1**, 59 (1964).

19. F. Prosser and S. Hagstrom, *Int. J. Quant. Chem.* **2**, 89 (1968).

20. B.T. Sutcliffe, *J. Chem. Phys.* **45**, 235 (1966).

21. G.G. Balint-Kurti and M. Karplus, *J. Chem. Phys.* **50**, 478 (1969).

22. M. Simonetta, E. Gianinetti and I. Vandoni, *J. Chem. Phys.* **48**, 1579 (1968).

23. H. Shull, *Int. J. Quant. Chem.* **3**, 523 (1969).

24. (a) G.A. Gallup, *Int. J. Quant. Chem.* **6**, 899 (1973); (b) G.A. Gallup, R.L. Vance, J.R. Collins and J.M. Norbeck, *Adv. Quant. Chem.* **16**, 229 (1982).

25. S.C. Leasure and G.G. Balint-Kurti, *Phys. Rev. A* **31**, 2107 (1985).

26. (a) J. Verbeek, *Ph.D. Thesis*, University of Utrecht, 1990; (b) J. Verbeek and J.H. van Lenthe, *J. Mol. Struc. (Theochem)* **229**, 115 (1991); (c) J. Verbeek and J.H. van Lenthe, *Int. J. Quant. Chem.* **40**, 201 (1991).

27. (a) V. Fock, *Z. Phys.* **61**, 126 (1930); (b) J.C. Slater, *Phys. Rev.* **35**, 210 (1930).

28. S.S. Shaik and P.C. Hiberty, *J. Amer. Chem. Soc.* **107**, 3089 (1985).

29. C.A. Coulson and I. Fischer, *Phil. Mag.* **40**, 386 (1949).

30. A.C. Hurley, J. Lennard-Jones and J.A. Pople, *Proc. Roy. Soc.* **A220**, 446 (1953).

31. (a) W.J. Hunt, P.J. Hay, W.A. Goddard, *J. Chem. Phys.* **57**, 738 (1972); (b) W.A. Goddard and L.B. Harding, *Ann. Rev. Phys. Chem.* **29**, 363 (1978); (c) F.B. Bobrowicz and W.A. Goddard, in *Methods of Electronic Structure Theory*, ed. H.F. Schaefer (Plenum Press, New York, 1977), pp. 79-127; (d) A.F. Voter and W.A. Goddard, *J. Chem. Phys.* **75**, 3638 (1981).

32. D.L. Cooper, J. Gerratt and M. Raimondi, *Adv. Chem. Phys.* **69**, 319 (1987); (b) D.L. Cooper, J. Gerratt and M. Raimondi, *Int. Rev. Phys. Chem.* **7**, 59 (1988); (c) D.L. Cooper, J. Gerratt and M. Raimondi, in *Valence Bond Theory and Chemical Structure*, eds. D.J. Klein and N. Trinajstic (Elsevier, 1990), p 287; (d) D.L. Cooper, J. Gerratt and M. Raimondi, *Top. Current Chem.* **153**, 41 (1990). (e) P.B. Karadakov, J. Gerratt, D.L. Cooper, M. Raimondi, *Chem. Rev.* **91**, 929 (1991).

33. M.B. Lepetit and J.-P. Malrieu, *J. Phys. Chem.* **97**, 94 (1993).

34. C.F. Jackels and E.R. Davidson, *J. Chem. Phys.* **64**, 2908 (1976).

35. (a) A.F. Voter and W.A. Goddard III, *Chem. Phys.* **57**, 253 (1981); (b) A.F. Voter and W.A. Goddard III, *J. Chem. Phys.* **75**, 3638 (1981); (c) A.F. Voter and W.A. Goddard III, *J. Amer. Chem. Soc.* **108**, 2830 (1986).

36. E. Hollauer and M.A.C. Nascimento, *Chem. Phys. Lett.* **184**, 470 (1991).

37. G. Ohanessian and P.C. Hiberty, *Chem. Phys. Lett.* **137**, 437 (1987).

38. P.C. Hiberty and J.M. Lefour, *J. Chim. Phys.* **84**, 607 (1987).

39. W.D. Laidig, P. Saxe and R.J. Bartlett, *J. Chem. Phys.* **86**, 887 (1987).

40. M. Merchan, J.-P. Daudey, R. Gonzales-Luque and I. Nebot-Gil, *Chem. Phys.* **141**, 285 (1990).

41. K.P. Huber and G. Herzberg, in *Molecular Spectra and Molecular Structures. IV. Constants of Diatomic Molecules* (van Nostrand Reinhold, New York, 1979).

42. C.W. Bauschlicher and P.R. Taylor, *J. Chem. Phys.* **86**, 887 (1987).

43. C.W. Bauschlicher, S.R. Langhoff, P.R. Taylor, N.C. Handy and P.J. Knowles, *J. Chem. Phys.* **85**, 1469 (1986).
44. S. Shaik, D. Danovich, A. Shurki and S. Shaik, manuscript in preparation.
45. L. Pauling, *J. Amer. Chem. Soc.* **53**, 3225 (1931)
46. S. Shaik and P.C. Hiberty, *Adv. Quant. Chem.* **26**, 99 (1995).
47. T. Clark, *J. Amer. Chem. Soc.* **110**, 1672 (1988).
48. P.M.W. Gill and L. Radom, *J. Amer. Chem. Soc.* **110**, 4931 (1988).
49. M.T.Nguyen and T.-K. Ha, *J. Phys. Chem.* **91**, 1703 (1987).
50. P. Maitre, personal communication.
51. W.R. Wadt and P.J. Hay, *J. Chem. Phys.* **82**, 284 (1985).
52. L.A. Curtiss, K. Raghavachari, G.W. Trucks and J.A. Pople, *J. Chem. Phys.* **94**, 7221 (1991).
53. L. Eberson, R. Gonzalez-Luque, J. Lorentzon, M. Merchan and B.O. Roos, *J. Amer. Chem. Soc.* **115**, 2898 (1993).
54. P. Archirel, personal communication, to be published.
55. F.X. Gadea, J. Savrda, I. Paidarova, *Chem. Phys. Lett.* **223**, 369 (1994).
56. P.C. Hiberty, S. Humbel, D. Danovich and S. Shaik, *J. Amer. Chem. Soc.* **117**, 9003 (1995).
57. J. Paldus and A. Veillard, *Mol. Phys.* **35**, 445 (1978), and references therein.
58. (a) D. Feller, E. Huyser, W.T. Borden, and E.R. Davidson, *J. Amer. Chem. Soc.* **105**, 1459 (1983).
59. A. Rauk, D. Yu and D.A. Armstrong, *J. Amer. Chem. Soc.* **116**, 8222 (1994).
60. A.D. McLean, B.H. Lengsfield III, J. Pacansky and Y. Ellinger, *J. Chem. Phys.* **83**, 3567 (1985).
61. (a) D. Feller, W.T. Borden and E.R. Davidson, *J. Amer. Chem. Soc.* **106**, 2513 (1984). (b) D. Feller, E.R. Davidson and W.T. Borden, *J. Amer. Chem. Soc.* **105**, 3348 (1983).
62. W.R. Wadt and W.A. Goddard III, *J. Amer. Chem. Soc.* **97**, 2034 (1975).
63. O.K. Kabbaj, M.B. Lepetit, J.-P. Malrieu, *Chem. Phys. Lett.* **172**, 483 (1990).
64. (a) M. Amarouche, G. Durand and J.-P. Malrieu, *J. Chem. Phys.* **88**, 1010 (1988). (b) M. Rosi and C.W. Bauschlicher Jr, *Chem. Phys. Lett.* **159**, 349 (1989). (c) F. Tarentelli, L.S. Cederbaum and P. Campos, *J. Chem. Phys.* **91**, 7039 (1989).
65. C. Lifshitz, *J. Phys. Chem.* **86**, 3634 (1982).

66. J. Hrusak, H. Friedrichs, H. Schwarz, H. Razafinjanahary and H. Chermette, *J. Phys. Chem.* **100**, 100 (1996).

67. D. Vogt, *Int. J. Mass Spectrom. Ion Phys.* **3**, 81 (1969).

68. C.M. Roehl, J.T. Snodgrass, C.A. Deakyne and M.T. Bowers, *J. Chem. Phys.* **94**, 6546 (1991).

69. M.A. Buntine, D.J. Lavrich, C.E. Desent, M.G. Scarton and M.A. Johnson, *Chem. Phys. Lett.* **216**, 471 (1993).

70. R. Benassi and F. Taddei, *Chem. Phys. Lett.* **204**, 595 (1993).

71. S. Humbel, I. Demachy and P.C. Hiberty, *Chem. Phys. Lett.* **247**, 126 (1995).

72. F. Grein and T.C. Chang, *Chem. Phys. Lett.* **12**, 44 (1071).

73. K. Raghavachari and J.B. Anderson, *J. Phys. Chem.* **100**, 12960 (1996) and references therein.

74. (a)Y. Apeloig, in *The Chemistry of Organic Silicon Compounds*, eds. S. Patai and Z. Rappoport (Wiley and Sons, Chichester, United kingdom, 1989), Chapter 2, p 57. (b) Y. Apeloig, *Stud. Org. Chem.* **31**, 33 (1987). (c)Y. Apeloig and O. Merin-Aharoni, *Croat. Chem. Acta* **65**, 757 (1992). (d) J.B. Lambert, L. Kania, and S. Zhang, *Chem. Rev.* **95**, 1191 (1995). (e) For a discussion of gas phase properties of silicenium ions, see: H. Acwarz, in *The Chemistry of Organic Silicon Compounds*, eds. S. Patai and Z. Rappoport (Wiley and Sons, Chichester, United kingdom, 1989), Chapter 7, Vol 1.

75. R.J.P. Corriu and M. Henner, *J. Organometallic Chem.* **74**, 1 (1974).

76. C. Eaborn, in *Organosilicon and Bioorganosilicon Chemistry*, ed. H. Sakurai (Ellis Horwood, United Kingdom, 1985).

77. G.A. Olah, L. Heiliger, X.-Y. Li and G.K.S. Prakash, *J. Amer. Chem. Soc.* **112**, 5991 (1990).

78. P.D. Lickiss, *J. Chem. Soc. Dalton Trans.* 1333 (1992).

79. G.K.S. Prakash, S. Keyaniyan, R. Aniszfeld, L. Heiliger, G.A. Olah, R.C. Stevens, H.-K. Choi and R. Bau, *J. Amer. Chem. Soc.* **109**, 5123 (1987).

80. (a) A.H. Gomes de Mesquita, C.H. MacGillavry and K. Eriks, *Acta Cryst.* **18**, 437 (1965). (b) G.A. Olah, *Top. Curr. Chem.* **80**, 19 (1979).

81. Z. Xie, D.J. Liston, T. Jelinek, V. Mitro, R. Bau and C.A. Reed, *J. Chem. Soc. Chem. Commun.* 384 (1993).

82. D. Lauvergnat, P.C. Hiberty, D. Danovich and S. Shaik, *J. Phys. Chem.* **100**, 5715 (1996).

83. (a) J.A. Pople, M. Head-Gordon, D.J. Fox, K. Raghavachari and L.A. Curtiss, *J. Chem. Phys.* **90**, 5622 (1989); (b) L.A. Curtiss, K. Raghavachari, G. Trucks and J.A. Pople, *J. Chem. Phys.* **94**, 7221 (1991).

84. (a) K.S. Pitzer, *J. Amer. Chem. Soc.* **70**, 2140 (1947); (b) K.S. Pitzer, *J. Chem. Phys.* **23**, 1735 (1955); (c) K.S. Pitzer, *Adv. Chem. Phys.* **2**, 59 (1959).

85. R.S. Mulliken, *J. Amer. Chem. Soc.* **72**, 4493 (1950); *ibid.* **77**, 884 (1955).

86. M.G. Brown, *Trans. Faraday Soc.* **55**, 9 (1959).

87. G.L. Caldow and C.A. Coulson, *Trans. Faraday Soc.*, **58**, 633 (1962).

88. (a) P. Politzer, *J. Amer. Chem. Soc.* **91**, 6235 (1969); (b) P. Politzer, *Inorg. Chem.* **16**, 3350 (1977).

89. R.T. Sanderson, in *Polar Covalence*, (Academic Press, New York, 1983).

90. D. Lauvergnat, P. Maitre, P.C. Hiberty and F. Volatron, *J. Phys. Chem.* **100**, 6463 (1996).

91. J.H. van Lenthe and G.G. Balint-Kurti, *J. Chem. Phys.* **78**, 5699 (1983).

92. For H_2N-NH_2 and HO-OH: D.F. McMillen, D. Golden, *Ann. Rev. Phys. Chem.* **33**, 493 (1982); for F-F, see ref. 41.

93. For leading references, see (a) K.-D. Asmus, in *Sulfur-Centered Reactive Intermediates in Chemistry and Biology*, eds. C. Chatgilialoglu and K.-D. Asmus (Plenum Press, New York and London, 1990) p 155. (b) B.C. Gilbert, *ibid.* p 135. See also refs 47 and 56 and references therein.

94. Y. Deng, A.J. Illies, M.A. James, M.L. McKee and M. Peschke, *J. Amer. Chem. Soc.* **117**, 420 (1995).

95. (a) N. Berthe-Gaujac, Ph.D. Thesis, Orsay, 1997; (b) P.C. Hiberty and N. Berthe-Gaujac, manuscript in preparation.

96. J. Verbeek, J.H. Langenberg, C.P. Byrman and J.H. van Lenthe, TURTLE-An ab initioVB/VBSCF/VBCI program: Theoretical Chemistry Group, Debye Institute, University of Utrecht, 1993.

INDEX